T0201918

Framboids

Framboids

DAVID RICKARD

OXFORD
UNIVERSITY PRESS

OXFORD
UNIVERSITY PRESS

Oxford University Press is a department of the University of Oxford. It furthers
the University's objective of excellence in research, scholarship, and education
by publishing worldwide. Oxford is a registered trade mark of Oxford University
Press in the UK and certain other countries.

Published in the United States of America by Oxford University Press
198 Madison Avenue, New York, NY 10016, United States of America.

Library of Congress Cataloging-in-Publication Data
Names: Rickard, David, 1943- author.
Title: Framboids / David Rickard.
Description: New York, NY : Oxford University Press, [2021] |
Includes bibliographical references and index.
Identifiers: LCCN 2020058329 (print) | LCCN 2020058330 (ebook) |
ISBN 9780190080112 (hardback) | ISBN 9780197571927 (epub)
Subjects: LCSH: Pyrites. | Ore deposits. | Sediments (Geology) |
Texture (Crystallography) | Aggregation (Chemistry)
Classification: LCC QE390.2.I76 R525 2021 (print) |
LCC QE390.2.I76 (ebook) | DDC 549/.32—dc23
LC record available at https://lccn.loc.gov/2020058329
LC ebook record available at https://lccn.loc.gov/2020058330

DOI: 10.1093/oso/9780190080112.001.0001

1 3 5 7 9 8 6 4 2

Printed by Integrated Books International, United States of America

For Simonne
who has been there since the beginning

Contents

Synopsis

This book describes what framboids are and how they are formed. It is consequently divided essentially into two sections: Chapters 1–9 describe framboids and Chapters 10–13 discuss how they are formed. This synopsis integrates the two parts into an integral whole which acts as a *vade mecum* for the volume. Section links are listed in parentheses.

The Basic Attributes of Framboids

Generally, framboids are ubiquitous microscopic, sub-spherical aggregates of often, geometrically arranged, pyrite microcrystals. The basic attributes of framboids are their microscopic size, their sub-spheroidal form, and their microcrystalline internal structure (1.1).

Framboid Mineralogy

Framboids are predominantly made of pyrite. This is due to the chemical properties of this mineral and its crystallography, as well as the relative natural abundances of iron and sulfide. The extreme dominance of pyrite means that term *framboid* generally refers to pyrite framboids unless otherwise qualified.

The question of whether other minerals form framboids directly is moot (8.2). Detailed examination of reports of other mineral framboids reveal microcrystalline material within and associated with framboids (e.g., greigite, Fe_3S_{4g}) and sub-spherical crystalline aggregates (e.g., marcasite, chalcocite-digenite, magnetite). In contrast, framboidal aggregates of, for example, polystyrene microspheres, have been widely synthesized, so there is no a priori reason for other minerals not forming primary framboids. The limitation is probably the requirements of crystal habit, solubility, and natural abundances of the constituent elements for framboid formation.

Framboids are sometimes observed replaced by other minerals. Pyrite framboids are often formed during the earliest stages of sedimentation or mineralization and therefore subject to further reactions with later fluids. Minerals such as copper, cobalt, zinc, and lead sulfides often display framboidal forms having replaced original pyrite framboids (8.2.3, 8.2.4). Likewise, oxidation of pyrite under some conditions can produce iron oxyhydroxide and iron sulfate framboids (8.3).

Distribution of Framboids in Time and Space

Framboids may be the most abundant mineral texture on Earth. There are around 10^{30} framboids on Earth, and they are forming at a rate of about 10^{21} per year or 10^{14} per second (1.2.1). They are found throughout the geologic record; the oldest framboid may be 2.9 Ga. They are abundant in rocks less than 600 Ma old, and decrease through the Proterozoic (>0.6 Ga) until they are rare in Archean rocks (>2.5 Ga) (1.3.2). Their absence in these ancient rocks may be related to the low concentrations of sulfate in contemporary seawater, and there is some suggestion that they tended to occur in systems where sulfate was locally enriched, such as through hydrothermal activity. They are far more abundant in sediments and sedimentary rocks than in hydrothermal deposits, and their presence in igneous rocks appears to be limited to hydrothermal veinlets within volcanic rocks (1.3.4).

History of the Study of Framboids

Historically, framboids have been known since at least 1885 and were defined in 1935 (1.2). Early interpretations suggested that they were fossilized microorganisms, and in the 1950s this idea was resurrected by the discovery of organic matter in some framboids (7.1). The pyrite-sulfur in sedimentary framboids is almost wholly sourced through the activities of sulfate-reducing microorganisms, so there is a direct relationship between sedimentary framboids and microorganisms (see p. 129). Subsequently, the limited distribution of organic matter in framboids, its absence in hydrothermal framboids, and inorganic framboid syntheses showed that organisms were not necessary for framboid formation.

Sizes of Framboids

One of the basic attributes of framboids is their microscopic size. Framboid size-frequency distributions are log-normal; the geometric mean size of framboids is 6.0 µm and 95% of framboids range in size between 2.9 and 12.3 µm in diameter (2.3). The largest framboids may be 250 µm in diameter, although spherical aggregates of framboids, known as polyframboids, may range up to

900 μm in diameter (2.7). Various spherical aggregates of nanoparticles have been described which are less than 0.2 μm in diameter. These do not form a continuum with framboids (2.1.1).

There is no evidence for any significant change in framboid diameters with geologic time (2.4.1), and the differences in mean sizes between hydrothermal and sedimentary framboids do not, at present, appear to be statistically significant (2.6). By contrast, it appears that the mean diameters of framboids from non-marine sediments are significantly larger (7.6 μm) than marine framboids (5.7 μm) (2.4.3). There is some evidence that framboids formed in the water column are smaller than those formed in sediments, but the non-critical use of this possible difference as a proxy for paleoenvironmental reconstructions is not robust (2.5.1).

Microcrystal Size

Framboids are constituted by microcrystals. Microcrystal size distributions are approximately log-normal and 95% of framboidal microcrystals are between 0.1 and 3.1 μm (4.1). Packing efficiencies range from cubic and hexagonal close packings (74 vol %) to random packings (56–64 vol %) (4.2.1).

Microcrystals Numbers

Framboids contain between less than 100 to over 500,000 microcrystals (4.2). The ratios of framboid diameters to microcrystal sizes show a clear bimodal distribution which reflects the populations of close packed ordered framboids and close packed randomly organized framboids (4.2). The average numbers of microcrystals in both disordered and ordered framboids are similar, which suggest that the organization of microcrystals is the result of an additional process.

Microcrystal Habits

Minerals that do not commonly produce equant crystals forms are unlikely to display the framboidal texture. This essentially limits the framboid texture to minerals, like pyrite, displaying isometric symmetry (4.4). The permutations of pyrite habits result in pyrite displaying the greatest variety of crystal shapes among the common minerals, which includes approximations to forbidden five-fold symmetries such as the pyritohedron and pseudo-icosahedron.

Framboid Sphericity

The original idea that framboids were generally spherical was found to be due to the limitations of the contemporary optical microscopic methods (1.2.2). Later scanning microscopic investigations showed that many framboids were at least partly faceted and some display polygonal icosahedral forms (3.3). This is significant since the assumption of framboid sphericity informed earlier explanations of how they could form. It cannot be assumed that framboids necessarily require a precursor template, such as a spherical space or spherical organic globule, to develop.

Framboid Microarchitectures

Framboids were originally studied in sections, and this is still a common method of investigation (5.1.1). Sections through framboids reveal that the microcrystals in pyrite framboids often show extremely regular arrangements. Framboids can be classified in terms of whether their constituent microcrystals are regularly arranged (ordered in a single domain), randomly arranged (disordered), or mixtures of both (partially ordered or multiple domains) (5.1.1). The relative proportions of these three types are unknown, but there may be a tendency for the proportion of ordered framboids to increase with geologic age.

Framboid Crystallography

Framboids are not mesocrystals or extreme skeletal varieties of single crystals, and single crystal X-ray diffraction analyses of even the most perfectly organized framboids show ring patterns indicative of dominantly randomly oriented particles (6.2). Further detailed studies of microcrystal crystallinity by electron backscatter diffraction showed that the microcrystals within a framboid are not crystallographically aligned (6.3.3). Both randomly packed and organized framboids show adjacent framboids with crystallographic orientations rotated 90°. Framboids are formed by the aggregation of pyrite microcrystals rather than the sequential growth of one microcrystal on another.

Free Energy Minimization and the Development of Framboid Sphericity

The fundamental driving for the self-assembly of framboid microcrystals is reduction in surface free-energy (13.1). The self-assembly of framboid microcrystals to form framboids is consistent with estimations based on the classical Derjaguin-Landau-Verwey-Overbeek (DVLO) theory, which balances the attraction between particles due to van der Waals forces against the interparticle electrostatic repulsive force (13.1).

Entropy Maximization and Self-Organization

Maximization of entropy is the fundamental process leading to the development of regular arrangements of microcrystals in framboids (13.3). The formation of framboids involves two distinct processes. First, pyrite microcrystals aggregate into sub-spherical groups through free energy minimization. Second, the microcrystals rearrange themselves into ordered domains through entropy maximization. Icosahedral symmetry tends to minimize short-range attractive interactions and maximize entropy (13.3.2). The physical processes that facilitate this rearrangement are Brownian motion and surface interactions.

Burst Nucleation

The formation of many thousands of equidimensional and equimorphic microcrystals in framboids is the fundamental evidence for burst nucleation (11.1). This is conventionally described by the LaMer model, which is characterized by (1) a lag phase before nucleation becomes significant, (2) burst nucleation where the rate of nucleation increases exponentially and may be completed in seconds, and (3) a short growth phase where nucleation becomes again insignificant (11.1). The growth phase is dominated by crystal growth, which is limited by the supply of nutrients, which is not replenished, and growth is rapidly extinguished. The result of LaMer kinetics is a large number of similarly shaped colloidal particles.

Pyrite Chemistry

The extreme insolubility of pyrite is one of the fundamental reasons for its particular involvement in framboid formation, as well as for the ubiquity of framboids (10.1.1). Pyrite solubility is determined by the activities of Fe(II) and S_2(-II). Aqueous Fe^{2+} does not appear to react directly with aqueous polysulfide species to produce pyrite, and the S-S bond in aqueous S_2(-II) is normally split by aqueous Fe^{2+} to produce aqueous FeS and sulfur (10.2.5). The mechanisms of pyrite formation include the polysulfide pathway and the H_2S pathway. FeS is the reactant moiety in both processes, either as a dissolved aqueous cluster or a surface-bound species. The polysulfide pathway involves the substitution reaction between aqueous $S_2(-II)$ and FeS with $[FeS.HS_2]^-$ as the reaction intermediate (10.2.2). The H_2S pathway involves the redox reaction between H_2S and FeS (10.3.1). Both are involved in pyrite nucleation from aqueous solution and pyrite crystal growth, and both processes have been proven by isotopic tracing. Both processes have been shown to produce framboids.

Critical Supersaturation

The supersaturation required for pyrite to nucleate is conveniently defined in terms of the critical supersaturation which defines a point at which nuclei begin to form at a measurable rate (11.3.1). The critical supersaturation is related to the solubility via the surface energy. The computed critical saturation for pyrite varies between 10^{11} and 10^{18}, which is consistent with experimental estimates.

Classical Nucleation Theory (CNT) and Framboid Formation

The formation of pyrite nuclei is extremely sensitive to the surface energies of pyrite nuclei, and small changes in supersaturation lead to changes in the rate of nuclei formation of several magnitudes. Assuming that the initiation of burst nucleation can be approximated to a rate of one nucleus per second, CNT suggests a mean surface energy of around 0.6 J m^{-2} for pyrite nuclei in aqueous solution at standard ambient temperature and pressure (SATP) (11.3). The critical radius for pyrite nuclei under these conditions is between 2.3 and 3.8 Å, which is similar to the dimensions of the pyrite unit cell (11.3.3).

Framboids and Euhedral Pyrite

By contrast with framboids, euhedral pyrite crystals evidence the formation of isolated nuclei (11.5.1). These isolated nuclei continue to grow in regimes where the nutrient supply is not depleted or restricted, resulting in the formation of euhedral pyrite crystals.

Iron (oxyhydr)oxides are the major source of reactant Fe in sedimentary environments. The reaction between aqueous $S(-II)$ and $Fe(III)$ (oxyhydr)oxides produces surface $=FeS$, polysulfides and surface disulfide. The resulting formation of surface FeS_2 moieties through the reaction between surface $=FeS$ and $S_2(-II)$ leads to the heterogeneous nucleation of pyrite (11.5.1). This reaction may the major route for producing individual pyrite crystals, rather than framboids, especially in sediments. The reaction with surface =FeS is, of course, not limited to iron (oxyhydr)oxides, but occurs with any iron mineral in a sulfidic environment, including the relatively scarce iron sulfides mackinawite and greigite, as well as pyrite itself. The reaction with surface =FeS sites on pyrite is a major route for euhedral pyrite crystal growth (12.4).

Framboid Crystal Growth

One of the major consequences of the relative insolubility of pyrite is the huge supersaturations with respect to pyrite that occur widely in low oxygen environments. The result is that the initial stage of pyrite microcrystal growth occurs in solutions with large supersaturations, and the less stable octahedral faces develop first (12.2.5). As the pyrite crystals grow, the solution becomes depleted in nutrients and the supersaturation begins to approach saturation and the most stable—or least soluble—cube faces develop. Truncated cubes and octahedra are thus common microcrystal habits.

The three basic processes in crystal growth in framboids—monomer addition, crystallization by particle attachment (CPA), and Ostwald ripening—are neither necessarily successive nor exclusive (12.2.4). End member monomer growth produces more extreme monodispersed populations of microcrystals. By contrast, CPA appears to result in quite irregular microcrystals which naturally lead to more disorganized geometries.

Rate of Framboid Formation

The rate of pyrite crystal growth is unknown. Various qualitative observations suggest that pyrite crystal growth is rapid and diffusion-limited. Since pyrite solubility is so low, diffusion-controlled growth can be closely modeled by linear approximations to the diffusion equations (12.3.3). These show that framboids take between a few hours to a few years to form. The average framboid takes 3–5 days to form whereas, because of the exponential nature of the relationship between framboid size and time, there is not much difference in the time taken for 80 μm (2.2 years) and the maximum observed 250 μm (3 years) diameter framboids to form.

Effect of Temperature

The effect of temperature on the rate of framboid formation is not well constrained. As a first approximation, it appears that rate of framboid formation is over a magnitude faster in hydrothermal systems than in sedimentary environments under similar monomer concentrations. In practical terms this means that an average 6 μm framboid forms in around 12 hours in hydrothermal systems rather than 3–5 days in sediments.

Effect of pH

Pyrite forms mainly through two routes: (1) the reaction between FeS species and polysulfides (10.2), and (2) the reaction of FeS species and H_2S (10.3). Both of these reactions produce framboidal pyrite. Although pyrite displays extreme stability in terms of pH space, variations in pH affect the rate of framboid formation through the effect of pH on the solubility of FeS_s, which limits the potential maximum concentration of dissolved Fe(II) and S(-II) in solution (12.4.2). At pH <7 the solubility of FeS_s is a function of the square of the proton concentration and thus its solubility increases rapidly with decreasing pH (10.2.5). Framboid formation is further constrained by pH since solutions with pH <5 produce marcasite rather than pyrite (8.2.1, 12.4.2). H_2S is the dominant dissolved sulfide species at pH <7, so the H_2S reaction is mainly limited to between pH 5 and 7. The dominant polysulfide species between pH 4 and 11 is HS_2^- and this would seem likely to be the major polysulfide reactant.

Organic Matter in Framboids

There is an intrinsic association between organic matter and sedimentary framboids since the sulfide in sedimentary pyrite is almost wholly the result of microbial sulfate reduction by mainly heterotrophic organisms (7.2). The exact nature of this organic material is unknown. However, it appears that microbial biofilm may be an important contributor (7.2.1). Likewise, the organic residues from framboids often appear similar to sulfur-rich organic geopolymers such as protokerogen (7.2.2). Most of the organic matter in framboids appears to be syngenetic with the framboids, and framboids seem to have grown in organic substrates.

The organic relicts tend to take on the form of the pyrite rather than vice versa (7.1.2). If organisms are involved as templates for framboid formation, they have not as yet been defined. Generally, known prokaryotic cells, such as giant sulfur bacteria, are too delicate to provide a scaffold for pyrite framboid formation (7.1.5). Eukaryotic organisms which might provide an internal template for framboid formation have not as yet been identified.

Framboid Geochemistry

At present, the composition of pyrite in framboids is unknown, although it would be interesting to know its stoichiometry (9.1). The trace element content of framboids has been widely reported since framboids usually constitute the earliest pyrite phase in a sediment and therefore are more likely to pick up trace element variations in contemporary seawater (9.3.1). However, the processes leading to trace element contents of framboids are complex and poorly understood.

No spatial variations in $\delta^{34}S$ compositions have been reported within individual framboids (9.3.3). This is consistent with the process of burst nucleation in framboids. It does suggest that framboids pick up a more accurate measure of the sulfur isotopic composition of the prevailing dissolved sulfide and the trace element composition of the contemporary environment and are likely to retain these over geologic time.

Acknowledgments

The pillars on which this monograph is constructed are three dissertations: my original undergraduate dissertation in 1965 and the PhD theses of my students Ian Butler (1994) and Hiroaki Ohfuji (2004). I thank Ian Butler and Hiroaki Ohfuji for allowing me to reproduce many of the images from their theses. They also reviewed many of the chapters and were involved in developing many of the ideas recounted in this book.

This monograph is multidisciplinary and could not have been completed without the dedicated work of a large number of reviewers. First and foremost was Rob Raiswell, Emeritus Professor of Sedimentary Geochemistry at Leeds University, United Kingdom, who read and commented on all 13 chapters. He also made a series of constructive comments on the structure of the volume based on his overview.

I thank the following colleagues who reviewed one or more chapters of the book and whose time and effort in helping to see the work on framboids published underlines the importance of the topic to a variety of research disciplines.

Ian Butler	*University of Edinburgh*
Michael Engel	*Friedrich-Alexander-Universität Erlangen-Nürnberg*
Daniel Gregory	*University of Toronto*
Stephen Grimes	*Plymouth University*
Miguel Angel Huerta-Diaz	*Universidad Autónoma de Baja California*
Ross Large	*University of Tasmania*
George Luther	*University of Delaware*
Hiroaki Ohfuji	*Tohuko University*
Rob Raiswell	*University of Leeds*
Kevin Rosso	*Pacific Northwest National Laboratory*
Zbigniew Sawłowicz	*Jagiellonian University, Krakow*
Nicolas Tribovillard	*Université de Lille*

A number of distinguished researchers also made comments on specific areas of the book and helped in clarifying particular areas.

David Bond   University of Hull
Julia Dshemuchadse   University of Michigan
Robert Finkelman   University of Texas, Dallas
Dimo Kashchiev   University of Sofia
Danlil Kitchaev   University of California, Berkeley
Paul Midgley   University of Cambridge
Raúl Merino Palomeres   University of Madrid
Juergen Schieber   Indiana University
Werner Stahel   ETH, Zurich
Gregor Trefalt   University of Geneva
Richard Wilkin   United States Environmental Protection Agency

The subject of framboids cannot be described without the inclusion of many images and, over the years, many colleagues have sent me images of framboids they have discovered or synthesized. Many of these images are included in this monograph and it could not have been completed without these contributions.

Robert Berner †   Yale University
Ian Butler   University of Edinburgh
Anders Elverhøi   Oslo University
Stephen Grimes   Plymouth University
Mizela Lay   University of Tasmania
George Luther   University of Delaware
Peter McGoldrick   University of Tasmania
David H. McNeil   Geological Survey of Canada
Christopher Morrissey †   Hagley, Worcestershire
Indrani Mukherjee   University of Tasmania
Edward B. Nuhfer   Niwot, Colorado.
Hiroaki Ohfuji   University of Tohoku
Heikki Papunen   University of Helsinki
Matías Reolid Perez   University of Jaén
Zbigniew Sawłowicz   Jagiellonian University, Krakow
Heide Schulz-Vogt   Leibniz-Institut für Ostseeforschung, Rostock
Nicolas Tribovillard   University of Lille
Laura A. Vietti   University of Wyoming
David Wacey   University of Western Australia

Many colleagues, in addition to the ones listed in the previous paragraphs, have been involved in discussions about framboids during the last 55 years, and these have helped mold the present volume.

G. Christian Amstutz†	Heidelberg University
Hubert Barnes	Pennsylvania State University
Dan Bubela†	Baas Becking Geomicrobiological Institute, Canberra
Jacques Jedwab †	University of Liege
George Kato	Kyushu University
Leonard G. Love †	University of Sheffield
Anthony P. Millman †	Imperial College, London
John Morse†	Texas A&M University
Marjorie D. Muir	Imperial College, London
Anthony Oldroyd	Cardiff University
Richard A. Read	Imperial College, London

The list is not complete, and I apologize to those colleagues I have inadvertently missed.

The monograph lists 469 references, often from ancient and esoteric sources. I thank Jacqueline Roach of the interlibrary loan team of Cardiff University for her dedicated assistance in sourcing many of these publications.

Finally, I thank my commissioning editor at Oxford University Press, Jeremy Lewis, for his unflagging support and encouragement.

1

Introduction

1.1. Definition of Framboids

Framboids are defined as:
Microscopic spheroidal to sub-spheroidal clusters of equant and equidimensional microcrystals.

The characteristics of the framboid texture need to be carefully defined since it is apparent that there are a large number of textures reported in the literature which are confused with it. This often refers to the more esoteric reports of framboids. For example, Garcia-Guinea et al. (1997) described large pyrite spheres in the remains of ink in the seams of 16th- and 17th-century books, which they described as framboids. In fact, these were smooth spherules of pyrite with no internal microcrystalline structure.

The key attributes of framboids are the microscopic size, the spheroidal to sub-spheroidal outer form, and the internal microcrystalline structure (Figure 1.1). It is these attributes which have made framboids so interesting to mineralogists, geologists, and geobiologists, as well as materials scientists. A particularly astonishing feature is the self-organization displayed by some pyrite framboids. However, the degree of self-organization is highly variable, and most framboids display only limited or even no obvious internal organization.

1.1.1. Microscopic Size

Pyrite framboids found in nature range from <1 to 250 μm in diameter and display a mean size of around 6 μm. This means that the number of average-sized framboids that can be placed on the period at the end of this sentence is 1850. The aggregates are made up of pyrite microcrystals, rarely more than 500 nm across, which sometimes display extraordinary ordering. The number of these microcrystals in any single framboid ranges between 10^2 and 10^7 (<0.1 ~20 μm in diameter). So probably more than 2 billion framboid microcrystals can be placed on a period. The point is that framboids are exceedingly small and not usually visible to the naked eye: they occur all around us without our being aware of them.

Framboids. David Rickard, Oxford University Press. © Oxford University Press 2021.
DOI: 10.1093/oso/9780190080112.003.0001

Figure 1.1. Group of pyrite framboids showing their major attributes: microscopic size, sub-spheroidal form, and discrete microcrystalline structure consisting of equant and equidimensional microcrystals.

Scanning electron micrograph by Stephen Grimes.

1.1.2. Spheroidal to Sub-spheroidal Form

Framboids had classically been assumed to be spherical, and the origin of the sphericity was a feature of earlier discussions regarding their origin. However, this proved to be an artifact of the limitations of optical microscopy: framboids were usually examined in sections under reflected light microscopy. A typical result is shown in Figure 1.4 later in the chapter. With increasing use of scanning electron microscope (SEM) techniques, it has become apparent that the outer surfaces of many framboids are faceted (Figure 1.1). Indeed, framboids may display polyhedron-like forms, such as icosahedra (Ohfuji and Akai, 2002). The original assumption of a spherical form had significant effects on theories regarding framboid origins, as discussed later, and the consequences echo in the literature even today.

1.1.3. Discrete Microcrystals

A key feature of the framboidal texture is that they are constituted of discrete microcrystals varying between approximately 0.1 and 2 µm in size. That is, the interior of the framboid is not solid, homogeneous pyrite. A number of other spheroidal pyrite forms are found in which the material is radiating, acicular pyrite, aggregates of pyrite crystals, and structureless spherules of massive pyrite. These are often described mistakenly as framboids but have different origins and distributions.

The microcrystals in framboids are often ordered to some degree. This ordering gave rise to the earlier suggestions that framboids were replacements of biological forms, as fossilized bacteria or discrete microorganic forms. It is clear, however, that this is not necessarily the case since framboids are found in high-temperature environments and may be synthesized without the involvement of microorganisms.

In general, the number of ordered framboids appears to be a small fraction of the total framboid population. However, some form of cryptic or localized, partial order may occur in a large number of framboids. Ordered or partially ordered microcrystalline pyrite in irregular, non-spheroidal, aggregates occurs widely, especially in recent sediments. But these lack the obvious spheroidal characteristics of framboids. However, the occurrence of this type of material in nature demonstrates that microcrystal ordering is unrelated to the outer form of the aggregate.

1.1.4. Equant and Equidimensional Microcrystals

A key feature of the framboid texture is that the microcrystals are all approximately equant. The microcrystals show a variety of the crystal shapes, including cubic, cuboidal, cuboctahedral, octahedral, and pyritohedral. The equant form of the microcrystals is important since it enables close packing arrays. It also means that minerals that do not commonly form equant habits cannot produce framboids. This generally limits the framboid texture to minerals that crystallize in the isometric system and amorphous materials that can form spheres.

Within any single framboid, the individual microcrystals all tend to have the same size. This enables close packing and contributes to the consequent stability of the aggregate. They are packed in a variety of ways, including cubic close packing, icosahedral packing, random packing, and mixtures of these, and consist of 10^2–~10^7 microcrystals (<0.1–~20 µm in diameter).

1.2. History of Framboid Research

Framboids were first described when optical microscopes became widespread in the mid-19th century. They are usually too small to be seen with the naked eye and, even with primitive microscopes, their internal structure could not be resolved. Gabriel Augustus Daubrée (1814–1891), professor of Mineralogy at the School of Mines in Paris (and later president of the French Academy of Sciences) described a potpourri of contemporary occurrences of pyrite in which he reported microscopic rounded pyrite nodules from the thermal spring at Bourbonne-les-Bains in northeastern France (Daubrée, 1875). He described one group from a Roman floor as having minute crystal faces. We cannot know whether these pyrite forms described by Daubrée were framboids, clusters of framboids, or simple pyrite nodules. However, he noted that his smallest pyrite crystals were 0.25 mm (250 µm) in size, and this is some 10 times larger than the normal size range of framboids. Similarly, Jakob Maarten van Bemmelen (van Bemmelen, 1866), who was to become one of the founders of colloid chemistry, described spherical particles of pyrite from polders near Groningen in the Netherlands, but did not observe their microstructure.

1.2.1. Discovery of Microcrystals in Framboids

Johann Jakob Früh (Figure 1.2) published the first description of framboids in 1885 (Früh, 1885). He found these in a moss peat near Gusev in Kaliningrad, Russia. At that time, he was a schoolteacher in the idyllic alpine village of Trogon in northeastern Switzerland. He published his description of pyrite framboids as part of his Habilitation submission to the University and Polytechnic of Zürich, which was to become ETH Zürich. Früh was appointed professor of Geography in Zürich in 1899 and published the monumental *Geographie der Schweiz* between 1930 and 1938. Früh's original contribution has been largely overlooked, and reference to Figure 1.3 explains why. The framboid illustration is just one on a page of 39 line drawings of objects he had observed in peat. His discussion of these objects is also limited: just one paragraph in 49 pages of text. The more widespread appreciation of framboids in the scientific community had to await almost 40 years until the work of Schneiderhöhn (1923).

Früh wrote (1885, p. 707):
These spherules are neither spores nor elm-leaf fungus perithecia. They do not react to potassium hydroxide or hydrochloric acid. Fungus or lichen spores (Fig. 4 and 29) are lightened by alkalis even if their membrane is highly humic. This is not the case with the spherules in question. Rather, as I have often convinced myself in

Figure 1.2. Johann Jakob Früh (1852–1938) in 1895.
Image courtesy of ETH Zürich.

various occurrences, they are aggregates of small pyrite crystals. They break down into numerous particles under pressure (Fig. 30), but spores do not. These particles are opaque, sharply delimited and, occasionally, show a brownish rim by the light reflected on their surfaces. Their diameter varies from 0.0007–0.001 millimeters. Despite this smallness, they mostly appear in 600 x magnification as sharp squares or short rectangles, i.e. usually as cubes, which can show Brownian molecular motion in isolation.[1]

[1] *Diese Kügelchen sind weder Sporen noch Ulmus-kügelchen. Sie reagiren nie auf Kalilauge oder Salzsäure. Pilz- oder Flechtensporen (Fig. 4 und 29) werden von Alkalien auch dann aufgehellt, wenn ihre Membran stark humificirt sein sollte. Mit den in Frage kommenden Kügelchen ist dies nicht der Fall. Es sind vielmehr, wie ich mich häufig an verschiedenen Vorkommnissen überzeugt habe, Aggregate von Schwefelkieskryställchen. Durch Druck zerfallen sie in zahlreiche Körnchen (Fig. 30), Sporen hingegen nicht. Jene Körnchen sind undurchsichtig, scharf begrenzt und höchstens vermöge des an ihren Flächen reflektierten Lichtes bräunlich umrandet. Ihr Durchmesser variirt von 00007—0·001 Millimeter. Trotz dieser Kleinheit erscheinen sie bei 600/1 meistens in scharfen Quadraten oder kurzen Rechtecken, d. h. in der Regel als Würfelchen, welche isolirt die Brown'sche Molekularbewegung zeigen können.*

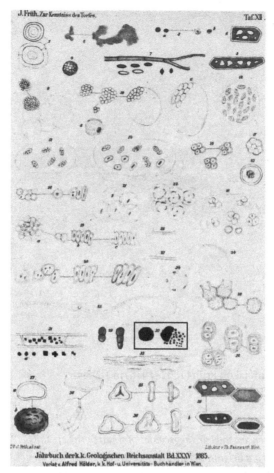

Figure 1.3. Johannes Jakob Früh's line drawings of microscopic objects he observed in peat with framboid images (no. 30) outlined. Früh's caption reads: 30. *Pyrite spherule. Disintegrating into individual cubes by pressure*[a] (Früh, 1885).

[a] *30. Schwefelkieskugel, z. Th. durch Druck in einzelne Würfelchen zerfallend.*

Früh observed the spherules in a transmitted light microscope at a magnification of about 600 ×. Even so, he was able to see the framboids themselves and their constituent microcrystals, which he reported had sizes between 0.7 and 1 micron. He does not estimate the sizes of the framboids themselves, but they appear from his figure to be between 10 and 20 μm in diameter. He carried out a series of chemical tests on these crystals, included demonstrating that they did not dissolve in hydrochloric acid but dissolved in nitric acid. Their reaction with carbon disulfide convinced him that they were constituted of pyrite and not marcasite. However, Früh was uncertain as to the origin of the sulfide.

In 1892 Ludvig Rhumbler, a research assistant at the Zoological Institute at the University of Göttingen, independently published a description of pyrite framboids from marine sediments (Rhumbler, 1892). Rhumbler reported that the spheres fragmented under gentle pressure from the cover glass into microscopic crystals. This was further confirmed by Potonié (1908), who crushed pyrite framboids from lake sediments between two glass slides to reveal their constituent microcrystals. In his original report on framboids, Rhumbler summarized most of the essential characteristics of these textures: (1) they are commonly pyritic; (2) they are aggregates of microcrystals; and (3) they are formed in sediments, with the iron being mainly provided from coastal sources and with the sulfide being produced by the reduction of sulfate. These characteristics have formed the basic pillars of framboid research over the last century and inform the structure of the present monograph.

1.2.2. Advances in Microscopy and Understanding Framboid Structure

The history of understanding of the internal microstructure of framboids is closely related to technical advances in observing and probing microscopic objects. It is important to understand this in appraising older work on framboids. For most of the history of framboid studies, the standard method of observation was through examining polished blocks of rock. Binocular microscopes generally did not have sufficient resolution to resolve their internal structure, and pyrite is opaque to conventional transmitted-light microscopes. Furthermore, the observation of materials through thin sections results naturally in some dispersion, and the size of the microcrystals approaches the wavelength of light. With the strictly two-dimensional surface of the polished blocks used in reflected-light microscopy, objects approaching the wavelength of light can be resolved optically (380 nm (violet) to 750 nm (red)). Normally, white light has been used to illuminate the specimen with a mid-wavelength of around 580 nm giving a maximum theoretical resolution of around 0.2 um with a plan apochromatic objective with 100 × magnification (Figure 1.4). In other words, bodies separated by 200 nm can theoretically be observed in reflected-light microscopy. Since the usual minimum size of the microcrystals approaches 200 nm, individual microcrystals can be resolved optically.

The key understanding of the optical properties of the opaque minerals was the development of the polarizing microscope and its application in reflected light microscopy by Koenigsberger (1901), which was systematized by Schneiderhöhn (1922). The introduction of the polarizing microscope to the study of opaque minerals was important since it enabled the mineral species to

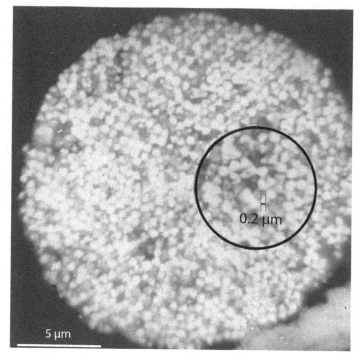

Figure 1.4. Reflected-light microscope image of a polished section of a framboid. This is imaged in white light and the insert shows the limit of resolution. The variable sharpness of the microcrystals is due to their varying heights relative to the plane of focus.

Reichert Zetopan microscope with 100 x oil immersion lens.

be determined and thus pyrite (cubic FeS_2), for example, could be distinguished from marcasite (orthorhombic FeS_2).

The first person to record the microarchitecture of pyrite framboids was Schneiderhöhn (1923) in his studies of the 255 Ma old Kupferscheifer ores at Mansfeld in Germany (Figure 1.5). Schneiderhöhn was a pioneer in the application of reflected light microscopy and helped develop the Leitz Erzmikroscop petrological microscope, which had both transmitted light and reflected light capabilities. The Leitz had a 95 × oil immersion lens, giving a linear magnification capability of over 800 × magnification and the possibility of resolving objects <1 μm in size. Schneiderhöhn himself claimed a resolution of 0.2 μm, which is at the limits of resolution under white light, as noted earlier. Whatever the actual resolution of his microscope, he was able to clearly discriminate the individual microcrystalline components of pyrite framboids.

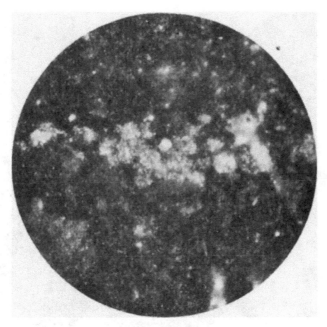

Figure 1.5. The first published image of framboids showing the constituent pyrite microcrystals in situ. The image was made by Schneiderhöhn (1923) using a 95 x oil immersion objective and the linear magnification was 800 x. The individual framboids are about 5 μm in diameter and the pyrite microcrystals about 0.5 μm in size.

The term *framboid* was first used for these pyrite bodies by American mineralogist George W. Rust of the University of Chicago in 1935. In a footnote to page 407, he explained his choice of the term laconically:

After framboise (Fr.) for "raspberry."

Rust was an assistant to Professor Edson Sunderland Bastin, professor of Economic Geology and head of the Department of Geology at the University of Chicago between 1922 and 1944. Bastin was the author of the paradigmatic publication *Interpretation of Ore Textures* (1950), which includes a section on framboids. Rust's original photomicrographs of the framboidal texture are shown, together with his original figure caption, in Figure 1.6. It is significant, in the light of the later overwhelming interest in framboids in sediments and sedimentary rocks, that these textures which were first defined as framboidal were from a hydrothermal copper ore in the Mississippi Valley mining district of southeastern Missouri.

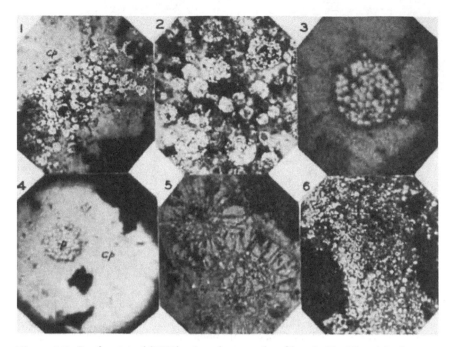

Figure 1.6. Rust's original (1933) microphotographs of framboids. His original caption read: *Cp, chalcopyrite; P, pyrite. 1. X 108. Solid spheroidal masses of pyrite in chalcopyrite. 2. X r 66. "Framboidal" spheroidal masses of pyrite in chalcopyrite due to metacolloidal crystallization. 3. X 580. "Framboidal" clusters of pyrite grains due to crystallization of a gel globule. Secondary minerals separate the discrete crystals of pyrite. 4. X 450. Same as 3 except that there has been no enrichment of the ore and hence chalcopyrite still forms the matrix of the pyrite cubes and grains. 5. X 400. Radial "bomb" type of texture developed in pyrite with a small "framboidal" nest of pyrite grains in the center of one pyrite area. Chalcopyrite matrix of "bomb" fragments. 6. X 64. Spheroidal globules of chalcopyrite, with pyrite core, and chalcocite (after bornite) matrix.*

Rust provided no details of the microscopic setup he used to prepare these images, although he probably used a Bausch and Lomb metallographic microscope. In 1931 M. N. Short of the United States Geological Survey (USGS) had published his account of the microscopic identification of the ore minerals. Since Bastin had arrived in Chicago from the USGS, it can be assumed that Short's techniques, which were state of the art in the United States in the 1930s, were used at Chicago by Rust. These systems involved huge bellows-type cameras mounted on top of metallographic microscopes, or even larger ones mounted horizontally. The images might require up to 12 minutes exposure time, and Short routinely developed them as 3.25 × 4.25 inch (8.3 × 10.8 cm) plates. The magnifications listed by Rust for the images in Figure 1.6 refer to the microscopic magnification of the objective lens multiplied by the eyepiece. The actual size of the framboids in these images is therefore unknown, and Rust merely refers to them as "tiny." In his paper, Rust does not refer to previous reports of occurrences of framboidal pyrite.

1.2.3. Discovering the Nature of Framboids

Framboids may represent the most abundant mineral texture in the natural world. In this case a *texture* is defined as the disposition or manner of union of the particles of a body or substance, using one of Webster's definitions. It differs from habit, which specifically refers to a crystal shape. Even with advances in microscopy, the question of what framboids actually are was not resolved until the early years of this century. The details are discussed in Chapters 3–6. Butler (1994) originally demonstrated that framboids were not single crystals. This precluded the idea that framboids were some extreme form of skeletal crystal. Ohfuji et al. (2006) showed, using enhanced computing techniques, that the microcrystals in organized framboids were arranged in sub-domains. Ohfuji et al. (2005) went further to demonstrate that the individual microcrystals in even the most perfectly organized framboids were not crystallographically aligned, even if they were arranged in regular geometric patterns. This proved that framboids were simple aggregates of microscopic pyrite crystals. Ohfuji and Akai (2002) demonstrated that some framboids, far from being spherical, showed icosahedral forms, which is a classically forbidden crystallographic symmetry. More detailed studies, recounted in Chapter 3, showed that framboids are commonly faceted. This meant that the problem earlier workers had in accounting for their extreme spherical shape was a chimera.

1.3. Distribution of Framboids

Framboids are ubiquitous. When the young and eager Polish framboid specialist Zbigniew Sawłowicz asked the great Professor G. Christian Amstutz of the University of Heidelberg where he could find some framboids, Amstutz replied, "Look in the Vistula." Amstutz's point was that framboids are all around us, in local rivers, such as the Vistula, and in your garden soils.

1.3.1. Framboid Numbers

Framboids are almost exclusively made up of pyrite. There are cogent reasons for this (detailed in Chapters 10–14), including the abundance of pyrite relative to other sulfide minerals, its extreme stability and low solubility, and its morphology. Other minerals also occur in framboids or in framboid-like forms, as is described in Chapter 8. However, when framboids are referred to loosely in this book, the reader can assume that pyrite framboids are being described.

The global quantity of pyrite is relatively well constrained in geochemical terms (Rickard, 2012b). The total amount of pyrite buried in sedimentary rocks is between 4 and 6 × 10^{21} g (Bottrell and Newton, 2004; Holser et al., 1988). Rickard (2015) estimated that between 1% and 10% of this might be in the form of framboidal pyrite. This estimate is relatively robust. Vallentyne (1963), for example, isolated 450 000 framboids per gram of dried lake sediments. Wilkin et al. (1996) reported that over 95% of pyrite from the Black Sea sediments occurs as framboids, irregularly shaped aggregates, and sub-micron sized microcrystals, and these textures dominated the pyrite fraction of all the sediments they examined.

The average framboid diameter is 6 µm (Rickard, 2019a) and the 0.64 packing density of the dominant randomly packed framboids (section 4.2 in Chapter 4) suggests that there would be 1.7 × 10^{30} framboids globally if framboids make up 10% of the global pyrite and 1.7 × 10^{29} framboids if they constitute 1%. This means that there are a billion times more framboids than sand grains on Earth, and a million times more framboids than stars in the observable universe.

These are minimal estimates since they do not consider pyrite framboids in igneous and volcanic rocks. The pyrite content of continental crust is <10^{22} g assuming all the sulfur is in the form of pyrite, the mass of continental crust is 2.17 × 10^{25} g (Peterson and Depaolo, 2007), and the average sulfur content is 0.4 wt % (Rudnick and Gao, 2003). The mass of pyrite in the oceanic crust is <10^{22} g assuming all the sulfur is in the form of pyrite, the mass of the oceanic crust is 0.6 × 10^{25} g and the average sulfur content is 0.096 wt % (Rickard 2012f). This suggests that the mass of pyrite in the crust is <2 × 10^{22} g. There is an element of double book-keeping here, since the continental and oceanic crust include a sedimentary component. However, these constitute a minor fraction (<<10% according to Peterson and Depaolo, 2007) and the error is insignificant with respect to the other approximations. The net result is that the mass of pyrite in the crust is approximately the same as that in sediments and sedimentary rocks. But the fraction that is in the form of framboids in igneous and volcanic rocks is relatively small and the contribution of continental crust to the total framboid inventory is insignificant.

By far the greatest amount of pyrite is contained in sediments and sedimentary rocks, and the pyrite sulfur is microbial in origin. Rickard (2015) estimated that volcanic sulfur makes up no more than 1% of the global sulfur flux. The estimation was based on a total global sub-aerial and oceanic volcanic S flux of about 10^{12} g year (computed from data in Andres and Kasgnoc, 1998; Brimblecombe,

2003; Canfield, 2004; Elderfield and Schultz, 1996) and assuming that no more than 25% of this is fixed as pyrite. This compares with microbial sulfate reduction which is estimated to be of the order of 30×10^{30} g S a^{-1} (e.g. D'Hondt et al., 2002; Turchyn and Schrag, 2004) with at least 80% of this re-oxidized (Jorgensen, 1977). This leads to a net burial of 6×10^{13} g S a^{-1} as pyrite in sediments. Within the limits set previously of between 1% and 10% of this pyrite to be framboidal in form, this suggests that between 3×10^{21} and 3×10^{20} framboids are formed in sediments every year, or between 10^{13} and 10^{14} framboids are formed every second. Again, the error in ignoring the volcanic and hydrothermal contributions is insignificant.

Apart from the enormous size of these numbers, the data show that the formation of the framboid texture is facile. It suggests that different pathways lead to framboid formation and the formation of pyrite microcrystals, their self-assembly into spherular forms, and the subsequent self-organization into organized internal arrangements are routine chemical and physical processes in Nature. That is, framboids are not especially remarkable but rather are a result of routine natural processes.

The strange anvil-shaped stability area of pyrite in pH-Eh space, first illustrated by Garrels (1960), is well known and is an artifact of the erroneous older National Bureau of Standards (NBS) free energy values for aqueous Fe^{2+} and S^{2-} in the calculations. Rickard and Luther (2007) showed that the pyrite stability region actually extended widely across pH space. A more detailed rendition is shown in Figure 1.7, where the pyrite stability region is illustrated for total dissolved sulfur activities of 10^{-3} and 10^{-6}, which covers the expected sedimentary concentration range between millimolar and micromolar sulfide.

Pyrite has an extensive stability region in sediments ranging from pH 2–10, both above and below the SO$_4$(-II) /S(-II) redox boundary. This is significant since this boundary is often used to distinguish "reduced" from "oxidized" environments. In fact, as pointed out by Rickard and Luther (1997) with respect to H$_2$S, *reduced* and *oxidized* are relative terms relating to electron exchange. Figure 1.7 shows that near the SO$_4$(-II) /S(-II) redox boundary, the calculated ion activity product (IAP) in equilibrium with pyrite changes extremely rapidly, up to 10^{14} within 50 mV (Butler and Rickard, 2000). The effect is significant in that supersaturations with respect to pyrite vary over very small regions of pH–Eh space. Because, as discussed in section 11.3.1 of Chapter 11, framboid nucleation requires extremely large supersaturations, this IAP gradient may affect whether framboids form or not in a particular sedimentary environment. The result is that framboid formation can vary over extremely small distances in a sedimentary or aquatic environment due to local spatial heterogeneities.

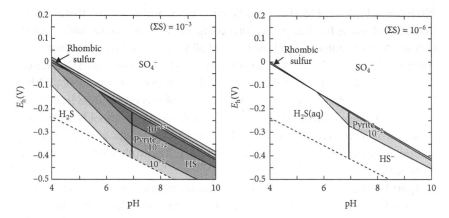

Figure 1.7. Pyrite stability in seawater at 25°C, 0.1 MPa total pressure and total dissolved sulfur activities of 10^{-3} ($(\Sigma S) = 10^{-3}$, left) and 10^{-6} ($(\Sigma S) = 10^{-6}$, right) for total dissolved Fe activities between 10^{-15} and 10^{-2}.

Reprinted from *Treatise on Geochemistry*: Second Edition, Chapter 9, The sedimentary sulfur system. Copyright (2013) with permission from Elsevier.

1.3.2. Distribution of Framboids in Sediments and Sedimentary Rocks

The early framboid workers could produce a list of sedimentary localities where framboids had been observed (e.g., Love and Amstutz, 1969). Such a list today would run to several volumes.

The distribution of framboids in sediments and sedimentary rocks mirrors the distribution of pyrite. Since the formation of sedimentary pyrite is a primary parameter in the biogeochemical cycle of the elements in Earth surface environments, its distribution through time can be modeled (e.g., Berner, 2006). Secular variations of pyrite contents of sedimentary rocks were discussed in some detail in Rickard (2012e, 2013) and are shown in Figure 1.8. The general trend is increasing sulfate in the Proterozoic with a possible peak around 2.0 Ga. Note the extreme fluctuations in oceanic sulfate concentrations in the Phanerozoic and the extremely low values in the Archean. The pyrite concentration in sediments is represented in terms of the balance between burial of reduced sulfur as pyrite and oxidized sulfur as evaporites. The Phanerozoic generally has seen the balance skewed toward evaporites in the prevailing oxygenated atmosphere.

The modeling of sedimentary pyrite formation during the Phanerozoic is more precise because more data are available. The models show that sedimentary pyrite formation has varied with time, with peaks in the Tertiary, Permo-Triassic, Devonian, Silurian, Ordovician, and Cambrian periods. These peak periods are

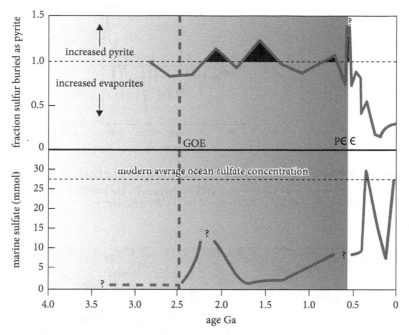

Figure 1.8. Variations in sedimentary pyrite formation with time (see text). GOE Great Oxidation Event. PЄ Cambrian-Precambrian boundary. Gray shading suggests increasing atmospheric O_2 during the Proterozoic.

Reprinted from *Developments in Sedimentology*, Vol. 65, Rickard, D., The evolution of the sedimentary sulfur cycle. Copyright (2012) with permission from Elsevier.

generally coincident with periods of widespread black shale development and extensive regions of anoxia in the contemporary oceans. The proportion of pyrite in the form of framboids may have been greater during these episodes, because of the prevalence of the stagnant conditions which facilitate framboid formation. Even so, the chart in Figure 1.9 roughly mirrors the history of sedimentary framboid formation during the Phanerozoic.

However, a similar relationship between peak sedimentary pyrite formation and framboid formation cannot be assumed for the Precambrian. Thus, the peaks in sedimentary pyrite formation relative to evaporites shown in Figure 1.8 at around 1.9 and 1.4 Ga do not appear to coincide with the occurrence of large numbers of framboids in contemporary sediments.

As described in section 7.1.6 of Chapter 7, in the vast deposits of sedimentary pyrite in 1.6 Ga Australian mid-Proterozoic shales, framboids are replaced by similar sized pyrite spherules and euhedra (Figure 1.10). Etching does not reveal any cryptic framboidal texture. The small pyrite euhedra were originally pyrite spherules which have subsequently been overgrown by euhedral pyrite. The

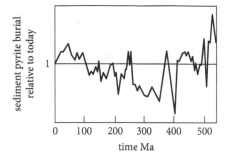

Figure 1.9. Variations in sedimentary pyrite deposition in the Phanerozoic.
Reprinted from *Developments in Sedimentology*, Vol. 65, Rickard, D., The evolution of the
sedimentary sulfur cycle. Copyright (2012) with permission from Elsevier.

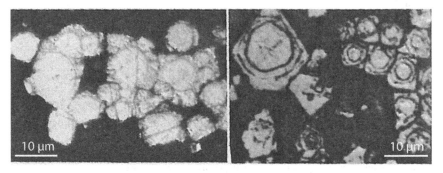

Figure 1.10. Photomicrographs of etched pyrite grains from the 1.6 Ga Mt. Isa
shales showing sub-spherical forms overgrown with successive generations of pyrite.
These grains represent the material from which the organic sacs were extracted by
Love and Zimmermann (1961).
Published by courtesy of the Society of Economic Geologists.

reason for the general absence of framboids in these sediments is unknown. In
view of the differential nucleation processes leading to pyrite euhedra formation
and framboid formation (section 11.5.1 in Chapter 11), it seems likely that the
extreme supersaturations leading to burst nucleation and framboid formation
were not achieved (except locally; see later discussion) This could be a result of
limited sulfate concentrations in the Proterozoic oceans at this time (Figure 1.8),
which would lead in turn to lower microbial sulfide reduction rates. Low oceanic
sulfate concentrations have been proposed to result from lower atmospheric oxygen concentrations at this time.

Framboids are occasionally reported from Proterozoic and even Archean
rocks. Rickard and Zweifel (1975a) reported framboids from 1.8 Ga massive

sulfide ore deposits in Sweden. Guy et al. (2010) reported framboids from 2.9 Ga carbonaceous shales from the Witwatersrand in South Africa (Figure 1.11) and noted that framboids had also been reported from 2.7 Ga rocks of the Rand Group. Liu et al. (2020) suggested that framboids formed in Archean sediments but were subsequently dissolved and the pyrite reprecipitated as nodules and dispersed fine-grains. It seems that in these environments, dissolved sulfide concentrations were locally sufficient to initiate framboid formation. It may be no coincidence that these environments were all associated with mineralized sediments, suggesting either a locally enhanced dissolved sulfate concentration or hydrothermal vents augmenting existing sulfide concentrations.

1.3.3. Hydrothermal Framboids

Framboidal pyrite is also associated with hydrothermal ore deposits, including many that are known or interpreted to have been formed at, or near, the sea floor (e.g., Chen, 1978; England and Ostwald, 1993; Halbach and Pracejus, 1993; Honnorez et al., 1973; Pichler et al., 1999; Rickard and Zweifel, 1975b). Schouten (1946) described framboids from tin veins from Cornwall, United Kingdom, and from the volcanogenic ores of Rio Tinto, Spain. The framboids at Rio Tinto, Spain, were the subject of a detailed study by Read (1968). Reedman

Figure 1.11. The oldest framboid? Photomicrograph of a 2.9 Ga framboid from carbonaceous shale from the Witwatersrand. The white specks are galena grains.
Modified from Guy et al., 2010. Published with permission of the Geological Society of South Africa (GSSA).

et al. (1985) described framboidal pyrite in quartz veins from north Wales which had been formed through hydrothermal processes at an estimated 2 km depth in the crust. Sassano and Schrijver (1989) described framboids in veins in ca. 490 Ma Cambro-Ordovician shales from Quebec. Annels and Roberts (1989) described framboids associated with a ca. 400 Ma old Au deposit. Scott et al. (2009) described framboids formed from hydrothermal solutions at 175°–220°C in Eocene (ca. 45 Ma) ore veins cutting the ca. 380 Ma sediments in the Carlin Trend, Nevada. Pyrite framboids have also been reported from probable hydrothermal breccias in two andesitic lavas (Love and Amstutz, 1969; Steinike, 1963). These textures were part of a wide spectrum of sub-spheroidal pyrite forms occurring in the andesites, which Love and Amstutz (1969) noted include some forms which might overlap with the framboidal texture *sensu stricto.*

1.3.4. Cryptic Framboids

One consequence of the electronic structure and extremely low solubility of pyrite is that the mineral is extremely tough. Recrystallization of pyrite is difficult to achieve, and pyrite withstands greenschist-facies metamorphism (<450 °C and <1 GPa). In order for pyrite to be recrystallized, temperatures and pressures in the amphibolite facies (>450 °C and <1.2 GPa) are required. At these temperatures the sulfur fugacity required to maintain pyrite stability becomes extremely high and pyrite tends to break down to pyrrhotite and sulfur. Depending on the confining pressure, subsequent cooling of the system leads to pyrite reformation. In these circumstances, pyrite retains no memory of its original state. Otherwise the original form of pyrite can be revealed by etching. Etching pyrite crystals in ancient rocks and ores commonly reveals a long and complex history, and no pyrite should be investigated microscopically without etching or observation of ultrathin (e.g., 100 nm) sections.

Rickard and Zweifel (1975b) originally showed that pyrite grains within 1.8 Ga massive sulfide ores, which had been metamorphosed to greenschist facies, contained remnant framboids. Wacey et al. (2015) reported apparently homogeneous pyrite grains from 560 Ma turbidites which had been metamorphosed to 350 °C and up to 0.7 GPa pressure. These reveal framboidal textures when ultrathin sections were observed with a transmission electron microscope (Figure 7.11 in Chapter 7).

The consequence of the cryptic nature of framboids within pyrite crystals in ancient deposits means that the frequency of framboid occurrences in the geologic record has probably been underestimated.

Since framboids are commonly formed during early diagenesis, the discovery of these cryptic framboids may also contribute to further understanding of the

history of the local environment. An example is shown in Figure 1.12 in a sample
from a mangrove swamp in Broad Sound, Queensland. Five events are recorded
in the sulfides. The first event was the formation of four pyrite framboids. These
were then oxidized, producing oxide rims. Then there was a further generation
of pyrite formation which produced euhedral pyrite crystals. At the same time,
the oxide rims were sulfidized. This was followed by a depositional hiatus when
the concentration of dissolved iron and sulfide was below the pyrite solubility
product. This resulted from acidification of the environment, which subsequently led to the formation of marcasite overgrowths. As a parenthesis here,
it is noted that optical microscopy readily identifies the marcasite through its
extreme anisotropy. Marcasite has the same composition as pyrite and thus it
cannot be distinguished through chemical analysis as, for example, by machine-
based energy dispersive X-ray analyses.

Pyrite framboids commonly act as nuclei for further pyrite growth and,
in the most picturesque examples, these were called "sunflower framboids"
by Sawlowicz (2000) (Figure 1.13). However, this term has subsequently been

Figure 1.12. Environmental history of a mangrove swamp revealed by sulfide
etching: (1) four framboids; (2) oxide rim; (3) pyrite overgrowth; (4) dissolution;
(5) marcasite overgrowth (see text).

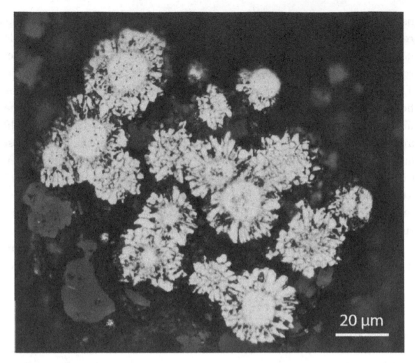

Figure 1.13. Classic "sunflower" texture: pyrite laths radiating around a central framboidal core.

Photomicrograph from the 700 Ma Tapley Hill Formation, South Australia, by Indrani Mukherjee.

used to describe a variety of textures (e.g., Merinero et al., 2017). Anyway, these textures only appear like sunflowers in section; in 3D they appear as small spheroids of radiating pyrite and would therefore not be reported as framboids.

Repeated overgrowths of successive generations of pyrite may mask evidence for original framboids. Successive overgrowths on pyrite may be a particular problem to the study of framboids in hydrothermal deposits. An example of this is shown in Figure 2.5 in Chapter 3, where a typical hydrothermal vein colloform pyrite texture is revealed to have numerous framboids which appear to act as sites for subsequent overgrowths as repeated pulses of hydrothermal solutions deposit successive layers of pyrite. Apart from the large framboids acting as cores to the developing texture, small framboids (e.g., 2 μm in diameter) are also scattered throughout this sample, including some within the outermost pyrite crystals of the texture.

2

Framboid Sizes

Framboids vary in size from less than 2 to over 200 μm in diameter (e.g., Love and Amstutz, 1966; Rickard, 1970; Rickard, 2019b; Sassano and Schrijver, 1989; Wignall and Newton, 1998). The mean size of framboids is 6.0 μm and 95% range between 2.9 and 12.3 μm in diameter with 99% between 2.0 and 17.7 μm in diameter.

The largest framboid may be the 250 μm framboid reported by Sweeney and Kaplan (1973) from a marine sediment off California. Various synthetic and natural sub-spheroidal aggregates of nanoparticulate pyrite have been reported (e.g., Butler and Rickard, 2000; Ohfuji and Rickard, 2005; Sawlowicz, 1993; Yücel et al., 2011) and have been called *protoframboids* or *nanoframboids* (Figure 2.1(a)). These aggregates are less than 0.2 μm in diameter and are constituted by particles with sizes as small as 2 nm. Large et al. (2001) used the term *protoframboid* for spherules between 3 and 17 μm in diameter consisting of particles with estimated sizes between 100 and 500 nm and indefinite iron sulfide compositions. These particles appear to have been at the limit of resolution of the scanning electron microscope (SEM) used by these investigators and may be nanoparticle aggregates. At the other end of the scale, spheroidal groups of framboids, called *Rogenpyrit* (Fabricius, 1961), *polyframboids* (Love, 1971) *multiframboids* (Soliman and El Goresy, 2012), which may range up to 900 μm in diameter, have been widely reported in nature (Figure 2.1(b)). Love and Amstutz (1966) also recorded *irregular aggregates* of pyrite microcrystals, and Wignall and Newton classified these as "type 2 framboids." These *irregular aggregates* do not fall within the formal definition of framboids, described earlier.

Much of the information about framboid sizes derives from framboids in sediments and sedimentary rocks: no systematic studies of higher temperature, hydrothermal framboids have been made, although these were the first framboids to be defined. Additionally, as discussed in Chapter 1, framboids in sedimentary rocks may not necessarily be formed at ambient temperatures, and hydrothermal fluids, such as those emanating from sub-aqueous vents, may have contributed to the framboid population of sedimentary rocks. This implies that sedimentary framboid size-frequency statistics may include framboids formed by higher temperature processes.

Framboids. David Rickard, Oxford University Press. © Oxford University Press 2021.
DOI: 10.1093/oso/9780190080112.003.0002

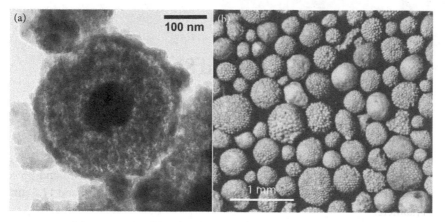

Figure 2.1. Extreme sizes of framboid-like objects: (a) sub-spherical cluster of pyrite nanoparticles from the Lau hydrothermal vent; (b) *rogenpyrit* from Rhaetic-Lias boundary mudstones in the German Alps.

(a) Reprinted by permission from *Springer Nature, Geoscience,* Hydrothermal vents as a kinetically stable source of iron-sulphide-bearing nanoparticles to the ocean. Yücel, M., Gartman, A., Chan, C. S., and Luther, G. W., 2011. (b) Reprinted by permission from *Springer Geologische Rundschau, Die Strukturen des Rogenpyrits (Kössener Schichten, Rät) als Beitrag zum Problem der Vererzten Bakterien.* Fabricius, F. 1961.

2.1. Framboid Size Distributions

2.1.1. Log-normal Distribution

Rickard (2019b) confirmed the original conclusion of Love and Amstutz (1966) that populations of sedimentary framboids commonly display a uni-modal log-normal distribution (Figure 2.2). There are obviously many statistical distributions that might fit framboid size-frequency measurements, but the log-normal distribution has a solid scientific heritage. It is based on the multiplicative hypothesis of elementary errors that states that if a random variation is the product of several random effects, a log-normal distribution must be the result. It seems that this sort of quasi-exponential relationship is fundamental to many natural processes (e.g., Heath, 1967). In the case of framboids, it might be reasonably assumed that a complex series of processes is involved in their formation and that this would be reflected in a log-normal size distribution.

One of the features of log-normal distributions is that multiplying log-normal random variables gives a log-normal product. Therefore, the geometric means and standard deviations are log-normally distributed and can be themselves summarized by global geometric means and standard deviations.

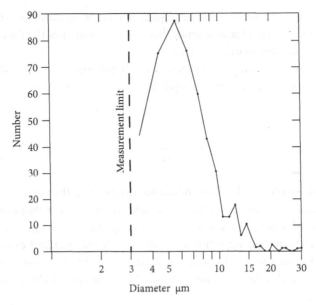

Figure 2.2. Size range of a typical framboid population measured by optical microscopy showing log-normal distribution.
Compiled from data in Love and Amstutz (1966).

A further consequence of the log-normal distribution is that apparently normal framboid size-frequency populations (e.g., Smolarek et al., 2017; Zhang et al., 2018) are expected to occur in a set of log-normal distributions (Limpert and Stahel, 2011) and these normal distributions are intrinsically log-normal.

A further test of the validity of the assumption of log-normal distributions for framboid size-frequency populations is the "95% range test" of Limpert and Stahel (2011). This is simply testing the minimum range sizes of two standard deviations (2σ) of the normal distribution. The 2σ range includes 95% of the observations and the application to reported framboid size-frequency populations shows that these have impossible negative diameters at the lower limit if the normal distribution is applied. That is, at least 5% of the framboids in the population display diameters <0 μm, which is obviously illogical. Limpert et al. (2001) go further to point out that additive statistics are generally flawed for natural objects such as framboids since continuing to subtract arithmetic standard deviations from the mean will usually end up with impossible negative dimensions. Thus, $\pm 3\sigma$ predicts the dimensions of 99% of the framboids, $\pm 4\sigma$ 99.9%, and so on. This suggests that additive statistics and, consequently, fitting the population to a normal distribution will ultimately predict a logical inconsistency. This also means that using additive statistics to predict positive dimensions for ± 1σ, or 68% of the population, is intrinsically flawed, since it incorporates the logical inconsistency that a small number of the framboids are predicted to have negative dimensions.

By contrast, the ranges predicted by geometric standard deviations are always positive since, in multiplicative statistics the range is given by Equation 2.1.

$$\frac{\bar{x}^x}{\sigma^n}$$ 2.1

where n refers to the number of standard deviations from the mean.

In fact, nearly all reports of framboid size-frequency distributions prior to Rickard's (2019b) meta-analysis are invalid and fail the 95% range test. It should be noted that framboid populations with a geometric standard deviation less than about 1.2 may appear to be normally distributed but are also intrinsically log-normal, and this remains the preferred interpretation of these populations (Rickard, 2019b).

2.1.2. Framboid Mean Diameters

The consequence of the log-normal size-frequency distribution for framboids is that their mean sizes and size ranges previously reported in the literature are mostly wrong. Framboid mean sizes are more accurately represented by the multiplicative (i.e., geometric) mean ($\bar{x}^*$) and standard deviation (σ^*). These geometric parameters are used throughout this book when referring to the average sizes of framboids and their size ranges. The various expressions for the log-normal and normal distributions are shown in Equations 2.2–2.5 in Table 2.1.

Table 2.1 Equations for the Statistical Parameters of the Normal and Log-normal Distributions

Arithmetic mean	$\bar{x} = (x_1 + x_2 + \ldots + x_n) / n$	2.2
Geometric mean	$\bar{x}^* = \sqrt[n]{(x_1 x_2 \ldots x_n)}$	2.3
Arithmetic standard deviation	$\sigma = \sqrt{[\Sigma (x_1 - \bar{x})^2 / (n-1)]}$	2.4
Geometric standard deviation	$\sigma^* = \exp{(\sqrt{[\Sigma(\ln{(x_1 / \bar{x}^*)})^2 / (n-1)]})}$	2.5

From Rickard (2019b).

The geometric mean is related to the arithmetic mean by Equation 2.6.

$$\bar{x}^* = \frac{\bar{x}}{\sqrt{\left(1+\left(\dfrac{\sigma}{\bar{x}}\right)^2\right)}}$$ 2.6

The geometric standard deviation is related to the arithmetic standard deviation by Equation 2.7.

$$\sigma^* = \exp\left(\sqrt{\ln\left(1+\left(\dfrac{\sigma}{\bar{x}}\right)^2\right)}\right)$$ 2.7

Equations 2.2–2.7 have several consequences for framboid size measurements. First, the geometric standard deviation is a ratio and therefore dimensionless. The geometric standard deviation is smaller than the arithmetic standard deviation (Equation 2.7). This means that the σ^* variation in framboid populations is low compared with σ. Indeed, Rickard (2019b) reported that values from some 13 published sets of sedimentary framboid–size frequency data form a narrow spread of around 2 (e.g., Figure 2.4). They are therefore nondiscriminatory and do not vary significantly between populations. The reason for the limited variation in σ^* values for framboid populations of all geologic ages is probably related to the framboid formation mechanism (see Chapter 13).

Second, the arithmetic mean, $\bar{x}$, is larger than the geometric mean, $\bar{x}^*$. This suggests that published estimates of average framboid sizes based on normal distributions are too high. The increase in x relative to x^* may be related to the magnitude of the geometric standard deviation, σ^* (e.g., Equation 2.8).

$$\sigma^* \sim 1.7, \bar{x} \sim 1.15\bar{x}^*$$ 2.8

That is, the mean diameter of a framboid population with a geometric standard deviation of 1.7 is some 15% lower than the reported values. The size ranges also differ. The lower limits of the 95% (e.g., 2σ) ranges turn increasingly negative with increasing σ, and the framboid size-frequency normal distribution fails the 95% free-range test. However, it remains positive for σ^* and represents a real value. The geometric upper limit of the 95% ranges exceeds the arithmetic limit by some 17% for $\sigma^* = 1.7$ and 25% for $\sigma^* = 2.5$ (Limpert and Stahel, 2011) giving a wider spread for the diameters of 95% of the framboids.

2.2. Measurement Errors

The definition of framboids (Ohfuji and Rickard, 2005) provides a limit on the minimum sizes of framboids. The constituent microcrystals of framboids are usually up to 1 μm in size. This means that a framboid, consisting of a number of microcrystals by definition, will not normally be less than 1 μm in diameter. This size limitation is close to the modal values of many measured framboid distributions of 5 μm or less (cf. Figure 2.2) since the chances of sectioning framboids through the maximum diameter are low: most sections of a 5 μm framboid will be < 5 μm in diameter. Electron microscopic measurements of framboids are robust enough to detect the sharp decrease in framboid diameters less than the modal value, which is a consequence of the log-normal distribution. This decrease is important since it distinguishes framboids from single crystals which show a continuous exponential increase in relative frequency to the nanoparticulate sizes characteristic of the constituent framboid microcrystals. This difference of the size distribution of framboids and single crystals has been used as an aid in discriminating populations using automated methods (e.g., Merinero et al., 2017). However, the smallest framboids may not be consistently detected by routine optical measurements since sections are likely to be less than 2 μm in diameter. The underreporting of the number of the smallest framboids will affect the shape of the framboid diameter-frequency distribution, leading to a greater uncertainty of the predicted size ranges and mean diameters (Farr et al., 2017). This is important in framboid studies since the distribution usually approaches log-normal and the smaller sizes dominate the populations. The sections through small framboids are likely to be even smaller, and these smallest sizes are also often missed or there is a lower limit to the observations.

Rickard (2019b) concluded that the previously reported uncertainties in the measurements of framboid diameters of within ± 10% (e.g., Wignall and Newton, 1998; Wilkin et al., 1996) are reasonable estimates of the precision of the measurements. Measurements made with the SEM microscope are more precise than the older measurements obtained by optical microscopy because of enhanced image resolution. As noted earlier, the effect is particularly marked at the lower end of the range of framboid measurements since the size-frequency measurements of framboid populations are skewed toward smaller sizes. Indeed, this contributed to an earlier discussion about whether framboid populations were merely exponentially distributed or were log-normal (e.g., Kalliokoski and Cathles, 1969; Love and Amstutz, 1966; Vallentyne, 1963) that was finally resolved by the demonstration that framboid numbers decreased at small sizes. The reason for the misconception was that framboids <3 μm in diameter were difficult to resolve accurately by the optical microscopic methods available at that time.

These measurement errors are random errors, which can be ameliorated by measuring large numbers of framboids. Rickard (2019b) concluded that at least 30 framboid diameters need to be measured and, ideally, 100 measurements is a good target.

A more serious source of error is the stereological error: the error in converting the measurements of 2D sections to 3D spheres. This is more serious because it is a systematic error. Nearly all framboid diameters are measured in section. Rarely are framboids separated from the rock and then measured, although Jack Vallentyne did this in his original study of framboids in lake sediments (Vallentyne, 1963). He filtered the sediments through a pillowcase, but warned that new pillowcases were not as effective as older ones. For a group of homogenous spheres, the average measurement is ca. 80% of the actual 3D diameter (Wicksell, 1925). However, framboids are not all the same size, and the ratio of the average measurement to the actual diameter varies according to the ratio of the standard deviations of the populations and the nature of the size distributions. Other methods of the measurement of size include (1) the volume-weighted mean diameter, in which random points are chosen in the section and the diameters of those framboids are recorded where they intersect a random point; and (2) the surface-related mean diameter, where the mean diameter is the probability proportional to their surface area. The correction factor in converting 2D diameters to 3D diameters is greater than 80% in both cases (88% and 85% respectively; Farr et al., 2017). Thus, the approximation of 80% suggested by the mathematics of monodisperse spheres encompasses the likely uncertainties in the measurements. In practical terms, this means that a framboid population with a reported mean size of 5 μm may display a mean diameter of 4 μm in section. The stereological error is rarely applied to published framboid size-frequency measurements. Wilkin et al. (1996) recognized the error but concluded that it was within the general uncertainty of the measurements. Other published reports rarely mention whether the correction factor was applied or not. The consequence is that, in the absence of any further information, the estimates of framboid diameters may be too low by a factor of up to 1.2.

The stereological error becomes even more significant when considering framboid volumes. Since the volume increases as the cube of the radius, the stereological error for estimates of framboid volume is close to 2. That is, volumes calculated on measurements of framboid diameters in sections need to be doubled for an estimate of the true volume.

2.3. Framboid Sizes

A summary of framboid sizes reported in the literature is presented in Table 2.2. Framboid distributions are overwhelmingly dominated by sedimentary framboids. As shown in section 1.3, sedimentary framboids constitute over 99% of the total framboidal pyrite on Earth. In terms of numbers, there are at least 10^{30} framboids on Earth, of which more than 10^{29} occur in sediments and sedimentary rocks. Against this background, the dominance of sedimentary framboids in determining the average size of framboids is clear. However, my statistical colleagues may hiccup when they see these numbers: even if I have data on the diameters of over 72,000 framboids, this is a wholly insignificant statistical sample of the entire population of framboids on Earth. Systematic measurements of framboid diameters are scarce or lacking for a number of framboid environments and, as discussed in the following, this seriously limits our understanding of the distribution of framboid sizes populations on Earth.

Some reported size ranges, lacking means or any other statistical analyses, are included to indicate the spread of framboid sizes that have been observed. The data are mainly derived from Rickard's (2019b) meta-analysis, where the geometric means have been calculated from the original arithmetic data via Equation 2.6. In the absence of reported σ values I have estimated these by means of the range rule (Equation 2.9).

$$\sigma = \frac{(max - min)}{4} \qquad 2.9$$

Where *max* and *min* refer to the maximum and minimum reported framboid diameters. This approximation works for normal distributions because 95% of the framboids fall with $\pm 2\sigma$. In some studies, I have extracted the statistical data by digitizing published histograms (using GRAPHCLICK™) and analyzing the data with XLSTAT™.

In Table 2.2, I have grouped the framboid measurements into two main groups and six broad categories, as described in the following.

2.3.1. Sedimentary Environments

- Sedimentary rocks
- Marine sediments
- Freshwater sediments.

As used throughout this book, a distinction is made between *sedimentary rocks*, being lithified or consolidated sediments, and *sediments*, being unlithified or unconsolidated sediments. Sedimentary rocks are uniformly ancient sediments, but unconsolidated sediments are not necessarily Holocene in age. Since the portmanteau terms *recent* and *modern* are now discouraged stratigraphically,

Table 2.2 Pyrite framboids sizes in different environments in terms of size ranges and computed geometric mean diameters (μm)

	Age	Ma	Location	Range (μm)	Mean (μm)	Source
Sedimentary rocks						
Black shales	Miocene	12	Japan	2–50	–	4
Siltstones	Cretaceous–Paleogene	65	James Ross Basin	1–31	4.3[a]	19
Shale	Upper Cretaceous	70	Egypt	60–200	–	22
Black shale	Upper Jurassic	145	Mexico	2–14	5.9	11
Mudstone	Upper Jurassic	155	UK	1–22	3.7	27
Mudstone	Lower Jurassic	180	Japan	1.5–35	5.1	3
Shale	Early Jurassic	185	UK	1–28	3.7	27
Black shale	Upper Triassic	215	Central China	3–38	12	31
Claystone and cherts	Lower Triassic	247	Central Japan	2–40	4.7[a]	25
Black shale	Early Triassic	250	South China	4–32	4.3	24
Carbonates	Early Triassic	250	South China	2–19	2.3	24
Breccias	Early Triassic	250	South China	2–16	3.4	24
Conglomerates	Early Triassic	250	South China	2–13	3.2	24
Carbonate	Permo-triassic	252	South China	1–18	6.3	7
Glaciomarine	Middle Permian	265	SE Australia	6–12	9.9[b]	1
Oil shales	Mississippian	330	Scotland	2–30	8.3[a]	8
Black shale	Late Devonian	375	New York	1–38	5.5	6
Shale	Devonian	390	Tennessee	2–39	5.9	8, 30
Black shales	Silurian	425	Poland	2–18	5.6	21
Black shales	Lower Cambrian	525	South China	1–15	4.6	33
Black shales	Lower Cambrian	525	South China	1–11	5.1	34
Marine sediments						
Methane zone	Holocene+		South China sea	20–40	–	32
Mud volcanoes	Holocene+		Gulf of Cadiz	4–44[e]	10.5[e]	12

(Continued)

Table 2.2 *Continued*

	Age	Ma	Location	Range (µm)	Mean (µm)	Source
Fjord	Holocene+		Framvaren Norway	3–10	5.9	30, 20
Nanofossil ooze	Holocene+		Angola basin	3–30	–	18
Sapropelic mud	Holocene+		Black Sea	2–9	4.8	30
Coccolithic ooze	Holocene+		Black Sea	2–19	4[b]	30
Continental slope	Holocene+		Peru margin	2–14	5.8	30
Freshwater						
Lake sediments	Holocene+			3–32	–	26
Lake sediments	Holocene+		Rhode Island	1–11	–	29
Peat bog	Holocene		Finland	3–45	–	14
Canal sediments	Holocene		UK	2–20	–	36
Salt marsh	Holocene+		Delaware	4–19[a]	9.1[a]	30
Salt marsh	Holocene+		Virginia	3–13[a]	6.7[a]	30
Estuarine muds	Holocene+		Rhode Island	3–8[a]	4.8[a]	30
Water column						
Euxinic water	Present day		Black Sea		~5[c]	13
Euxinic water	Present day		Pettaquamscutt estuary, RI	8.5	~5[c]	29
Hydrothermal deposits						
Shallow marine vent	Holocene+		West coast Mexico	1–34	5.7	15
Lead-zinc veins	Jurassic	180	Central Wales	1–40	6.8[c]	16
Lead-zinc ore	Carboniferous	340	Ireland	1–37	11.2	35
Mineralized sediments						
Mineralized peat	Holocene		Cornwall, UK	15–67	43.2[c]	2
Ore mud	Tertiary	10	Tynagh Ireland	<40	–	17

Table 2.2 *Continued*

	Age	Ma	Location	Range [e] (μm)	Mean (μm)	Source
Ore shale	Permian	255	Mansfeld Germany	3–20	8.3[c]	9
Ore shale	Devonian	380	Rammelsberg Germany	3–17	4.3[a]	8, 30
Ore shale	Lower Proterozoic	1650	Mt. Isa Australia[d]	1–15	~2	10

[a] computed from reported arithmetic means and standard deviations values (Equation 2.6.).
[b] geometric mean estimated from reported arithmetic means and ranges using range rule (Equation 2.9) and Equation 2.6.
[c] see text.
[d] pyrite spherules (see text).
[e] computed directly from machine measurements. All other data from computations in Rickard (2019b). Ages are listed in Ma as approximate guides. Holocene refers to the present era (11500–0 a).

Sources: 1. Gong et al. (2007); 2. Hosking and Camm (1980); 3. Izumi et al. (2018); 4. Kato (1967); 5. Large et al. (2001); 6. Lash (2017); 7. Liao et al. (2017); 8. Love (1957); 9. Love (1962); 10. Love and Zimmermann (1961); 11. Martinez-Yanez et al. (2017); 12. Merinero et al. (2017); 13. Muramoto et al. (1991); 14. Papunen (1966); 15. Prol-Ledesma et al. (2010); 16. Raybould (1973); 17. Rickard (1970); 18. Schallreuter (1984); 19. Schoepfer et al. (2017); 20. Skei (1988); 21. Smolarek et al. (2017); 22. Soliman and El Goresy (2012); 23. Suits and Wilkin (1998); 24. Sun et al. (2015); 25. Takahashi et al. (2015); 26. Vallentyne (1963); 27. Wignall and Newton (1998); 28. Wilkin and Arthur (2001); 29. Wilkin and Barnes (1997b); 30. Wilkin et al. (1996); 31. Yuan et al. (2017); 32. Zhang et al. (2014); 33. Zhang et al. (2018); 34. Zhou and Jiang (2009); 35. Graham (1971); 36. Large et al. (2001).

I have referred to these as Holocene+, suggesting that the samples may come from sediments of Quaternary age or even older (e.g., >2.5 Ma): with the slow sedimentation rates (e.g., ≤ 1 cm 10^{-3} a) that characterize the fine-grained sediments which often include abundant framboids, the age of the "modern" sediment rapidly increases with depth below the sediment-water interface.

2.3.2. Other Environments

- Water column
- Hydrothermal
- Mineralized sediments

Framboids from the water column refer specifically to framboids recovered from contemporary waters and do not include framboids recovered from sediments, both ancient and modern, which may have originally been formed in the water column. I have distinguished mineralized sediments from ordinary sediments because they contain relatively large concentrations of metals and are often the sites of base metal mines. These mineralized sediments include both sediments and sedimentary rocks, according to the definitions noted earlier. The potential origin of the framboids in mineralized sediments ranges from end-member sedimentary processes, formed at ambient temperatures, to hydrothermal processes, formed at more elevated temperatures, usually distally with respect to hydrothermal vents.

2.4. Sedimentary Framboids

2.4.1. Framboids in Sedimentary Rocks

Authors have used various sedimentological terms to describe the rocks hosting the framboids and, in Table 2.2, I have used the broad descriptions provided in the reports. Sedimentary rocks characterized by large framboids are included here for illustration even though statistical data are not provided (e.g., Kato, 1967; Soliman and El Goresy, 2012). Table 2.2 shows that there is no obvious relationship between rock type and framboid size. For example, framboids in the ca. 220 Ma upper Triassic black shales of central China described by Yuan et al. (2017) have a mean diameter of 12 μm compared with the 5.1 μm mean size of the other six occurrences listed in Table 2.2.

Table 2.3 lists framboid diameters in sediments and sedimentary rocks from various geologic eras. The data are extracted from several publications where the stratigraphic assignment of the samples is defined. The data are non-critical. Holocene+ refers to framboids in modern unconsolidated sediments which may extend to older sediments than simply Holocene, and samples from various sedimentary environments are included. For example, the Holocene+ samples also include framboids from cold seeps and lake sediments. The table is based mainly on Rickard (2019b), and several further examples have been added so that the total number of measurements is 67 106. The analysis shows that the mean diameter of framboids in sediments and sedimentary rocks is 5.7 μm and 95% range between 2.6 and 12.3 μm in diameter. There is no systematic change with geologic age: the mean size of Holocene+ framboids is 6.0 μm, and they range from 2.6 to 14.2 μm in diameter. The slightly higher mean value is due to the larger proportion of measurements of framboids in freshwater Holocene+ sediments: 50% of the Holocene+ framboids are from freshwater sediments which, as noted in the following, have greater mean diameters. Framboids in Holocene+ marine sediments have mean diameters of 5.9 μm (see section 2.4.2), which is indistinguishable from that of ancient sedimentary rocks.

Table 2.3 Framboid sizes in sedimentary rocks of various ages: age in millions of years before present Ma, geometric mean size $\bar{x}^*$, geometric standard deviation σ^*, 95% confidence interval min and max, number of measurements, n. Data from Rickard (2019b) complemented by additional reports.

Age Ma	Era	$\bar{x}^*$ μm	σ^* μm	main μm	max μm	n	Reference
	Holocene+	6.0	1.5	2.6	14.2	10339	Merinero et al., 2017, 2008; Suits and Wilkin,1998;Vallentyne, 1963; Wilkin et al., 1996.
	Paleogene	4.8	1.3	2.8	8.5	415	Schoepfer et al., 2017.
100	Cretaceous	5.0	1.3	2.8	9.1	1473	Martinez-Yanez et al., 2017; Schoepfer et al. 2017.
200	Jurassic	5.0	1.4	2.5	9.9	2652	Izumi et al. 2018; Martinez-Yanez et al., 2017; Wignall et al. 2010; Wignall and Newton, 1998.
	Triassic	6.6	1.4	3.2	13.9	9522	Bond and Wignall, 2010; Liao et al., 2017; Sun et al., 2015; Takahashi et al., 2015; Wignall et al. 2010; Yuan et al., 2017.
300	Permian	6.6	1.6	2.6	16.2	19582	Bond and Wignall 2010; Wei et al., 2015,2016; Wignall et al. 2010.
	Carboniferous	8.6	1.3	4.8	15.6	157	Love, 1957.*
400	Devonian	6.0	1.3	3.5	10.3	3827	Lash, 2017; Love and Amstutz, 1969; Marynowski et al., 2011; Wilkin et al., 1996.
	Silurian	5.0	1.3	2.8	8.8	2069	Smolarek et al., 2017.
500	Cambrian	4.7	1.3	2.6	8.4	6865	Zhang et al., 2018; Zhou and Jiang, 2009.
600	Ediacaran	5.2	1.3	3.3	8.2	10205	Wang et al., 2012.
	Total	5.7	1.5	2.6	12.3	67106	

There is no significant difference between the mean diameters of framboids in ancient and modern sediments. Thus 95% of Holocene+ framboids have mean sizes of 2.6–14.2 μm and 95% of framboids in ancient sediments have mean diameters of 2.6–12.3 μm.

The lack of any significant difference between the mean diameters of ancient and modern framboids constrains the processes involved in framboid formation. Once formed, framboids do not grow, although they may be infilled, experience mineral overgrowths, and, ultimately, recrystallize under high PT conditions. The result is that ancient framboids are indistinguishable from framboids in modern sediments and that the physical and chemical characteristics of modern framboids can be interpolated to ancient sedimentary environment conditions.

2.4.2. Framboids in Marine Sediments

Systematic measurements of framboids in marine sediments have been reported by Wilkin et al. (1996) and Merinero et al. (2017). Wilkin et al. (1996) measured the diameters of over 5000 framboids from the Black sea, the Framvaren fjord, and the Peru margin. The mean size of framboids from marine sediments listed by Wilkin et al. (1996) is 5.3 μm, and 95% ranged between 2.4 and 11.5 μm. Merinero et al. (2017) reported the diameters of 378 framboids associated with methane-derived carbonate pipes in the Gulf of Cadiz. They reported a mean diameter of 10.5 μm, with a 95% spread between 6.2 and 17.6 μm. Similarly, large size ranges have been reported from a methane-related environment in the South China Sea by Zhang et al. (2014) (Table 2.2). Schallreuter (1984) also reported a wide range of framboid diameters from a nanofossil ooze from the Angola basin but, by contrast to the methane-related framboids, the reported minimum size was just 3 μm (Table 2.2). The conclusion is that framboid mean diameters vary considerably in different modern marine environments. Wilkin et al. (1996) related the variations at the lower end of the range to variations in the oxygenation state of the contemporary waters. Framboids related to methane seeps may have an altogether different average size distribution, and it appears that framboid populations in calcareous oozes in the deep sea and nearshore marine systems may also vary.

2.4.3. Framboids in Non-marine Sediments

I have classified a diverse group of sedimentary environments as non-marine where they do not occur in normal marine systems. These environments include lakes, estuaries, and wetlands including salt marshes.

Some of the earliest systematic measurements of framboid diameters were made in lake sediments (Vallentyne, 1963). These measurements also avoided the stereological error since they were made on framboid concentrates rather than sections. However, they were made with a routine transmitted light microscope and the measurements of the smallest framboids consequently suffered. The mean size of framboids in non-marine sediments is 6.9 μm. Papunen (1966) reported an arithmetic average diameter of 30 μm for framboids from a peat bog in southeastern Finland. Recomputing these data suggests a geometric mean diameter of 28.4 μm and a 95% range of 14.6–55 μm. This value is large and distinctly different from most other framboid populations, where the average size is usually <10 μm. However, similar mean large framboid diameters were reported in the systematic study of peat associated with ore mineralization by Hosking and Camm (1980), as discussed later. Indeed, as described in section 1.2.1 of Chapter 1, the original description of framboids came from peat samples (Früh, 1885) probably because they were relatively large and more easily discernible by the microscopic technology available at that time. The significantly larger diameters of pyrite framboids in peat is predicted by the framboid formation model of Rickard (2019a), which relates framboid size in part to the concentration of Fe and S in solution. In acidic peat waters, the concentration of dissolved Fe and S is not limited by the solubility of mackinawite, metastable FeS_m, which constrains the dissolved Fe and S in ambient systems with pH >~6, such as seawater, to ~10^{-6} M.

Wilkin et al. (1996) reported the mean diameters of 2068 framboids from salt marsh sediments of the eastern United States. Their data suggest a geometric mean diameter of 7.9 μm and a 95% range of 2.8–22.7 μm. These values are indistinguishable from the recomputed values for framboids from lake sediments, which show a geometric mean value of 7.6 μm and a 95% range of 2.7–21.2 μm for 1601 measurements.

The mean diameter of framboids in freshwater sediments is significantly larger than that of framboids in marine sediments. This is consistent with the framboid formation model outlined in Rickard (2019a), which predicts that framboids formed in freshwater environments will be larger than those formed in marine environments. The availability of dissolved Fe and S species in sulfidic waters at ambient temperatures is controlled ultimately by the solubility of FeS_m, mackinawite. This is pH dependent at pH <~6 (Rickard, 2006) which is characteristic of many freshwaters and lower than the average pH of seawater. Therefore, there is more dissolved Fe and S available in freshwater systems and the framboids can grow to larger sizes in the same time period. The results from more extreme acidic systems, such as peats, is consistent with this conclusion.

2.5. Framboids in Other Environments

2.5.1. Framboids in Water Columns

The idea that framboids could be less dense than water and float was first proposed by Rickard (1970). However, there has been no systematic study to date of framboid size measurements from framboids occurring in any water column. Degens et al. (1972) first described framboids suspended in a water column in Lake Kivu in East Africa. Subsequently, framboids averaging 10 μm in size were reported from suspended solids in the water column of Kau Bay (Middelburg et al., 1988) and from the euxinic waters of the meromictic Powell lake in British Columbia (Perry and Pedersen, 1993). Muramoto et al. (1991) originally described pyrite framboids in the anoxic waters of the Black Sea and suggested that they formed primarily near the oxic-anoxic interface. They reported that 100 pyrite framboids in the water column of the Black Sea had an average diameter of 6 μm. Skei (1988) reported that pyrite framboids in the anoxic waters of Framvaren Fjord ranged in size between 3 and 10 μm and also suggested that they formed near the redoxcline. Wilkin and Barnes (1997b) reported the diameters of 27 framboids from 3 samples of sulfidic water from the lower basin of the Pettaquamscutt River (Rhode Island).

There are two potential sources for framboids in the water column: (1) framboids formed in the water column (e.g., Muramoto et al., 1991), and (2) framboids re-suspended from sediments (e.g., Middelburg et al., 1988). Most interest has been focused in framboids forming in the water column. In order for framboids to be formed within the water column, the water must be sulfidic. The redoxcline, separating sulfidic and non-sulfidic water, is usually located at or near the sediment-water interface. In some environments it rises into the water column itself and euxinic conditions prevail. The word *euxinic* derives from the Latin name for the Black Sea, *Pontus Euxinus* (see discussion in Rickard, 2015, p. 215) and the Black Sea has been a major target for studies of framboids formed in the water column. In the most radical interpretation, Lyons and Berner (1992) concluded that almost all the pyrite in Black Sea sediments is largely formed at or above the sediment-water interface.

In the absence of direct systematic studies of framboids formed in the water column, investigations have been centered on framboids found in the sediments which may have been formed in the water column. These framboids are described as syngenetic (Raiswell and Berner, 1985) since they are formed at the same time as the sediment. Wilkin et al. (1996) originally proposed that framboids formed in the water column were smaller on average than those formed in sediments. They correlated the sizes of framboids in Black Sea sediments with contemporary periods of euxinia. Recalculation of Wilkin et al.'s (1996) data for syngenetic framboids from Black Sea sediments shows that the geometric mean of the framboid diameters formed in euxinic waters is 4.7 μm (Rickard, 2019b). Wilkin et al. (1996) also noted that the sedimentary framboid populations which correlated with euxinic conditions in the Black Sea appeared to have a small size range. They originally calculated this in terms of additive statistics, but recalculation by Rickard (2019b) showed that the geometric standard deviations in both syngenetic and diagenetic framboid populations were relatively constant and non-discriminatory. The predicted size range for 95% of syngenetic Black Sea framboids is 2.4–9.1 μm (Rickard, 2019b).

By contrast, Schieber and Schimmelmann (2007) found framboid size distributions typical of euxinic systems in Santa Barbara Basin sediments under sub-oxic bottom waters. They concluded that framboid size distributions in modern sediments and black shales have no diagnostic value for paleo-water column oxygenation: if a sub-population of these framboids were formed in the water column, they left no remnant size signature. They further showed that the variation in framboid sizes between successive varves in the sediments did not correlate with historic changes in water column oxygenation. They showed that framboid formation tends to completion after five years of sedimentation and concluded that the framboids formed entirely within the sediment. Smolarek et al. (2017) suggested that there was also a problem time-averaging samples in environments where euxinia were brief. They suggested that this effect is particularly notable in environments subject to high-amplitude redox changes. For example, the Salton Sea, a hypereutrophic lake in southern California, experiences euxinic conditions in summer and oxic conditions in winter. The framboids in its sediments show a narrow size-frequency distribution around 5 μm that apparently records the euxinic conditions but not the oxic conditions (de Koff et al., 2008).

2.5.2. Framboid Size-Frequency Distributions and the Oxygenation State of Paleo-Water Columns

Wilkin et al. (1996) proposed that framboid size-frequency measurements of framboid populations of sedimentary rocks could be used to probe the distribution of euxinia in paleo-waters. The original framboid size-frequency proxy (Wilkin et al., 1996) suggested than euxinic and non-euxinic environments could be discriminated in terms of a locus on a plot of the arithmetic mean diameter versus the standard deviation. As mentioned earlier, Rickard (2019a) showed that a consequence of the application of multiplicative statistics to framboid size-frequency distributions was the decrease in spread of the geometric standard deviation relative to the arithmetic standard deviation. The consequence is that the plot of the framboid geometric mean diameter versus the geometric standard deviation shows considerable overlap between euxinic and oxic-dysoxic populations and it is therefore not an effective discriminator for the oxygenation conditions of paleo-water columns (Figure 2.3).

Wignall and Newton (1998) suggested that further insights into framboid size-frequency distributions in ancient environments might be provided by comparing the framboid data with paleoecology. That is, information on bottom water oxygenation conditions could be determined by the contemporary nektobenthos. This approach has been expanded by comparing framboid size-frequency distributions with whole-rock pyrite-S isotope analyses and trace element data as further proxies (e.g., Hammarlund et al., 2012; Marynowski et al., 2011; Smolarek et al., 2017; Sun et al., 2015; Takahashi et al., 2015; Wei et al., 2015; Zhang et al., 2018). Even so, these extensive investigations have hitherto provided little information about any variations in sizes of framboids potentially formed within ancient water columns since the putative ancient syngenetic framboids have been primarily identified on the basis of Wilkin et al.'s Holocene+ Black Sea data.

Since Wilkin et al.'s (1996) original contribution various loci have been suggested between increasingly refined oxygenation conditions, including oxic, various low-oxygen categories, and anoxic as well as euxinic (e.g., Bond and Wignall, 2010; Liao et al., 2017; Martinez-Yanez et al., 2017; Wei et al., 2015; Wei et al., 2016). The effectiveness of framboid size-frequency measurements

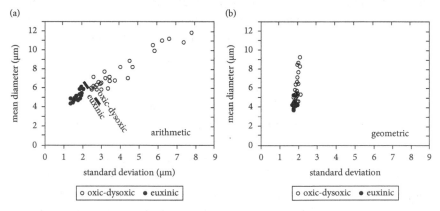

Figure 2.3. (a) Wilkin et al.'s (1996) original data distinguishing diagenetic framboids formed in sediments under oxic-dysoxic water from syngenetic framboids formed in a euxinic water column on a plot of arithmetic mean diameter versus standard deviation. (b) The same data on a plot of geometric mean diameter versus geometric standard deviation.

in discriminating between these more detailed oxygenation conditions has been criticized by Smolarek et al. (2017), who showed that distinguishing between euxinic and dysoxic paleo-water column oxygenation states was possible, whereas discriminating between euxinic and sub-oxic conditions is less clear-cut.

Rickard (2019a) noted that the apparent oxic trends observed on arithmetic mean diameter versus standard deviation plots may be partly a consequence of their log-normal distributions with relatively constant geometric standard deviations (Figure 2.4).

The results of framboid size-frequency measurements provide a statistical probability that the framboid population includes a significant fraction of syngenetic framboids. In simple terms, for example, the calculated size range of 95% of syngenetic Black Sea framboids of 2.4–9.1 µm predicts that 5% of syngenetic framboids will have a mean diameter greater than 9.1 µm. That is, there is a 1 in 20 chance of a framboid population with a mean diameter of >9.1 µm being formed under euxinic conditions. But there is still a 1 in 20 chance, which means that, on average, 1 in 20 framboid size-frequency distributions will give a contrary (or "wrong") result. Unfortunately, there is no way of determining when these events will occur and, since they are likely to be random, they will cluster at times. The effects of probability help explain the conflicting evidence for framboid size distributions as an indicator of euxinic conditions.

The basic problem, however, remains that there have been no systematic analyses of framboid size-frequency distributions formed in the water column itself. Although framboids have been collected from water columns, as noted earlier, the number of observations is small. All the data from Wilkin et al. (1996) derive from framboids sampled from within the sediments themselves, which were then recalibrated to historical data on water column conditions.

2.6. Hydrothermal Framboids

Framboids are commonly observed in hydrothermal veins, volcanogenic deposits, and are frequently described from their modern equivalents: marine hydrothermal vents, including black smokers. Although framboids were originally defined from an ore deposit (i.e., the Cornwall copper mines in the Mississippi Valley Pb-Zn district by Rust, 1935), few size measurements have been reported from ore deposits where the framboids are distinctly hydrothermal in origin. The problem is that framboids in these systems are commonly embedded in more massive pyrite, often described as colloform, or have major overgrowths (e.g., Figure 2.5). This makes the analysis of framboid size distributions both difficult and complex. However, Graham (1971) reported the size distributions of 1124

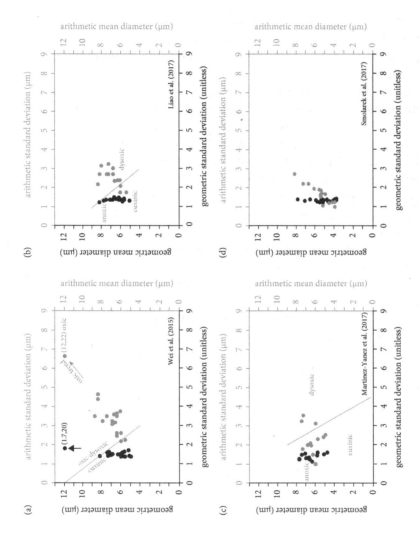

Figure 2.4. Plots of the geometric mean diameters (μm) versus geometric standard deviations (unitless) superimposed on the grayed-out plots of arithmetic mean diameters (μm) and standard deviations (μm) for a selection of data sets. (a) 250 Ma Permian, China (Wei et al., 2015). (b) 252 Ma, Triassic, China (Liao et al., 2017). (c) 145 Ma Cretaceous, Mexico (Martinez-Yanez et al., 2017). (d) 430 Ma Silurian, Poland. (Smolarek et al., 2017).

Reprinted from *Palaeogeography, Palaeoclimatology, Palaeoecology*, 522, 62–75. Sedimentary pyrite framboid size-frequency distributions: A meta-analysis. Rickard, D. Copyright (2019) with permission from Elsevier.

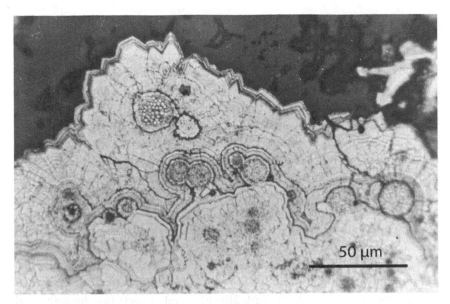

Figure 2.5. Hydrothermal framboids in colloform pyrite from Silvermines, Ireland. Etched polished section, reflected light image.

framboids from the hydrothermal ores of the Silvermines district in Ireland. These deposits commonly show framboids embedded in colloform textures (Figure 2.5), and measuring framboids within pyrite overgrowths is notoriously difficult (e.g., Wacey et al., 2015). Recalculating Graham's data from published histograms (and including the stereological correction) shows the mean size of the Silvermines framboids was over 11 μm with a standard deviation of 7 μm, and the framboids ranged in size from 1 to over 37 μm in diameter.

The sizes of hydrothermal framboids have also been recorded in systems where they are not embedded in massive or colloform sulfides. Sizes of framboids in shallow ocean hydrothermal vents have been reported by Honnorez (1969) and Prol-Ledesma et al. (2010). Honnorez (1969) reported framboids between 20 and 40 μm in diameter from the vent deposits at Vulcano, Italy. Prol-Ledesma et al. (2010) reported framboid sizes ranging between 1 and 35 μm for over 100 framboids from a shallow submarine vent system from Banderas Bay off the western coast of Mexico. These hydrothermal vent populations appear at first sight to be more uniform than usual, but in fact fail the 95% range test, suggesting a distribution which more nearly approximates to log-normal with a geometric mean of 7.4 μm and a geometric standard deviation of 2. This suggests that 95% of these vent framboids range between 2.0 and 29.3 μm.

Raybould (1973) reported the size distribution of framboids in ca. 180 Ma hydrothermal veins in Wales. He found a range of sizes from 1 to 40 μm in diameter with a mode around 5–6 μm. Most of the framboids were less than 12 μm in size. Recalculating Raybould's data shows that the mean framboid diameter in these veins was 6.8 μm and the geometric standard deviation was 1.48. This suggests that 95% of these framboids were in the range of 3.1–14.9 μm. Ostwald and England (1977) reported framboids in chalcedonic veins and amygdales in a ca. 250 Ma Permian andesite from Allendale in New South Wales, Australia. The environment is classified as hydrothermal since these veins and amygdales are likely to have been formed by groundwaters that were above ambient temperatures and the framboids represent an early stage of the paragenesis. They reported the diameters of 235 freestanding framboids in the chalcedony with a computed geometric mean diameter of 16.4 μm and a geometric standard deviation of 1.6. They also reported the diameters of framboids in pyrite aggregates within agates. These appear to have a similar, if slightly large mean size, but were not included here since they were enclosed in colloform pyrite.

It might be worth making a more systematic study of the sizes of hydrothermal framboids to see if any specific size signature could be extracted. The factors controlling framboid size (Rickard, 2019b) might suggest that hydrothermal framboids would tend to larger diameters since the rate of pyrite crystal growth is temperature dependent and many hydrothermal fluids are acidic.

2.6.1. Mineralized Sediments

Mineralized sediments where framboid size measurements have been reported include the 250 Ma Permian Kupferschiefer where Schneiderhöhn (1923) made his original observations, the ca. 380 Ma Devonian Rammelsberg deposit (e.g., Love and Amstutz, 1966), the late Tertiary (possibly ca. 10 Ma) muds from the Tynagh Ag deposit (e.g., Rickard, 1970), and the Holocene peats of west Cornwall (Hosking and Camm, 1980).

Hosking and Camm (1980) made a systematic study of framboids in peats associated with mineral veins in Cornwall, United Kingdom. They measured the diameters of 1,558 framboids which they had separated from the peat. These measurements, as in the case of Vallentyne (1963), do not suffer from the stereological error of measurements of framboids in section. By contrast, Hosking and Camm (1980) remark that the optical microscopy at their disposal was limited, and it is possible that smaller framboids were missed or not collected. I analyzed their published graphs with GRAPHCLICK and XLSTAT. In both sample sites, the peats were associated with tin placers. In the case of the Red River Valley peats, this included copper sulfides such as chalcocite, Cu_2S, and chalcopyrite, $CuFeS_2$, and the peat was sulfide-rich with H_2S gas being produced. The two sample sites analyzed by Hosking and Camm (1980) were about

10 km apart and associated with different rivers. However, the geometric mean sizes of the samples were similar (41.9 and 44.4 μm) and the geometric standard deviations were closely coincident ($\sigma^* = 1.2$). The data suggests that 95% of these Cornish peat framboids range between 29.9 and 62.2 μm in diameter. As discussed earlier, the large mean size of these framboids in peat has less to do with the associated mineralization and is related to the acidic pH of the peat waters. From the point of view of mineralized sediments, therefore, larger diameter framboids are predicted to occur in systems where the amount of dissolved Fe and S is greater than in normal marine sediments, due to either more acid solutions or higher temperatures, or both. The obverse of this conclusion is that populations of smaller framboids in mineralized sediments probably evidence framboid formation under ambient temperatures at pH conditions more generally associated with marine environments.

The occurrence of spheroidal pyrite that does not exhibit the internal framboidal texture has been widely noted, especially in Precambrian mineralized sedimentary rocks. Careful distinction must be made here between pyrite spherulites exhibiting a radiating internal structure and small massive pyrite spherules. The radiating spherulites derive from an entirely different process and are distinct from framboids (Ohfuji, 2004; Ohfuji and Rickard, 2005; Rickard, 2015).

The pyrite grains in the 1.6 Ga lower Proterozoic Mount Isa Shale are discussed in sections 1.3.2 (Chapter 1) and 7.1.6 (Chapter 7). They are different (Figure 2.6) from the framboids in the 250 Ma Permian Kupferschiefer, although

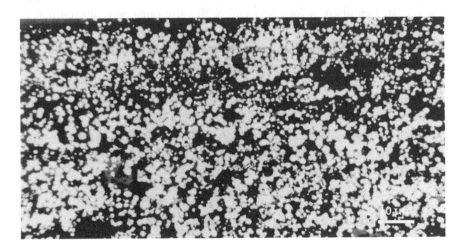

Figure 2.6. Typical view of bedded pyrite particles in ca. 1.6 Ga Proterozoic sedimentary rocks from Mt. Isa, Australia.
Reflected light micrograph by Peter McGoldrick.

they show the similar size ranges (e.g. <5–~40 μm; Painter et al., 1999) and, as discussed in section 7.1.6, are commonly associated with organic envelopes (Love and Zimmermann, 1961; Schouten, 1946). They are part of a spectrum of similarly sized pyrite forms ranging from euhedral crystals to spherules. The spherules themselves consist generally of a homogeneous core of rounded to euhedral pyrite, surrounded by a rim of inclusion-rich pyrite (Grondijs and Schouten, 1937; Painter et al., 1999). Deep etching of some of these textures does not reveal any convincing cryptic framboidal structures (e.g., Love and Zimmermann, 1961). Although they could be original framboids in which the individual microcrystals have coalesced with age and metamorphism (England and Ostwald, 1993; Painter et al., 1999; Scott et al., 2009), there is little evidence to support this idea. On the other hand, the occurrence of these spheroidal pyrites in place of framboids in mineralized sediments >1.5 Ga old might reflect changing conditions in the Earth system.

2.7. Polyframboid Sizes

There are two significant size measurements for polyframboids: the size of the constituent framboids and the size of the aggregates as a whole. Neither has been systematically measured. Love (1971), in his paper originally defining polyframboids, reported that polyframboids from the 430 Ma Wenlockian of north Wales were between 35 and 900 μm in diameter. However, many of the polyframboids had been crushed during compaction of the sediment so that the reported numbers at the lower end of this size range were overestimated. Alternately, I suppose that this "crushing" could have occurred during late diagenesis by disaggregation when an enclosing biogenic casing dissolved. The framboids in the polyframboid pictured in Figure 3.14 in Chapter 3 vary between 18 and 25 μm in diameter. This seems at the high end of the general framboid size distribution and is consistent with Love's measurements, but the sampling is obviously too low to draw any conclusions. However, I get the general impression that the individual framboids of polyframboids are indeed larger than the global mean. If so, it has some implications for the formation of polyframboids and might be worth examining further.

2.8. The Maximum and Minimum Sizes of Framboids

It is reasonable to ask what limits the size ranges of framboids. The evidence that framboid size-frequency distributions do not show simple exponential relationships demonstrates that framboids have a finite minimum size. This size appears to be somewhat less than 2 μm diameter. At that size, the microcrystals must be less than 100 nm and are, by definition, nanoparticles.

Naturally occurring clusters of nanoparticles, so-called nanoframboids, have been reported as being around 200 nm in diameter (e.g., Yücel et al., 2011) with the individual microcrystals being replaced by nanoparticles often as small as 4 nm in size. The properties of nanoparticles cannot be predicted by merely extrapolating the properties of microcrystals to a smaller size. At nanoparticle dimensions the surface area to volume ratios of the particles become extremely high and surface properties dominate nanoparticle behavior. Spherical clusters of FeS_2 nanoparticles are regularly synthesized experimentally (Butler and Rickard, 2000; Farrand, 1970) and are reported from, especially, hydrothermal natural environments (Gartman and Luther, 2014; Yücel et al., 2011).

The lack of a continuum of framboid sizes is consistent with the conclusion that framboids do not significantly increase in diameter with time: the final framboid size is similar to the original pyrite nucleation volume. So-called micro- or nanoframboids are chemically labile and tend either to dissolve, oxidize (e.g., Gartman and Luther, 2014; Yücel et al., 2011), or are involved in the aggregative growth (i.e., crystallization by particle attachment) of pyrite crystals (Chapter 13). Indeed, spherical aggregates of inorganic nanoparticles need to be kept apart experimentally with organic ligands, otherwise they rapidly fuse together (e.g., De Yoreo et al., 2015, see Chapter 13). Clusters of nanoparticles are not produced by the same processes as those involved in framboid formation, nor do they behave in the same way. They are more akin to atomic clusters, which form similar constructs. Microframboids and nanoframboids are discrete entities which are distinct from framboids.

At the other end of the scale, the largest framboids are around 250 μm in diameter, but framboids are rarely found much larger than 80 μm in size. The factors ultimately controlling the maximum size of framboids are related to the limitations of the supply of dissolved Fe and S to the site of framboid nucleation and crystal growth (e.g., Raiswell et al., 1993; Rickard, 2019a). The process of framboid formation requires the buildup of extreme supersaturations of dissolved Fe and S with respect to pyrite and the sudden nucleation of all of the microcrystals (see section 11.1 in Chapter 11). The maximum concentration of dissolved Fe and S in the bulk solution is determined by the solubility of metastable FeS (see 10.2.5 in Chapter 10). In order to achieve similarly shaped and sized microcrystals, the nucleation volume needs to be homogeneous. The larger the nucleation volume, the less probable it is to remain stable enough for sufficient time for enough dissolved Fe and S to be transported to the framboid site for framboids to nucleate. The requirement then to maintain a constant environment over the years that it takes large framboids to form (Rickard, 2019a) places a natural constraint on the maximum framboid size. The decreasing abundance of framboids with size and the scarcity of very large framboids are consistent with this conclusion.

Large framboids are also more sensitive to gravitational forces (section 3.2.3 in Chapter 3). Although gravity is a relatively weak force, it becomes significant for larger volumes of relatively dense pyrite/fluid mixes and competes with the surface free energy and interparticle van der Waals forces that are ultimately responsible for framboid shape and stability. In fact, large volumes of nanoparticulate pyrite form irregular masses rather than spheroidal framboids (e.g., Figure 3.15 in Chapter 3). Gravity also has a significant effect on framboid shape and leads to framboidal constructs reminiscent of those commonly attributed to soft-sediment deformation.

3

Framboid Shapes

A key attribute of framboids is that they are sub-spheroidal in shape (Ohfuji and Rickard, 2005). The term *sub-spheroidal* may appear, at first sight, to be a tautology since a spheroid by definition is an object that is approximately spherical in shape. However, framboids diverge significantly from ideal sphericity, with the occurrence of facets and quite irregular rounded shapes. Most display a degree of symmetry, but many are asymmetric. *Sub-spheroidal* is used to encompass all these forms.

Originally framboids were described as spherical (e.g., Love and Amstutz, 1966) and they were described as spheres by Love (1957). Schneiderhöhn (1923) and Neuhaus (1940) had described them as *Kügelschen*, the diminutive of *Kügel* (a solid ball of the type used in skittles) and the term used by both Früh (1885) and Rhumbler (1892) in their original descriptions. Vallentyne (1963) called them *spherules*, a more-or-less direct translation: spherules are simply small spheres, and the word derives from the Latin diminutive of *sphaera*, a sphere. Rust (1935) in his original definition of framboids called them *spheroids*.

The assumption that framboids were spherical colored the early discussions concerning their origin. In particular, the sphericity was suggested to have been inherited from preexisting structures such as microorganisms (e.g., Love, 1957; Schneiderhöhn, 1923), organic coacervates (e.g., Papunen, 1966; Rickard, 1970), spherical cavities in sediments caused by gas bubbles (Rickard, 1970) and preexisting iron sulfides (Berner, 1969; Wilkin and Barnes, 1997a). Much of this debate was predicated on the assumption that most framboids were in fact spherical, and this assumption was based on the limitations of the microscope technology available at that time.

Understanding of framboid shape has basically followed technical developments in microscopy since the mid-19th century. Individual framboids are too small to be seen with the naked eye, and they were originally studied with biological microscopes using transmitted light microscopy. They were added to a liquid medium, usually water, in a shallow well in a glass slide, covered with a thin glass lid and illuminated by a sub-stage light source.

The problem with using transmitted light microscopy to study framboids is that pyrite is opaque, and framboids appear as small black spheres (Figure 3.1). The effects of light dispersion around the edges of these three-dimensional objects, as well as microscopic resolution, resulted in the framboids appearing as

Framboids. David Rickard, Oxford University Press. © Oxford University Press 2021.
DOI: 10.1093/oso/9780190080112.003.0003

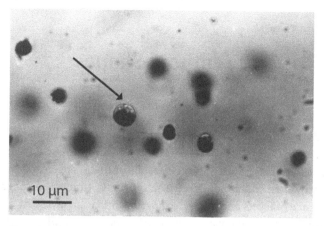

Figure 3.1. Framboids in transmitted light microscopy. The light can catch the edges of the framboids (arrowed) revealing the metallic luster of pyrite.

more-or-less perfect spheres. The effect was often enhanced by the birefringence caused by the water in which the framboids were often originally immersed for study. If any facets were present, they were not discernible, and small deviations from sphericity were also not easily spotted.

The development of reflected light microscopy at the beginning of the 20th century provided enhanced optical resolution, and much of the history of framboid studies have been based on this technique. The recognition of facets or deviations from sphericity for framboids was generally dependent on extreme examples, such as framboids which were clearly pentagonal or hexagonal in section (e.g., Morrissey, 1972) or elliptical in section (e.g., Read, 1968). The first clear account of faceted framboids was provided by Skripchenko and Berber'yan (1976).

The introduction of the scanning electron microscope (SEM) in the second half of the 20th century enabled framboid shapes to be detailed. Even so, it was not until the early years of the 21st century that framboids with regular polyhedral shapes were described (e.g., Butler, 1994; Ohfuji and Akai, 2002).

3.1. Spheroidal Framboids

It is convenient to define spheroidal framboids as framboids with curved outer faces, which encompasses those sub-spheroidal forms that do not display facets. Truly spheroidal framboids are possibly less abundant. The fraction of framboids that are more-or-less spheroidal, rather than perfectly spherical, in any given framboid population is probably less than 20%, although statistics on this are lacking.

3.1.1. Sphericity

Classical sphericity is a shape factor that defines the roundness of an object. It was originally defined by Wadell (1935) as the ratio of the surface area of a sphere with the same volume as the object to the surface area of the object itself. In this case, a perfect sphere has a sphericity of 1. Regular polyhedra have sphericities between 0.8 for cubes and >0.9 for icosahedra. The polygonal forms exhibited by framboids are multifaceted and therefore have sphericities >~0.9. The sphericity of an ellipsoid, ψ, is given by Equation (3.1).

$$\psi = \left(\frac{d_s}{d_l} \right)^{\frac{1}{3}}$$

(3.1)

where d_s is the shortest diameter and d_l is the longest.

Sphericity estimates have been used to attempt to estimate the numbers and sizes of framboids using machine-based techniques (e.g., Merinero et al., 2017). Most framboids have a sphericity >0.8 and this cutoff was used to discriminate framboids from other mineral forms in automatic analyses of framboid size-frequency measurements.

3.1.2. Deviations from Perfect Sphericity

Most framboids do not display sphericities approaching 1. The problem is often intrinsic and results from the complex interplay of history and the psychology of shape perception where the brain interprets shapes differently depending on context and association. The result is that sub-spheroidal framboids are often perceived as being spherical. This is compounded by the lack of sphericity being obvious when pointed out and observers stating that they knew this all along—which is actually probably true. However, in the literature framboids are often reported to be spherical, and measurements are presented assuming that this is a close approximation.

Figure 3.2 shows a typical field of view of framboids in section. Of the ca. 50 framboids captured in this image, fewer than 10 actually display approximate circular sections. Many of the framboids are elongated. Some appear to be deformed in contact (e.g., Figure 3.2a). In others, contact appears to produce facets (e.g., Figure 3.2b). Even framboids which are apparently circular in section often display one or more facets (e.g., Figure 3.2c). And there are also quite irregular, asymmetric shapes (e.g., Figure 3.2d).

Figure 3.3 shows a selection of sub-spheroidal framboids in section. The framboid in Figure 3.3(a) appears to be perfectly spherical. Interestingly, it is

Figure 3.2. Shapes of framboids from Holocene sediments (see text).
Reflected light photomicrograph.

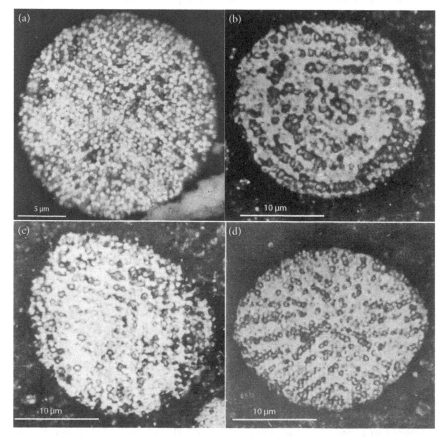

Figure 3.3. Framboid shapes: (a) spherical; (b) oblate; (c) teardrop; (d) almond.
Reflected light photomicrographs.

dominated by mainly randomly organized microcrystals, and Skripchenko and Berber'yan originally proposed that more spherical, less faceted framboids normally lacked extensive internal microcrystal organization. The oblate sphere in Figure 3.3(b) is a common framboid form. However, many examples of this form actually display one or more facets when examined in 3D under SEM. Another common form is the teardrop, which has one end more pointed (Figure 3.3(c)). Again, this may result from a facet developing at one side. Finally, the almond shape (Figure 3.2d) has developed apices at opposite sides of the framboid.

3.2. Ellipsoidal Framboids

3.2.1. Ellipsoidal Framboids Produced by Deformation

Ellipsoidal framboids are in a separate class since several researchers have related these to deformation. This relates to the geometrical definition of an ellipsoid as a form that may be derived from a spheroid by directional scaling. The idea is that original sub-spheroidal framboids have been deformed by applied unidirectional stress resulting from tectonism. Figure 3.4(a) shows ellipsoidal framboids from the massive pyrite ore body at Rio Tinto, Spain. Read (1968) concluded that these were formed from sub-spheroidal framboids during deformation. He provided experimental data to demonstrate that these ellipsoidal framboids were produced by partial recrystallization of the pyrite under metamorphism to greenschist facies conditions (300°–450°C and 0.2–1 GPa pressure). Under these conditions in massive pyrite deposits, the fugacity of sulfur was so high that pyrite remained stable and did not decompose to iron monosulfide and sulfur. The ellipsoidal framboids in Figure 3.4(b) have been produced by simple shear stress during deformation of mudrocks at the Fosterville Au deposit, Victoria, Australia. The original sub-spheroidal framboids have been mechanically distended to lesser or greater amounts during the imposition of cleavage. The associated organic matter was graphitized and also extended parallel to the cleavage.

3.2.2. Primary Ellipsoidal Framboids

Ellipsoidal framboids (e.g., Figure 3.5(a)) are commonly found in unconsolidated sediments and in association with contemporary spheroidal framboids (Figure 3.5(a)). That is, that ellipsoidal framboids are formed in the absence of the imposition of an external force. These are not deformed sub-spheroidal framboids and the development of the ellipsoidal form is a primary feature.

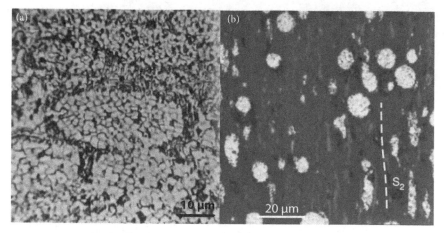

Figure 3.4. Ellipsoidal framboids produced through deformation of originally spherical framboids. (a) Reconstituted framboids paralleling foliation.
Reflected photomicrograph from a volcanogenic ore deposit, Rio Tinto, Spain (Read, 1968).
(b) Elongated framboids and graphite (light gray) paralleling cleavage. From the Fosterville Au deposit, Victoria, Australia (Scott et al., 2009).
Reflected light photomicrograph published by courtesy of the Society of Economic Geologists.

    In fact, there appears to be a continuum in framboid shapes between ellipsoids, oblate spheroids, prolate spheroids, and spheroids that supports this conclusion (Figure 3.6). This is consistent with the geometric definition of ellipsoids which include oblate spheroids in which the c-axis is shorter and prolate spheroids in which the c-axis is longer. The key feature from the point of view of framboids is that every cross-section of these forms is an ellipse. This means that the elliptical sections through framboids illustrated in Figure 3.5 may have been derived from any of these 3D forms.

    The ellipsoidal sedimentary framboid shown in Figure 3.5(a) has an axial ratio of 0.625 and a sphericity of 0.85. Soliman and El Goresy (2012) reported framboid populations with minimum axial ratios of 7:10. This would imply sphericities ≥ 0.89 according to Equation 3.1. The accepted cutoff of ≥ 0.8 for the minimum framboid sphericity appears to be a robust limit for framboid form.

    Figure 3.7 shows an SEM image of an ellipsoidal framboid. In 3D this form approaches a prolate spheroid with an axial ratio ca. 0.8, implying a sphericity of 0.93 (Equation 3.1). Skripchenko and Berber'yan (1976) suggested that spheroidal forms in highly organized framboids displayed disorganization in the outermost layers, but I have not been able to confirm this suggestion generally. Locally, however, this seems to be true. For example, Figure 3.7 shows well-organized microcrystals visible in the layer beneath the surface

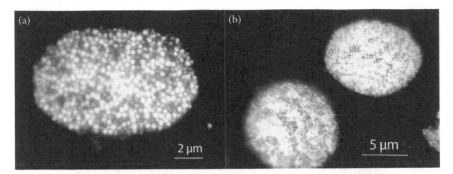

Figure 3.5. Primary ellipsoidal framboids from unconsolidated sediments:
(a) ellipsoidal framboid; (b) coexisting ellipsoidal and spheroidal framboids.
Reflected light micrographs.

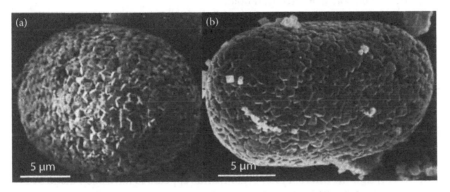

Figure 3.6. Spheroidal framboids: (a) oblate spheroid; (b) prolate spheroid.
Scanning electron microscope images by Schallreuter (1984) published courtesy of the IODP.

layer. The surface layer itself shows patches of organization partly separated
by clear dislocations as the surface layer accommodates to the curved outer
surface.

3.2.3. Effect of Gravity on Framboid Sphericity

Pyrite has a density of around $5g\,cm^{-3}$, which is around five times greater than
water. The result is that larger pyrite particles sink in water. Deviations from
sphericity have long been observed in groups of framboids, and this has been
widely interpreted as a result of the framboids growing into their neighbors. In
fact, as shown in section 11.2 (Chapter 11), framboid nucleation requires that

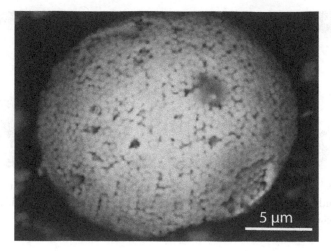

Figure 3.7. Ellipsoidal ordered framboid.
Scanning electron microscope image by Hiroaki Ohfuji.

the framboids originally form in a volume similar to the final framboid and their diameter does not increase significantly with time. The growth of framboids is limited to the growth of the individual microcrystals.

An alternative explanation for divergences from sphericity is gravitational effects, and this appears more obvious when examining framboid groups such as those in Figure 3.2. Various points of contact are arrowed in Figure 3.2. Arrow **a** shows impingement resulting in a clear depression in one framboid. It is also noteworthy that the microcrystal organization appears to have been disturbed at the point of impact. This is consistent with the conclusion (see Chapter 13) that microcrystal organization develops later in framboid development. Arrows **b** and **c** show linear contacts which appear to affect both framboids equally. My interpretation here is that, in contrast to the framboids indicated by arrow **a**, these framboid pairs were in similar states of development when the impact occurred. Framboid **d** has been squeezed on all sides by its neighbors, resulting in a complex final form. Close inspection of the framboids in Figure 3.2 reveals that virtually all the framboids in this cluster have been deformed more or less from the ideal spherical form.

Similar effects to these are well documented in sedimentology and are described as soft-sediment deformation. It appears that partially completed framboids have impacted each other simply under the effects of gravity, leading to the equivalent of soft-sediment deformation of the framboids. For this to occur, the framboids need to be still fluid-rich when they are impacted, as is the case in sediments. It also implies that the final exceptionally robust shapes of framboids are not acquired until microcrystal growth is completed, analogous to lithification of sediments during diagenesis. It is also consistent with the general conclusion that the volume occupied by framboids does not change greatly after initial nucleation. As with diagenetic changes in general, the crystals grow and water is expelled, leading to the formation of a highly stable entity.

The implication is that some deviations from sphericity in individual framboids (e.g., Figure 3.3) are a result of gravitational forces on the neo-framboid. The consequence of this is that the preservation of spheroidal forms for individual framboids reflects further attributes, such as small size (where gravitational forces do not dominate), rapid microcrystal growth (which leads to rapid lithification of the framboid), or the presence of a con-fining medium.

3.3. Faceted Framboids

3.3.1. Polyhedral Framboids

The clear identification of polyhedral framboids had to await advances in micro-scope technology as well as obvious examples (e.g., Morrissey, 1972; Skripchenko and Berber'yan, 1976).

Figure 3.8 shows typical examples of framboids showing polygonal shapes in section: Figure 3.8(a) shows a symmetrical pentagonal framboid and Figure 3.8(b)

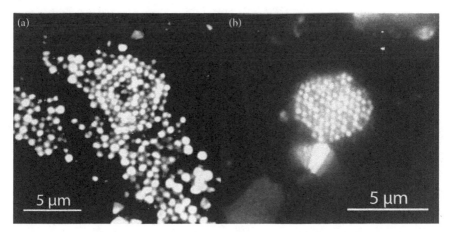

Figure 3.8. Polygonal sections through framboids: (a) pentagonal; (b) hexagonal.
Reflected light photomicrographs by C. J. Morrissey.

a hexagonal framboid. However, it is not clear which 3D polyhedron gave rise to these shapes since they might be derived from various solid forms (Zimmerman and Amstutz, 1973). The framboids showing hexagonal sections cannot have been derived from icosahedra since hexagons cannot be obtained from sectioning solids with pentagonal symmetry. The external morphology of these framboids is likely to be an octahedron or cube.

One feature of both of these polygonal framboids is the organization of their constituent microcrystals. Skripchenko and Berber'yan (1976) originally suggested that framboids with highly organized arrangements of microcrystals were multifaceted.

The introduction of the scanning electron microscope permitted detailed images of 3D polyhedral framboids. Figure 3.9(a) illustrates an early example. The facets of this polyhedral framboid might be interpreted as pentagons, triangles, and squares (Figure 3.9(b)). Because pyrite exhibits possibly the greatest range of habits in the mineral world, this polyhedral form could be interpreted as a crystalline form, a modified pyritohedron, for example. However, this would imply that framboids were extreme examples of single crystals with concomitant consequences for their origin and formation. As shown in Chapter 6, this is not the case: framboids are crystal aggregates.

The problem was finally solved by Ohfuji and Akai (2002), who showed that some framboids were icosahedral. Regular icosahedra (Figure 3.10) are interesting forms since they are a classically forbidden fivefold crystallographic symmetry. The problem is fundamental since the packing of regular pentagonal forms must necessarily leave gaps. There are two ways these can be accommodated: (1)

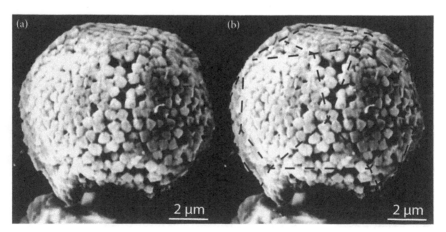

Figure 3.9. (a) Polyhedral framboid. (b) Possible interpretation of facet geometry.
Scanning electron microscope image.

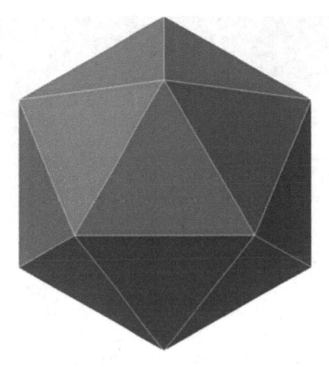

Figure 3.10. Regular icosahedron.

through dislocations and (2) through deviations from equilateral triangular faces. This latter process is observed in pyrite crystals which show, relatively rarely, icosahedra with varying triangular crystal faces (Figure 3.11).

Reference to the detailed scanning electron micrograph of an icosahedral framboid in Figure 3.12 shows that, at first sight, the faces appear to be regular equilateral triangles and that the space-fitting problem is resolved by dislocations between microcrystals, shown by the black zigzag lines of spaces between some of the surface rows. In fact, more detailed observation reveals that the faces are not all of equal size and the icosahedral space problem is accommodated by differently sized triangles. This is analogous to the relationship between pentagonal dodecahedra and pyritohedra. The pyrite pseudo-icosahedron in Figure 3.11 results from a combination of an octahedron and pyritohedron.

3.3.2. Framboids with Partial Facets

The introduction of the SEM to framboid studies revealed that many apparently spheroidal framboids displayed facets as well as curved surfaces (Figure

Figure 3.11. Natural pseudo-icosahedron of pyrite from the Caknakkaya mine, Murgul Cu-Zn-Pb deposit, Turkey. The crystal is about 2.5 cm across.

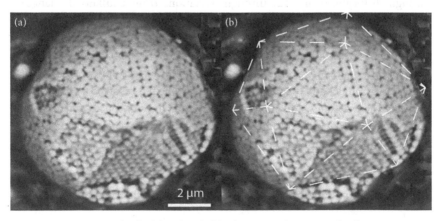

Figure 3.12. (a) Icosahedral framboid. (b) The 20 regular triangular faces.
Scanning electron microscope image by Hiroaki Ohfuji.

3.13). Such combined framboid shapes were difficult to discern under the relatively low resolution of optical microscopy, and the psychological factors in
seeing sphericity also contributed to a lack of reporting of these combined
shapes.

An alternative suggestion regarding the formation of the combination forms,
as shown in Figure 3.13, is that the facets are actually fracture planes and the
original framboid displayed a uniform spherical shape. Certainly, the curved
faces of that framboid are composed of incomplete microcrystals, whereas the
faceted faces are constituted by complete microcrystals. Furthermore, the outer
coating of incomplete microcrystals which define the curved surface on the
right-hand side of the framboid clearly ends in a disturbance consistent with a
fracture. However, reference to other framboid images, such as Figure 3.2 earlier in this chapter, clearly shows framboids with both curved and planar faces.
Some of these planar margins are obviously the results of contacts between adjacent framboids, but others appear to be on free surfaces. There also appears to
be no clear relationship between the curvature of the faces and the development
of packing geometries in the framboids. The obverse may not be the case, however: the development of planar facets appears to be related, at least in part, to the
packing arrangements of the constituent microcrystals, as originally proposed
by Skripchenko and Berber'yan (1976).

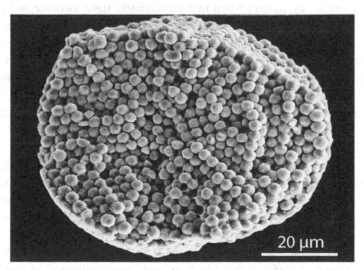

Figure 3.13. Framboid displaying both rounded and faceted margins from the 2 Ma
Nishiyama Formation, Japan.

Scanning electron microscope image by Hiroaki Ohfuji.

The development of combinations of curved and planar facets in framboids has implications for the development of framboid forms. It cannot be assumed that framboids necessarily require a precursor template, such as a spherical space or spherical organic globule, to develop. Framboid shape must be an intrinsic property of the growth mechanism of framboids.

3.4. Polyframboids

Although framboids commonly aggregate, as defined in this book, *polyframboids* refer specifically to spheroidal to sub-spheroidal aggregates of framboids (Figure 3.14 and section 3.1.1). *Polyframboids* have also been used to describe clusters of framboids which have various shapes (e.g., Tribovillard et al., 2008) and which can form in a variety of ways. But polyframboids *sensu stricto* are exceptional in that they consist of roughly equidimensional framboids packed into a sphere.

As with discrete framboids, the shapes of the framboids within a polyframboid only approximate to perfect sphericity. Close examination of the polyframboid in Figure 3.14 shows that the constituent framboids are often oblate spheroidal to ellipsoidal in form and they display abundant facets. These facets appear where framboids are in contact and suggest that the framboids were deformed by contact when in a more plastic state, as described earlier. However, the outer framboids of the cluster also show curved faces, and there is no suggestion of flattening against some outer curved surface. The origin of polyframboids is discussed in section 7.1.7 (Chapter 7). There it is noted that discoid framboid clusters resulting from infilling of diatoms have been clearly shown to retain spheroidal framboid forms at the margins of the clusters, against the outer walls of the diatom skeleton. Indeed, discoid clusters where the diatom shell has dissolved maintain the curvature of the outer framboid layer.

The packing style in the polyframboid shown in Figure 3.14 is cubic close packing with each framboid apparently situated over the depressions be-tween adjacent framboids in the lower layer. Assuming a packing density of $\pi/3\sqrt{2}$ and an average framboid volume of 4×10^{-9} cm^3 for 20-μm framboids, then a polyframboid of 130 μm diameter, as shown in Figure 3.14, would contain about 200 framboids. Love (1962), in his paper originally defining polyframboids, suggested that 130 framboids were contained in one of his Devonian examples.

Nanocrystalline pyrite refers to pyrite crystals less than 100 nm or 0.1 μm in size; crystals larger than this are described as *microcrystalline* and are the usual

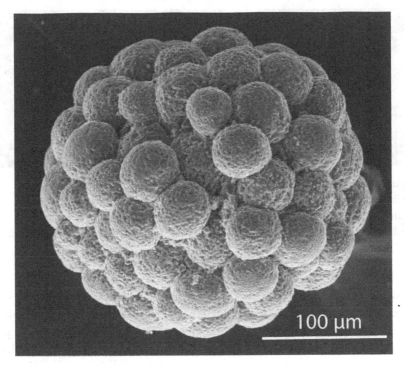

Figure 3.14. A polyframboid from 180 Ma Jurassic sediments, Spain.
Scanning electron microscope image by Matías Reolid Pérez

constituents of framboids. One of the most common textures obtained in the experimental syntheses of pyrite in aqueous solutions, both at ambient and hydrothermal temperatures, is nanocrystalline pyrite, which we call "pyrite dust" (Gartman and Luther, 2013; Rickard, 1997; Rickard, 1969). It is not surprising, therefore, that this form of pyrite is commonly encountered in Holocene sediments. Indeed, it has been suggested that nanoparticulate pyrite debouching from deep ocean vents is a major source of oceanic iron (Yücel et al., 2011). What is perhaps surprising is that the nanoparticles in the natural materials often show distinct areas of ordering.

Figure 3.15 shows images of typical masses of nanocrystalline to microcrystalline pyrite from 1Ma Pleistocene and Holocene+ sediments. Regular patches of organized pyrite nanocrystals are shown in Figure 3.15(a). These patches seem to occur in roughly circular areas, suggesting a gradation between these clusters of pyrite nanocrystals and discrete framboids. Figure 3.15(b) shows that these irregular clusters of pyrite particles are not restricted to nanocrystals, but grade into pyrite microcrystals. There also appears to be a complete gradation in form

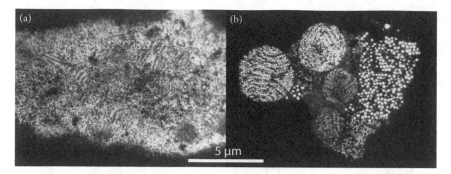

Figure 3.15. Optical microscope images of irregular masses of nanoparticulate pyrite showing (a) areas of partial organization (Holocene sediments, mangrove swamp, Queensland), and (b) associations with discrete pyrite framboids and gradation into microcrystalline pyrite (ca. 1 Ma Pleistocene sediments DSDP hole 262 Timor Trough).

between the amorphous mass of pyrite, spheroidal pyrite areas (Figure 3.15(b) lower right), and discrete framboids. The conclusion is that the spheroidal shapes of framboids are developed through reorganization of the constituent pyrite microcrystals, rather than due to some sort of preexisting spherical template.

4

Microcrystal Morphology

4.1. Microcrystal Sizes

The microcrystal is the fundamental unit of framboids. I refer to these as *microcrystals* since they show systematic differences from pyrite macrocrystals. Furthermore, they are generally of colloidal or microscopic size, rather than nanoparticulate (i.e., ≤ 100 nm) although, as discussed in the following, this may be partly a result of the limiting resolution of conventional microscopic techniques. Various compilations of microcrystal sizes have been made. Love and Amstutz (1966) described microcrystals up to 12 µm in diameter, but admitted that the vast majority of microcrystals were less than 2 µm in size. Kalliokoski and Cathles (1969) also described large microcrystals, up to 5 µm in size. Microcrystal sizes between 0.1 and 1.4 µm in size were reported by Wilkin et al. (1996) and between 0.5 and 1.5 µm by Large et al. (2001). Ohfuji (2004) measured a set of very large framboids with microcrystal sizes ranging between 1 and 10 µm. Indeed, since the early accounts, reported microcrystal size measurements have been more conservative, with averages around 1 µm in size or less (Love, 1971; Rickard, 1970; Schallreuter, 1984; Skei, 1988; Vallentyne, 1963; Wilkin et al., 1996). However, more recently, microcrystal size measurements have been relatively rarely compiled since the simple measurement of framboid diameters has supplanted the relative size of framboids and their constituent microcrystal as a potential environmental indicator (cf. Wignall and Newton, 1998; Wilkin et al., 1996).

It is fairly obvious that since framboids are made up of microcrystals, the microcrystals must be considerably smaller than the framboids. The ratios of framboid diameters, D, to microcrystal diameter, d, were originally reported by Wilkin et al. (1996) for modern sediments and were shown to vary mainly between 5 and 30, although ratios as high as 118 were reported.

One of the more recent compilations by Ohfuji (2004) is summarized in Table 4.1. This is significant since it records over 300 measurements made by scanning electron microscopy rather than optical microscopy. This set of measurements avoids the stereological error (section 2.2 in Chapter 2) and provides far greater precision, especially at smaller microcrystal sizes. Recalculations of the size distributions of microcrystals in over 300 framboids of all ages measured by Ohfuji (2004) show that they approximate

Framboids. David Rickard, Oxford University Press. © Oxford University Press 2021.
DOI: 10.1093/oso/9780190080112.003.0004

Table 4.1 Size of Framboids and Microcrystals Measured with a Scanning Electron Microscope by Ohfuji (2004)

			Ma	Framboid Diameter μm	Microcrystal Diameter μm
Lake Harutori	Japan	Present		<5–15	0.7–2
Mitarase Lagoon	Japan	Present		<5–20	1–2
Sakata Lagoon	Japan	Present		<5–40	0.2–2
Septiba Bay	Brazil	Present		<5–10	0.5–1
Shirone Drill Core	Japan	Holocene		5–120	0.5–7
Kanai Drill Core	Japan	Holocene		20–110	1.5–2
Udenaha mudstone	Japan	Pleistocene	0.5	20–145	3–10
Uonuma F.	Japan	Pleistocene	1	30–50	2–3
Nishiyama F.	Japan	Miocene	2	15–140	1–10
Teradomari F.	Japan	Miocene	10	5–70	0.5–4
Hawsher Bottoms	UK	Lower Jurassic	190	3–20	0.3–2
Mid Clare Shale	Ireland	Lower Carboniferous	335	4–20	0.3–1
Chattanooga Shale	US	Upper Devonian	350	3–30	0.2–1
Geneseo Shale	US	Upper Devonian	385	5–30	0.2–2.5

log-normal distributions. The geometric mean size of the microcrystals is 0.5 μm. The geometric standard deviation is 2.5, giving a range in sizes for 95% of framboidal microcrystals of between 0.1 and 3.1 μm. As discussed in section 2.1.1 of Chapter 2, the log-normal distribution is expected for natural systems and reflects a complex series of interdependent processes involved in microcrystal size determination.

4.2. Numbers of Microcrystals in Framboids

Rickard (1970) originally calculated the potential number of microcrystals in a framboid by assuming spherical microcrystals in cubic close packing (ccp). The number of microcrystals, N, could then be estimated by rearranging Hartley's (1949) formula (Equation 4.1):

$$N = \varphi \left(\frac{D}{d} \right)^3 \qquad\qquad 4.1$$

where φ is function describing the packing efficiency, D is the diameter of the framboid, and d is the diameter of the microcrystal. A similar formula was used by Wilkin et al. (1996). The D/d ratio itself is a measure of the relative volume of the framboid, V, to the average relative volume of the microcrystals, v, assuming that these are spheres. As discussed later, although spherical microcrystals might occur, framboidal microcrystals are usually modified octahedral and cubes. Even so, the spherical assumption is a relatively good approximation of their volume. The average value is also a robust measure since, by definition, framboids are necessarily constituted by roughly equidimensional microcrystals.

The maximum packing efficiency for spheres is displayed by cubic and hexagonal close packing where the packing coefficient, φ, is given by Equation 4.2:

$$\varphi = \sqrt{2\pi} / 6 \approx 0.74 \qquad\qquad 4.2$$

All other packings have packing functions less than 0.74. Random close packing of spheres, which may be used as an approximation to disorganized framboid microstructures, has no precise packing function. The closest random packing of spheres has been estimated to have a packing function of around 0.64, but loose random packings may have packing functions as low as 0.56 (Dullien, 1991). The packing function for icosahedral packing tends to 0.69 (Mackay, 1962).

Plots of the maximum numbers of microcrystals for framboids of various sizes between 5 and 150 μm, calculated according to Equations 4.1 and 4.2, are shown in Figure 4.1. These calculations are for the real diameters of framboids and need to be corrected for the stereological error for measurements of framboid diameters from sections. As discussed in section 2.2 of Chapter 2, the stereological error is substantial because the consequent error in the computation of framboid volume is equal to the cube of the error, which is close to 2. This means that the actual framboid volumes computed from measurements of framboid diameters in sections are approximately half of the uncorrected figures, so that, for example, a D/d value of 10 is equivalent to a V/v ratio of 5.

As pointed out by Rickard (1970), over 1 000 000 microcrystals could be contained in a framboid. The maximum reported number of microcrystals in a framboid may be the 14 μm framboid with 0.1 μm microcrystals reported by Wilkin et al. (1996, p 3904, Figure 4b) which I estimate could contain a maximum 1 100 000 microcrystals if they were ideally cubic close packed and up to

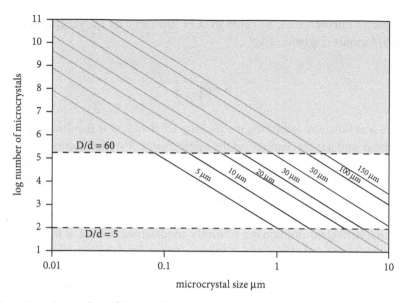

Figure 4.1. Logarithm of the number of spherical microcrystals in cubic close packing against microcrystal size (μm) for framboids between 5 and 150 μm in diameter. The limiting values of the ratios of framboid diameter to microcrystal diameter (D/d) of 5–60 reveal that framboids contain between about 100 and 500 000 microcrystals.

980 000 if they were randomly packed. Correcting this for the stereological error would suggest between 550 000 and 490 000 microcrystals in this framboid. Measured D/d values for framboids vary between about 5 and 120 (Ohfuji, 2004; Skipchenko and Berber'yan, 1976; Wilkin et al., 1996). The upper limit is computed for measurements in section, and correction for the stereological error suggests that it is more likely to approach 60. This is more compatible with estimates of a maximum D/d between 40 and 50 and the estimate of 50 for framboids measured by SEM, which consequently did not involve the stereological error (Large et al., 2001; Ohfuji, 2004). The result is that the number of microcrystals in framboids ranges between less than 100 to over 500 000, and the limiting minimum size for the microcrystals themselves is close to 0.1 μm. Below this size, the microcrystals become nanoparticles and are subject to different physical effects, as noted in section 2.8 of Chapter 2, more akin to atomic clusters. Although microframboids and nanoframboids, with diameters less than 1 μm and constituted by nanoparticles, are observed experimentally and in contemporary, especially hydrothermal, environments, there is no continuum in size between these microframboids or nanoframboids and framboids generally, as shown in section 2.1.1 of Chapter 2. The "protoframboids" described by

Large et al. (2001) form a distinct grouping on a plot of spherule diameter versus particle size and display a correlation between the apparent particle diameter and the diameter of these non-pyritic spherules. This suggests a different growth process for these objects and is consistent with the conclusion that they are not framboids but nanoparticle clusters.

There appears to be no simple relationship between framboid diameter and microcrystal diameter. For example, recalculation of the Wilkin et al. (1996) data set for Holocene+ framboids shows no correlation between framboid sizes and the number or sizes of their constituent microcrystals. Large et al. (2001) suggested such a correlation, but closer inspection reveals that this depended on their essentially structureless "protoframboids" (see Chapter 2); the framboids themselves showed a scattered distribution. However, there is a weak correlation between the logarithm of the number of microcrystals and the microcrystal diameters in a set of very large framboids measured by Ohfuji (2004). This is discussed further in section 4.2.2.

4.2.1. Packing Densities

The consequence of the variable packing coefficient in Equation 4.1 is that disordered framboids contain fewer microcrystals than ordered framboids. The relative efficiency of close packing is illustrated in Figure 4.2. The D/d values for over 300 framboids of all ages measured by Ohfuji (2004) show a clear bimodal distribution, with framboids containing disorganized microcrystals having distinctly lower D/d values than framboids with organized microcrystals microstructures. The populations of framboids with organized microarchitectures (*org* on Figure 4.2) fit closely to log-normal distributions. The hypothesis for a log-normal distribution for the D/d values for the disorganized framboids (*dis* in Figure 4.2) also cannot be dismissed, but the fit is relatively poor.

If it is assumed that the population of ordered framboids has an average packing density function, φ (Equation 4.2), approaching 0.74, then the estimated φ value for the disordered framboids with the same number of microcrystals averages 0.64, which closely approaches the estimated value of this function for the closest random packing of spheres. The agreement between the computed and theoretical packing functions suggests that the packing of disordered framboids is tight and is similar to the closest random packing of spheres. This is consistent with observations about the relative robustness of framboids, which are essentially aggregates of microcrystals. It also implies that even the packing of disordered microcrystals in framboids involves forces which hold the microcrystals together and bind them tightly in the framboid (section 13.1 in Chapter 13).

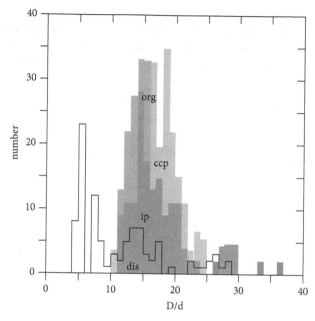

Figure 4.2. Plots of the ratio of framboid diameter to microcrystal diameter (D/d) for over 300 framboids of all ages versus number measured by Ohfuji (2004). The framboids with an organized microstructure (org, icosahedral packing, ip, and cubic close packing, ccp; see section 5.2 in Chapter 5) have greater D/d values that framboids where the microcrystals are randomly packed and show no ordering (dis).

Wang et al. (2018) demonstrated that spheroidal clusters of colloidal particles showed number-dependent free energy minima analogous to the well-established magic numbers of atomic aggregates. Wang et al. (2019) replaced the D/d ratio with the confinement ratio, which they defined as R/d, where R is the radius of the spheroid. The confinement ratio is then half the D/d ratio. They found that plots of the computed free energy versus the confinement ratio showed distinct minima which coincided with the number of particles needed to build ideal icosahedra with increasing numbers of shells or layers of particles. The interpolation of the Wang et al. (2019) free energy computations to framboids is not possible directly since their calculations were based on spherical rather than crystalline particles in relatively water-rich ($\varphi = 0.52$), rather than anhydrous, framboidal, spherical clusters. However, qualitatively, their data are consistent with the relationship between framboid packing densities and free energy minimization.

The result further suggests that the average numbers of microcrystals in both disordered and ordered framboids are similar, and that the reason for the bimodality displayed by the D/d function is the intrinsic variation of the packing density. The consequence of this is that framboid microcrystal formation is not dependent on microcrystal packing or organization. Microcrystals form in framboids and are subsequently subject to organization, or not, as the case may be. This further suggests that the organization of microcrystals in framboids is the result of an additional process which does not occur in a majority of framboids.

4.2.2. Ostwald Ripening

Although most microcrystals in framboids are necessarily small, Ohfuji (2004) studied a set of 25 very large framboids ($\geq$80 μm diameter) from the 2 Ma Neogene of Japan (Table 4.2). As shown in Figure 6.8 in Chapter 6, there is a weak correlation ($R^2 = 0.8$) between the logarithm of the number of microcrystals in a framboid and the microcrystal diameters for this set of framboids. The larger the microcrystals, the fewer there are in the framboid, and this is independent of the framboid volume. In particular, the observations may evidence Ostwald ripening in microcrystal growth. That is, the larger microcrystals have a lower solubility than the smaller ones and grow at the expense of the smaller crystals. In the context of large framboids, this may explain why there is a weak negative correlation between the number of microcrystals in the framboid and the microcrystal size. The lack of an arithmetic proportionality, which might be expected, is interesting. The weak exponential relationship suggests that the processes controlling microcrystal numbers in framboids are complex but are ultimately related to the processes involved in microcrystal growth itself, such as Ostwald ripening (see section 12.2 of Chapter 12). The limiting size of framboid microcrystals is related to the fundamental processes of nucleation and diffusion-controlled crystal growth and the increasing relative importance of gravitational forces in larger pyrite crystal aggregates, as discussed in section 3.2.3 (Chapter 3) and section 13.3 (Chapter 13).

At the other end of the scale, the smallest microcrystals in framboids appear to be approximately 0.1 μm (100 nm) in size. In fact, the statistics suggest that the lowermost size limits of 95% of framboid microcrystals approaches 0.08 μm (80 nm) and the 99% range is defined by a lower limit of 0.03 μm (30 nm). As discussed in section 2.8 of Chapter 2, nanoparticle aggregates are chemically labile and may rapidly fuse to form pyrite crystals via aggregative growth. This appears to constrain the lower limit of framboid microcrystal sizes.

Table 4.2 Microcrystal Sizes (d μm) and Numbers for a Set of 25 Very Large Framboids of Diameter D μm, from the 2 Ma Neogene of Japan Arranged in Order of Increased Packing Efficiencies.

D (μm)	d (μm)	D/d	Number of Microcrystals	Microcrystal Packing Structure
119	6	21	6877	Disordered
140	5	30	20726	Disordered
108	3	35	31596	Disordered
105	3	37	38199	Disordered
77	1	54	114468	Disordered
84	7	13	1576	Icosahedral
94	7	14	2142	Icosahedral
144	10	14	2223	Icosahedral
120	8	15	2277	Icosahedral
90	6	15	2372	Icosahedral
107	7	15	2379	Icosahedral
98	7	15	2405	Icosahedral
119	8	15	2742	Icosahedral
98	5	19	4771	Icosahedral
92	4	23	9515	Icosahedral
117	4	29	18658	Icosahedral
117	3	34	27881	Icosahedral
107	10	11	880	ccp
110	10	11	1003	ccp
96	8	12	1228	ccp
79	7	12	1277	ccp
92	7	13	1507	ccp
83	6	13	1803	ccp
80	4	18	4600	ccp
103	3	37	36444	ccp

From Ohfuji (2004).

4.3. Microcrystal Size and Geologic Age

There seems to be no correlation between the sizes of microcrystals in framboids and their geologic age. This is illustrated, for example, in Table 4.1 where the microcrystals in ca. 375 Ma old framboids have sizes ranging from 0.2 to 2.5 μm, which is similar to the range displayed in modern lake sediments. This means that generally microcrystals remain the same over geologic time period and do not, as might be expected, grow larger with time. The Ostwald ripening suggested by the very large Neogene framboids discussed earlier is therefore a result of continuous crystal growth of these microcrystals and is part of their formation process. Framboid microcrystals appear to remain stable over geologic time periods.

4.4. Pyrite Crystal Habits

The microcrystals comprising a framboid are basically all equant, equidimensional, and equimorphic by definition (see section 1.1 of Chapter 1). Thus, minerals that do not commonly produce equant crystals forms are unlikely to display the framboidal texture. This essentially limits the framboid to minerals, like pyrite, displaying isometric symmetry.

The three common fundamental habits of pyrite are shown in decreasing order of abundance in Figure 4.3. The cube is by far the most abundant simple habit displayed by pyrite, followed by the less common pyritohedron and the rarer octahedron. All three of these forms resemble Platonic solids, although the pyritohedron is necessarily a modified pentagonal dodecahedron.

The other regular forms that satisfy the symmetry requirements of pyrite are the diploid, rhombic dodecahedron, trisoctahedron, and trapezohedron, and these are described together with the cube, pyritohedron, and octahedron in Table 4.3.

Figure 4.3. The basic crystal habits of macroscopic pyrite in order of abundance: (a) cube (42 mm side length); (b) pyritohedron (42 mm height); (c) octahedron; (20 mm side length).

Table 4.3 The End-Member Regular Crystal Forms of Pyrite in Decreasing Order of Abundance

Name	Description	Form Symbol	
Cube	4 rectilinear faces	$\langle 100 \rangle$	
Pyritohedron	12 irregular pentagonal faces	$\langle 201 \rangle$	
Octahedron	8 triangular faces	$\langle 111 \rangle$	
Diploid	24 irregular 4-sided faces	$\langle 321 \rangle$	
Rhombic dodecahedron	12 rhombic faces	$\langle 110 \rangle$	
Trisoctahedron	24 triangular faces	$\langle 221 \rangle$	
Trapezohedron	24 similar 4-sided faces	$\langle 211 \rangle$	

However, these simple habits themselves are uncommon, and most pyrite crystals display combinations of these forms. In his quite amazing compilation, Birkholz (2014) displayed 691 different pyrite habits. Topological crystallography reached its zenith in the 1920s, and by 1931 Goldschmidt (1920) needed to make the case for its continued value in the face of the great advances made by X-ray crystallography. A sample of one of the 42 pages of his line drawings of pyrite crystal habits is shown in Figure 4.4. In this last hurrah of topological crystallography, Tokody (1931) reported that 449 different crystal forms had been described for pyrite by 1930, of which around half were *sicher* or proven forms. The permutations of the pyrite forms result in pyrite displaying the greatest variety of crystal shapes among the common minerals. This variety has an added significance in that pyrite is able to approximate forbidden fivefold symmetries such as the pentagonal dodecahedron, but with asymmetric pentagonal faces, and the icosahedron, again with different sized triangular faces, as a combination of the octahedron and pyritohedron.

Studies of framboids were originally based on studies of polished sections using optical microscopy. Since this resulted in random sections through the constituent microcrystals, an extensive scheme based on the probability of a particular habit producing a defined shape in two-dimensional sections was developed by Tokody (1931).

4.5. Framboid Microcrystal Habits

It was earlier suggested that the microcrystalline habit determined the organization of the framboid (cf. Wignall and Newton, 1998; Wilkin et al., 1996). However, as discussed in the following, small variations in the form of packed polyhedra can produce significant changes in the structure of the resulting aggregate, and there is no way, at present, of predicting aggregate structure from the shape of the constituent polyhedra (Love and Amstutz, 1966).

Reports on the habits shown by pyrite microcrystals in framboids have varied considerably over the years as technology has improved. Thus, early investigations were with optical microscopy, using reflected light on polished sections of framboids. Reflected-light microscopy has some advantages over conventional transmitted light microscopy since the object being observed is two dimensional and its image suffers less degradation by dispersion. Even so, the resolution is limited by the wavelength of light in the visible spectrum to around 0.2 µm, as discussed in section 1.2.2 of Chapter 1. The other problem with the optical microscopic approach was that framboidal microcrystals were observed in random section. Some considerable effort was made to provide guides to the probability of a particular 2D image being related to a specific 3D habit (e.g., Zimmerman

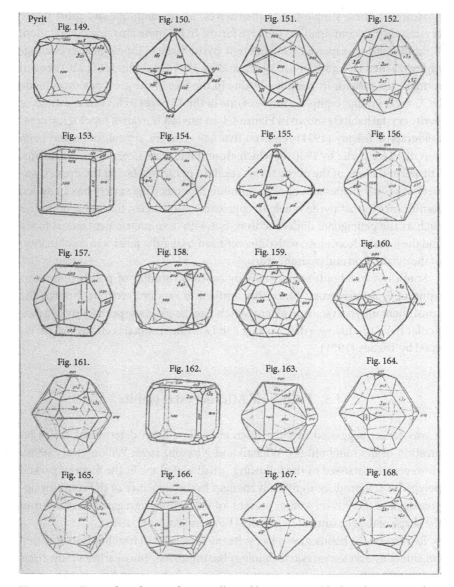

Figure 4.4. Examples of pyrite forms collected by Victor Goldschmidt in 1920. This is one page of 42 pages illustrating almost 700 different pyrite forms.

and Amstutz, 1973), but this approach was limited by the large variety of potential pyrite crystal forms. There was a brief period when transmitted electron microscopy was applied to investigating framboid microcrystals, but this approach used images from collodion peels and suffered from the same 2D problem as polished sections. The introduction of scanning electron microscopy initiated a revolution in framboid microcrystal mineralogy. This technique provides views of whole framboid microcrystals and not just sections, permitting a more robust assignment of crystal forms to be applied. However, it should be recalled that SEM resolution has also improved with time: the early Stereoscan Mark 1 of 1965 had a resolution of 10 nm, whereas a more modern instrument might approach 0.8 nm resolution. The significance here can be seen in the relative sharpness of the SEM images comparing, for example, the images from 1974 in Figure 8.7 in Chapter 8 with those from 2010 in Figure 1.1 of Chapter 1, which were collected on different generations of SEMs. Although the general forms of the framboid microcrystals were resolvable on the earlier SEMs, the detailed forms were only visible with some certainty on the later models.

4.5.1. Cubic Microcrystals

Simple cubes have not been widely reported from framboids since the introduction of electron microscopic techniques. The original identification of microcrystals as simple cubes was made by Love and Amstutz (1966) in framboids from the 390 Ma Devonian Chattanooga Shale in the United States. Kalliokoski and Cathles (1969) also showed an image of pyrite microcrystals with square cross sections. Love et al. (1971) argued that both these were "convincing evidence" of cubes in framboids. They based their conclusion on the statistical improbability of such square cross sections being derived from forms other than cubes. Figure 4.5 shows the problem that early workers had with identifying the shapes of sections of microcrystals with optical microscopy.

Although at first glance the microcrystals look cubic, in detail there are few symmetric crystals and even fewer 90° interfacial angles. The microcrystals in the left side of the image of Figure 4.5 are clearly octahedra, and a layer of octahedral vertices occurs just below the main surface of the section, visible as a row of points of light in the image. Similarly, the image presented by Kalliokoski and Cathles (1969) also showed microcrystals with square cross sections admixed with other polygonal shapes, including clear octahedra.

When cubes are obvious as the form of pyrite microcrystals, the subsequent crystal packing is often highly disordered and becomes a simple cluster of interlocking pyrite microcrystals (Figure 4.6). These forms have close similarity to some of the synthetic framboids described in Chapter 11. There is an obvious gradation between these clusters and framboids (see section 10.4 in Chapter 10). Apparent spheroidal clusters of cubic microcrystals can also be produced by overgrowths on structureless pyrite spherulites (Wang and Morse, 1996), and there is no reason to assume that similar overgrowths could not be produced on framboids.

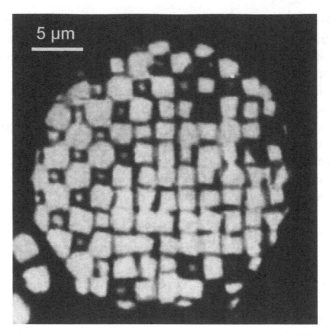

Figure 4.5. Possible cubic microcrystals from an optical microphotograph by Love and Amstutz (1966). Detailed study suggests that, at best, the microcrystals are irregular cubes and they may actually be sections through octahedra. The scale is estimated.

The higher resolution offered by the SEM provided more detailed insight into apparent cubic microcrystals in framboids. The image in Figure 4.7(a) looks at first sight to include cubic forms, but detailed imaging (Figure 4.7(b)) reveals that the microcrystals are quite irregular and appear to be octahedra in different orientations. Even those few microcrystals with square sections appear to display a faint diagonal cross, suggesting that they may be truncated octahedra.

4.5.2. Pyritohedral Microcrystals

The idea that pyritohedral microcrystals occurred in framboids originally derived from elementary theoretical considerations of the origin of some packing geometries observed by Love and Amstutz (1969).

Love et al. (1971) used electron microscopy to demonstrate that pyritohedral forms did occur in framboids (Figure 4.8). However, even here the forms were mostly identified as individual pentagonal faces admixed with cubes $\langle 1\ 0\ 0 \rangle$.

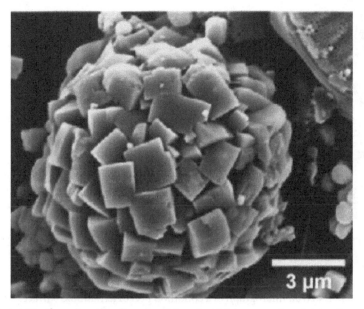

Figure 4.6. Pyrite cubes in a disordered framboid from a modern lake sediment.
Scanning electron microscope image by Hiroaki Ohfuji.

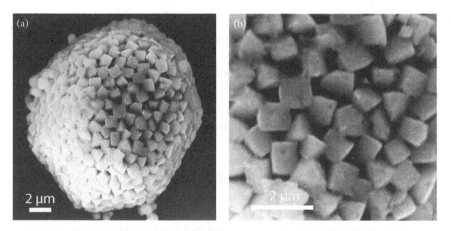

Figure 4.7. (a) Framboid appearing to display cubic microcrystals. (b) Detail
showing that the few square sections may be truncated octahedra.
Scanning electron microscope images by Anders Elverhøi.

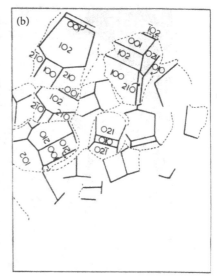

Figure 4.8. Electron micrograph of a collodion peel of a framboid surface.
(b). Interpretation of the crystal forms including pyritohedral faces (2 1 0), (1 0 2)
and (0 2 1).

From Love, L. G., et al. 1971. Framboidal pyrite: Morphology revealed by electron microscopy of
external surfaces. *Fortschritte der Mineralogie*, 48(2): 259–264. Fig.1a and b.

Some of these microcrystals are reminiscent of the pyrite habit described by
Goldschmidt (Figure 4.4 # 166) of a pyrite crystal found in Traversella, an ancient mine in the Piedmontane Alps north of Turin.

4.5.3. Octahedral Microcrystals

Sweeney and Kaplan (1973) reported an early investigation using SEM of
framboids from sediments off the Californian coast. They noted that the majority of microcrystals were octahedra. With improved electron microscope
techniques, Ohfuji (2004) and Butler (1994) were able to study the habits of
the microcrystals in pyrite framboids in detail and found that high-resolution
analyses of framboids showed that a vast majority of microcrystal forms were
actually octahedra or modified octahedra. This is an important observation,
since octahedra are the least common of the three basic forms of macroscopic
pyrite. This suggests that the processes forming the pyrite microcrystals
in framboids are different from those forming the macrocrystals. The
differences in abundance of octahedra and cubes in pyrite framboids and

macrocrystals are obviously related to crystal size. Pyrite requires extreme supersaturations to nucleate, and thus the early formed crystals are likely to possess the least stable, most soluble form: the octahedron. As crystallization proceeds, the supersaturation decreases and cubic faces form. Long-term crystal growth under relatively low supersaturation conditions leads to the formation of larger cubic crystals. These observations are consistent with the physico-chemical model of microcrystal growth described in section 12.2.6 of Chapter 12.

It is rare for simple octahedral microcrystals to occur, and detailed analyses of most reveal cube terminations, reflecting the decreased supersaturation at the end of microcrystal growth (e.g., Figure 4.9). Most octahedral microcrystals are modified by cube terminations leading to the formation of cuboctahedra or, more commonly, truncated octahedra. The series has been identified by Ohfuji (2004) (Figure 4.10).

The morphology varies normally between simple cubes to octahedra via cuboctahedra to truncated octahedral shapes as the {1 1 1} faces develop. These forms include three Archimedean polyhedra, the truncated cube, the truncated octahedron, and the cuboctahedron. The significance of these Archimedean solids is that they are highly symmetric and are composed of regular polygons, hence facilitating efficient packing geometries.

Figure 4.9. (a) Octahedral microcrystals, some with cube terminations, from a framboid in the 360 Ma, upper Devonian Gowanda shale, New York.
Scanning electron microscope image by Ian Butler.
(b) Octahedral microcrystals from Holocene sediments, Japan.
Scanning electron microscope image by Hiroaki Ohfuji.

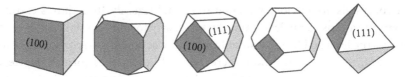

Figure 4.10. Common pyrite microcrystal habits in framboids varying between simple cubes and octahedra with combinations of these.

Figure 4.11. (a) Rhombic dodecahedra. (b) Trisoctahedra (triakis octahedra).
Scanning electron microscope images of pyrite framboids in Holocene sediments in Japan by Hiroaki Ohfuji.

4.5.4. Higher Order Forms

Ohfuji (2004) reported a series of higher order pyrite microcrystals forms from framboids in Holocene+ sediments in Japan. The growth of triangular pyramids on the faces of the original octahedra leads to the formation of trisoctahedra or triakis octahedra (Figure 4.11(b)). Trisoctahedra have Miller indices {2 2 1} and are known forms in macroscopic pyrite.

The trisoctahedra appear to develop as space-filling forms between octahedra in ccp. In this model, the octahedra formed and aggregated into organized ccp, and then growth continued in the void spaces. Alternatively, and more probably, the process was continuous. As discussed in section 12.2.1 of Chapter 12, the form probably developed as triangular dislocation hillocks on the octahedral face. Ohfuji (2004) also observed rhombic dodecahedra in the microcrystals of pyrite framboids (Figure 4.11(a)). Rhombic dodecahedra are composed of 12 rhomb faces and are sometimes simply referred to as dodecahedra. The rhombic dodecahedra are geometrically related to the trisoctahedra and can be envisioned as (1 1 0) faces infilling the spaces between the trigonal pyramids. Rhombic dodecahedra were also described from framboids in coal from northeastern Spain by Querol et al. (1989).

It is interesting to note that both of these forms—the rhombic dodecahedron and the trisoctahedron—can pack regularly to produce organized arrays of microcrystals. In Figure 4.11 the trisoctahedra appear to pack in hcp (hexagonal close packing) and the rhombic dodecahedra in ccp.

4.5.5. Spheroidal Crystals

Under the optical microscope, pyrite microcrystals in framboids often appear to be spheroidal. However, it has been difficult to determine whether or not these are actual curved or spheroidal crystals because of the limitations of optical resolution. Akai et al. (2004) found that the individual faces on some microcrystals could not even be resolved in the SEM and reported that the microcrystals appeared to be rounded cuboids. An example of apparently spheroidal microcrystals is shown in Figure 4.12. Even under this resolution, the microcrystals appear to be similar to the spherical polystyrene latex spheres used

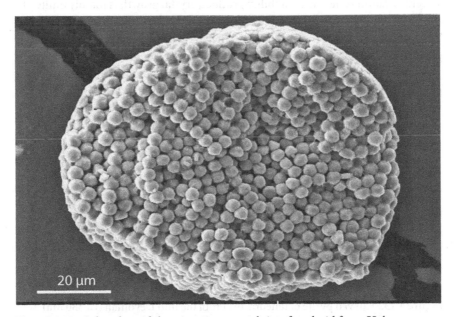

Figure 4.12. Sub-spheroidal pyrite microcrystals in a framboid from Holocene sediments in Japan. Detailed examination reveals them to be cuboctahedra. Note the icosahedral packing.

Scanning electron microscope image by Hiroaki Ohfuji

in calibrating SEM magnification. However, in detail they appear to be cubes modified by octahedra. The nascent crystallinity of these microcrystals leads to more diffuse edges and a more spheroidal appearance.

However, clearly curved pyrite microcrystals do exist in framboids. Butler (1994) discovered pyrite crystals with obvious curved faces in a 380 Ma pelite at Rammelsberg, Germany (Figure 4.13(a)). Ohfuji (2004) presented a high-resolution image of a Holocene+ framboid from Japan which clearly displays curved microcrystals (Figure 4.13(b)). These curved crystals would appear virtually spherical in section. There has been some recent theoretical interest in the development of curved pyrite crystals (Andersson, 2005). The curvature may be a result of incomplete development of higher order crystal forms, such as the pyritohedron or rhombic dodecahedron, on the original octahedral faces.

Alternatively, they may reflect differential surface free energies across a crystal face, known as the Berg effect (Berg, 1938). The conventional Berg effect explains the growth of skeletal or hollow crystals during rapid crystal growth as due to the supersaturation at the center of a crystal faces being much lower than that at the higher energy sites at the crystal edges and apices (see section 4.5.6). The reverse effect may result from extremely low supersaturations. That is, when the supersaturation is near the solubility product, crystal growth is mainly limited to the low energy sites in the centers of the crystal faces, resulting in curved faces. That is, whereas the conventional Berg effect is generally related to rapid crystal growth, the formation of curved faces may reflect slower growth.

Butler (1994) presented an alternative origin for rounded microcrystals in pyrite framboids. He showed that rounded microcrystals in framboids within fossilized wood were coated with organic material derived from biofilm

Figure 4.13. Curved pyrite octahedra. (a) From the mid Devonian (ca 380 Ma) Rammelsberg deposit associated with framboids.
Scanning electron microscope image by Ian Butler.
(b) From a framboid in the Holocene Shirone formation.
Scanning electron microscope image by Hiroaki Ohfuji.

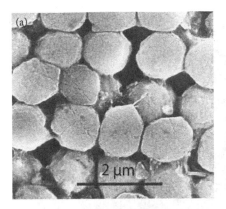
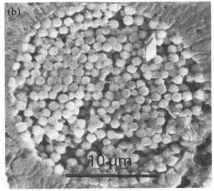

Figure 4.14. (a) Rounded pyrite microcrystals coated with organic material. (b) Framboid with rounded microcrystals in ca. 50 Ma Tertiary wood.
Scanning electron microscope images by Ian Butler.

(Figure 4.14), either relicts of the original plant material in which the pyrite developed or, as discussed in more detail in section 7.2.1 of Chapter 7, microbial biofilm. The microcrystals appear to be essentially octahedral in various states of impeded growth by the biofilm.

4.5.6. Hollow Microcrystals

Cevales (1961) reported small depressions on the faces of individual microcrystals (Figure 4.15) using a thermionic emission electron microscope—a primitive forerunner of the modern photo-emission electron microscope. Cevales (1961) was examining framboid sections and thus the small depressions he reported were actually hollow microcrystals. He also noted that optical microscopic observations of these polished sections of framboids did not reveal the depressions in the center of the microcrystals.

One of the more dramatic discoveries by Butler (1994) was that apparently perfect octahedral microcrystals in some framboids were hollow. Hollow microcrystals had been reported in framboids before (e.g., Schallreuter, 1984). However, these were at the framboid surface. Butler's discovery was significant since it showed that apparently pristine microcrystals within framboids could be hollow. He also showed that the hollow microcrystals were not a preparation artifact since these samples had not been subject to any chemical treatment and that it was improbable that there was originally some material in the cores of the crystals that has since disappeared. His findings explained the original report of Cevales (1961).

Figure 4.15. Original emission electron microscope image of hollow microcrystals in framboids from the Italian Alps by Cevales (1961).

The fact that these apparently perfect octahedra are hollow further raises the possibility that other pyrite crystals which seem to be solid are also hollow. Of course, once someone has noticed a form such as that marked A on Figure 4.16, it is easy to find more examples. However, spotting the first one is the key. These hollow crystals are varieties of skeletal crystals which develop through rapid growth in high supersaturation environments where molecular pyrite moieties are added more rapidly to the high energy edges and apices of the rapidly growing crystals (Berg, 1938).

In pyrite framboids, hollow microcrystals are more commonly expressed as depressions or holes in microcrystal faces, especially those at the framboid exteriors (Figure 4.17). Of course, it is unknown whether these skeletal forms are indeed concentrated at the exterior since the apparently whole octahedral microcrystals in the interior might in fact be hollow. The discovery of similar forms in 275 Ma middle Permian sediments from Australia were interpreted as fossil microbial colonies (Gong et al., 2007). In fact, these framboids are indeed associated with copious amounts of probable microbial biofilm within worm burrows. Gong et al. (2007) erroneously stated that these framboids with hollow

Figure 4.16. Hollow octahedral microcrystals in a framboid from the ca 280 Ma Cafana deposit in Turkey. The broken crystal at A clearly shows that the pyrite octahedra are hollow.

Scanning electron microscope image by Ian Butler.

microcrystals have not been observed outside the worm burrows, and cite this as the key reason why they thought that these were fossil microbial colonies. The conclusion follows in the great tradition of some workers considering that these complex structures cannot be formed by physico-chemical processes, but must be molded on a biologic template.

Ohfuji (2004) presented even clearer evidence for the Berg effect in framboid crystal growth with the discovery of hollow-faced octahedral microcrystals in the outer layer of a framboid from Eocene sediments in Japan (Figure 4.18). The centers of the crystal faces have clearly not grown at the same rate as the edges

Figure 4.17. (a) Hollow surface microcrystals in framboids from the ca. 370 Ma Upper Devonian Gowanda shale New York. (b) Detail showing the incomplete forms at the surface.

Scanning electron microscope image by Ian Butler.

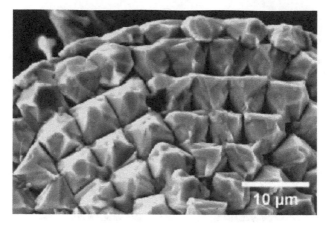

Figure 4.18. Hollow faces on octahedral microcrystals from a framboid in ca. 45 Ma Eocene sediments.

Scanning electron microscope image by Hiroaki Ohfuji.

and apices, leading to this odd dimpled effect. The depressions on the faces can be seen to be stepped in places.

4.5.7. Incomplete Microcrystals

Incomplete microcrystals at the surface of pyrite framboids were first noted by Sweeney and Kaplan (1973) in framboids from the San Diego trough and the Santa Barbara basin off the Californian coast. They considered that these microcrystalline growth forms were caused by impeded growth against an unidentified surface membrane which enclosed the framboids and which dissolved readily in HCl. The nature of this surface membrane was not defined. Images of framboids untreated with HCl were blurred, and Sweeney and Kaplan attributed this to the presence of the surface membrane. Surface membranes on framboids have been observed since and generally consist of organic material, usually biofilm. However, the membrane reported by Sweeney and Kaplan dissolved in HCl and therefore organic composition is improbable. In the absence of subsequent reports of surface membranes of this type, it is possible that they were formed artifactually during sample preparation.

Many framboids do not appear to grow into free space but seem to be limited by either minerals in the hosting sediments or organic materials in which the framboids have grown. This appearance of impeded growth of the surface microcrystals is enhanced by the relatively smooth outer surfaces of some framboids (Figure 4.19).

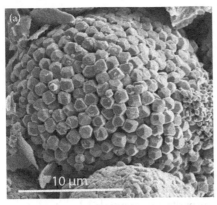

Figure 4.19. (a) Smooth surfaced framboid from a Holocene sediment, Japan.
Scanning electron microscope image by Hiroaki Ohfuji.
(b) Section of a smooth surfaced framboid in ca. 50 Ma Tertiary wood, UK, showing evidence of impediment to growth of outer microcrystals.
Scanning electron microscope image by Ian Butler.

In fact, closer observation of the surface microcrystals reveal that their poorly developed habits have not been caused by impeded growth. An example is shown in Figure 4.20, where framboids in the 390 Ma Chattanooga shale occur in a clay mineral matrix typical of formation in muddy sediments. Growth in the outermost crystals of the framboids was severely curtailed, leading to the formation of small octahedra with large outer cubic {0 0 1} faces. A similar effect was observed by Sweeney and Kaplan (1973) in framboids in modern clayey sediments. However, it is clear that the framboids in the ancient shales are not growing against the clay minerals, and a clear gap occurs between the enclosing minerals and the framboid surface (Figure 4.20(b)). Of course, it could be argued that subsequent dewatering of the sediment led to differential shrinkage of the clay and pyrite producing the gap. Even so, the fact that the clay minerals are deformed around the framboid (Figure 4.20(a)) proves that the framboid—and its outermost crystals—were formed before the framboid was sheathed in clay. The impediment to microcrystal growth was not, then, the clay minerals. This is consistent with the early diagenetic formation of the framboids in muddy sediments with >80% water content and thus high porosity. The framboids grew in an aqueous solution and the origin of the partially formed outer crystals must be sought elsewhere.

The association of framboids with organic material may show a different aspect of the relationship of the outer layer of microcrystals with the encasing material. In the simplest case, framboids growing within the conductive tissue of fossil wood appear to show clear examples of impeded growth of their outer

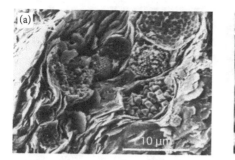
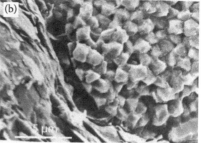

Figure 4.20. (a) Framboids sheathed in clay minerals from the ca 370 Ma Chattanooga shale. (b) Detail of edge of framboid showing incomplete microcrystals and the gap between the clay minerals and the pyrite microcrystals.
Scanning electron microscope image by Ian Butler.

Figure 4.21. Detail of apparent impeded growth of microcrystals at the margins of a framboid in 50 Ma fossil wood.
Scanning electron microscope image by Ian Butler.

microcrystals (Figure 4.19(b)). However, detailed analysis (Figure 4.21) shows that the microcrystals have developed the characteristic outer cubic faces, but that the relationship with the organic material is not simple. In particular the microcrystals appear to have grown into (replaced) organic matter, rather than grown against it, as in wall growth. In this case, the matrix in which the framboids occur is lignified organic material which occupies the original fluid-filled xylem of the tree (Grimes et al., 2002).

It appears that the framboids originally formed in organic matter and have subsequently been encased in more kerogenous material during diagenesis and dewatering (see section 7.2 of Chapter 7). Comparison of framboids in modern

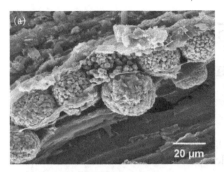 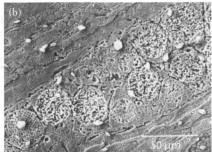

Figure 4.22. Comparison of framboids: (a) in wood in Holocene muddy sediments from the Sakata lagoon, Japan.
Scanning electron microscope image by Hiroaki Ohfuji.
(b) ca. 50 Ma old framboids in fossil wood from London Clay.
Scanning electron microscope image by Ian Butler.

wood with the 50 Ma old Tertiary material (Figure 4.22) seems to support this interpretation.

The original suggestion for impeded growth by Sweeney and Kaplan (1973) was for framboids sheathed in a diffuse material which they thought was organic matter. The fact that this material was soluble in HCl makes this identification less likely. However, it did reflect the discovery by Love (1957) of organic sacs encasing framboids. Love used classical micropaleontological techniques to separate these organic sacs. These were insoluble in HCl, HNO_3, and HF. A typical organic sheath around a framboid is shown in figure 7.13 in Chapter 7. The material is similar in appearance to that shown by Sweeney and Kaplan (1973) but, by contrast, is insoluble in HCl. The origin of these sacs is discussed in section 7.1.2 (Chapter 7). However, the key feature was that pyrite microcrystals grew with these organic bodies, and dissolving out the pyrite with HNO_3 left an organic meshwork. In other words, these organic sacs did not impede microcrystal growth, and the cause of any limitations of crystal growth of facets of their surface microcrystals must be sought elsewhere.

Butler (1994) concluded that the reason for limited growth of microcrystals on the outer surfaces of framboids was a steep negative concentration gradient in dissolved nutrients. He thought that the enclosing mineral or organic material produced a partially closed environment that enhanced this gradient. As discussed in some detail in Chapter 11, Butler's conclusion was essentially correct. However, the concentration gradient was produced by rapid pyrite nucleation and growth in a diffusion-limited regime. The diffusion-limited regime was, of course, at least partly due to the fine-grained mineral and organic matter in which the growing framboid was embedded.

5

Framboid Microarchitecture

The extraordinary microcrystal ordering displayed by some pyrite framboids led to considerable interest in the later decades of the 20th century because processes leading to self-organization in nature were at the cutting edge of international research at the time. Some workers considered that these remarkable microarchitectures could not arise through inorganic processes, but required the prior presence of a biological template, either in the form of colonies of microorganisms or single multicellular bodies. Another group thought that a preexisting mineral template was required.

Since that time, understanding of the processes leading to assembly of nanoparticles and colloidal particles into microscopic structures has become a significant research theme in the chemical, biological, and material sciences. The attention has changed to a focus on understanding the formation processes of nano- and micro-crystals in order to synthesize specific forms. Pyrite framboids are obvious targets here because of the industrial potential of manufacturing organized microscopic arrays of doped semiconductor particles. However, the only systematic modern work on the nature of framboid microstructures has been carried out by Ohfuji and his coworkers at the beginning of the century (Ohfuji, 2004; Ohfuji and Akai, 2002), and this preceded the surge in studies of self-assembly process in the materials science literature.

It is tempting to compare these organized microstructures with *superlattices*. Classically, superlattices were defined in terms of semiconductor superlattices as man-made periodic structures with a period larger than the solid unit cell usually consisting of two or more materials. Since that time, superlattices have expanded to include periodic structures composed of nanocrystals and natural superlattices, such as opal, are well known. Indeed, since 1985 there has been a *Journal of Superlattices and Microstructures* dedicated to research into synthetic heterostructures. In this context, a framboid with an organized (i.e., periodic) microstructure might be thought of as a natural superlattice. However, as is shown in section 6.3 of Chapter 6, the individual microcrystals in framboids are not in crystallographic continuity. Thus, although the morphologic arrangement of the microcrystals may be periodic, the crystallographic ordering is not. Additionally, the term *superlattice* might be confusing in this context, since in crystallography a *lattice* is a mathematical abstraction which expresses the periodicity of the crystallographic structure.

Framboids. David Rickard, Oxford University Press. © Oxford University Press 2021.
DOI: 10.1093/oso/9780190080112.003.0005

Although a number of earlier studies described the microstructures of pyrite framboids, most of them used reflected-light microscopy, and therefore they contributed to understanding framboid microarchitecture in two dimensions. The detail of the 3D packing structure of microcrystals was not fully understood until the work of Ohfuji and his colleagues at the beginning of this century, cited earlier. I discuss the earlier work since many investigators observe framboids in rock sections and it is therefore important to discuss the relationship between these two-dimensional slices through framboids and the three-dimensional reality.

5.1. Framboid Sections

The studies of sections through framboids revealed that the microcrystals in pyrite framboids commonly showed extremely regular arrangements (e.g., Kalliokoski and Cathles, 1969; Love and Amstutz, 1966; Rickard, 1970; Skipchenko and Berber'yan, 1976). These two-dimensional studies reached their acme with the work of Love and Amstutz (1966). They described a variety of two-dimensional (2D) microcrystal organizations observed in pyrite framboids in two 350 Ma Carboniferous black shales. They classified the ordered structures into four types: (I) square patterns, (II) hexagonal patterns, (III) irregular six sided patterns, and (IV) more complex patterns. Kalliokoski and Cathles (1969) also studied the 2D ordered structure in framboids and assumed that it exclusively reflected cubic close packing (ccp). Later studies reported that in most cases such 2D domain structures displayed symmetrical patterns, such as threefold (Morrissey, 1972; Ohfuji and Akai, 2002; Skipchenko and Berber'yan, 1976), fivefold (Large et al., 2001; Love and Amstutz, 1966; Morrissey, 1972; Ohfuji and Akai, 2002; Skipchenko and Berber'yan, 1976), and sixfold (Morrissey, 1972; Skipchenko and Berber'yan, 1976), suggesting that framboids might be polyhedral rather than simply spherical. Indeed, Large et al. (2001) and Ohfuji and Akai (2002) demonstrated that the framboids displaying threefold and fivefold domain structures in sections are pseudo-icosahedral in both external shape and internal structure (see Figure 5.8 later in the chapter).

Rickard (1970) noted that the framboid microstructures are not always uniform in a framboid, but are sometimes made up of several sub-domains which display various geometries including disordered sub-domains. He classified framboids into three basic levels according to the packing state of their constituent microcrystals: (1) ordered, (2) partially ordered, and (3) disordered structures. Ohfuji (2004) added a fourth group which displayed different packing geometries within the same framboid. He referred to these as *multiple ordered structures* and showed that these explained some of the complex patterns displayed by framboids observed in 2D sections.

The relative abundances of ordered and disordered framboids were studied by Ohfuji (2004). As shown in Table 5.1, Ohfuji observed a tendency for ordered framboids to be more abundant in ancient sedimentary rocks, but the data are insufficient to establish this. I also get the impression of a greater proportion of disordered framboids in recent sediments, but I have no explanation for this phenomenon if it is real.

5.2. Single Domain Microarchitectures

Apparent single domain microarchitectures have been widely reported in framboids (e.g., Butler, 1994; Kalliokoski and Cathles, 1969; Love and Amstutz, 1966; Rickard, 1970; Skipchenko and Berber'yan, 1976). The constituent microcrystals are exceptionally uniform in both size and morphology and are arranged into a single array, that is, visually at least, coherent through the entire volume of the framboids. In 2D sections, they show a variety of arrangement patterns, which can be classified into three basic geometries: cubic (Figure 5.1(a)), hexagonal (Figure 5.1(b)), and linear arrays (Figure 5.1(c)).

Love and Amstutz (1966) proposed that the different internal geometries observed in 2D sections of framboids reflected different packing types of their constituent microcrystals. Thus, cubic geometries reflected cubic close packing (ccp), and hexagonal geometries reflected hexagonal close packing (hcp). Linear geometries were then the result of incomplete microcrystal packing, reflecting varying microcrystal habits. This was an important proposal since it informed contemporary understanding about the nature of the physical processes responsible for the packing.

5.2.1. Cubic Close Packing

Kalliokoski and Cathles (1969) questioned the Love and Amstutz (1966) interpretation and proposed all organized framboids were ccp and that hcp rarely, if ever, existed. The discussion of the occurrence of alternative packing geometries for framboids was continued by Love et al. (1971) and Wilkin and Barnes (1997a), who proposed that both ccp and hcp occurred. The problem was that Kalliokoski and Cathles never actually described how the various microcrystal geometries observed in 2D sections could all derive from simple ccp.

In fact, as shown in Figure 5.2, various sections through ccp can display all the uniform 2D arrays observed by Love and Amstutz (1966). Thus, if a section is cut

Table 5.1 Properties of Pyrite Framboids: Framboid Diameter (D μm), Microcrystal Size (d μm), Packing Geometry

			Ma	Framboid D (μm)	Microcryst d (μm)	Packing Geometry		
						R	PO	O
Sedimentary rocks								
Udenaha mudstone	Okinawa	Japan	Pleistocene	1	20–145	3–10	+	+	+++
Uonuma Formation.	Niigata,	Japan	Pleistocene	1	30–50	2–3		+++	+++
Nishiyama Formation	Niigata,	Japan	Miocene	2	15–140	1–10	+	+++	+++
Teradomari Formation	Niigata,	Japan	Miocene	10	5–70	0.5–4	++	+++	+++
Cleveland ironstone	Yorkshire	UK	Lower Jur.	185	3–20	0.3–2	+	+++	
Mid Clare shale	Kerry	Ireland	Lower Carb.	335	4–20	0.3–1	+	+++	+++
Chattanooga shale	Tennessee	US	Upper Dev.	350	3–30	0.2–1	+	+++	+++
Geneseo shale	New York	US	Upper Dev.	350	5–30	0.2–2.5		+	+++
Holocene Sediments								
Shirone Drill Core	Niigata	Japan			5–120	0.5–7	+	++	+++
Kanai Drill Core	Niigata	Japan			20–110	1.5–2	+	++	
Lake Harutori	Hokkaido	Japan			<5–15	0.7–2	+++		
Mitarase Lagoon	Niigata	Japan			<5–20	1–2	+++	+	+
Sakata Lagoon	Niigata	Japan			<5–40	0.2–2	+++	+	+
Septiba Bay	Septiba	Brazil			<5–10	0.5–1	+++		

R: totally disordered; PO: partially ordered; O: ordered. Relative abundance: +++: frequent; ++: occasional; + rare; blank: not observed. Ages in Ma are indicative.
Holocene refers 11500–0 a.
Modified from a compilation in Ohfuji (2004).

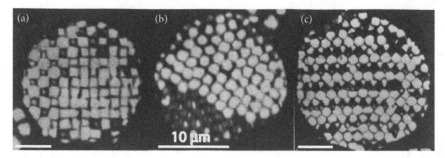

Figure 5.1. Love and Amstutz's (1966) original classification of framboids in 2D into three types: (a) Type 1 (ccp), (b) Type 2 (hcp), and (c) Type 3 (linear arrays). The scale on each image is 10 μm; this is perhaps exaggerated since the images were presented with linear magnifications only.

Reflected light microphotographs from Love, L. G., and Amstutz, G. C., 1966. Review of microscopic pyrite from the Devonian Chattanooga Shale and Rammelserg Banderz. *Fortschritte der Mineralogie*, 43: 277–309. Plate 1, figures 7 and 9; plate 2, figure 1.

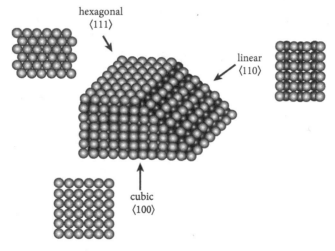

Figure 5.2. Derivation of various 2D geometries of microcrystal arrangements from simple cubic close packing. Spheres are shown for illustration. A cubic arrangement is obtained when the section is at 55° to the packing plane, a hexagonal arrangement is observed parallel to the packing plane, and a linear arrangement vertical to the packing plane.

Redrawn from Ohfuji (2004).

parallel to the cubic closed packed layers (i.e., parallel to ⟨1 1 1⟩), a 2D hexagonal array is observed. A cubic array is displaycd in a scction cut approximatcly 55° to the closed packed layers (i.e., parallel to ⟨1 0 0⟩). Linear arrays are obtained if the ccp array is observed at 90° to the closed packed plane (i.e., parallel to ⟨1 1 0⟩).

A single layer of microcrystals packed in ccp is identical to that in hcp since both are close packed. This means that it is impossible to distinguish between ccp and hcp on the basis of the geometry of a single layer of microcrystals, as might be obtained by examining a framboid externally in 2D. Likewise, the presence of an hexagonal array in 2D sections is impossible to assign to ccp or hcp. Consequently, although cubic and more extreme linear arrays in 2D section are characteristic of ccp, hexagonal arrays may be a result of either ccp or hcp. The conclusion is that, although most organized framboids are ccp, simple observations may not preclude the presence of hcp.

The uncertainty in discrimination between ccp and hcp is exacerbated by the fact that the second layer in close packed microcrystals is also identical in both packing types. The microcrystals in the second layer in both ccp and hcp are situated in the wells caused by the holes between the hexagonally arranged microcrystals in the underlying layer. It is only in the third layer that differences between ccp and hcp occur. Detailed observations of the microcrystal arrangements in organized framboids by scanning electron microscopy may only rarely penetrate more than two layers. Random sections thus produce the same geometries in scanning electron microscopy as in optical microscopy.

5.2.2. Close Packed Spheres

The thermodynamically stable crystalline geometry for hard spheres is the cubic close packed system which maximizes entropy at high densities. This arrangement has the densest packing ($\varphi = 0.74$) and does not exhibit five-fold symmetry. In a sphere it would appear that all angles of view of close packing should be observed. In other words, a sphere of cubic close packed microcrystals such as a framboid should display all the apparent geometries described earlier: cubic arrays, hexagonal arrays, and linear arrays. Models of packings in a sphere have been developed for a surprising number of applications, including materials science, biology, and astrophysics, as well as interesting pure mathematicians.

Figure 5.3 shows a hemisphere of cubic close packed spheres computed by the BALSAC (Build and Analyze Lattices, Surfaces, and Clusters) algorithm by Klaus Herman of the Fritz Haber Institute, Berlin. The image is described as a *spherical fcc tip* as, I suppose, might be encountered in the metal probe of an atomic force microscope. However, the hard sphere model is equally applicable to framboids, where the spheres approximate constituent microcrystals. The close packed layers are tiled vertically and the ⟨1 1 1⟩ and ⟨0 0 1⟩ crystallographic directions are indicated. It is interesting to see

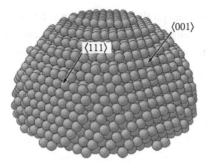

Figure 5.3. Hemisphere of cubic close packed spheres.

Computed by Balsac visualization and analysis software, Version 4, by K. Hermann (FHI), Berlin, 2020; see also http://www.fhi berlin.mpg.de/KHsoftware/Balsac/index.html.

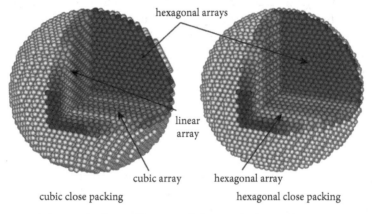

Figure 5.4. Spheres of cubic and hexagonal close packed particles.

Computed by Diehl, S., Rockefeller, G., Fryer, C. L., Riethmiller, D., and Statler, T. S., *Generating Optimal Initial Conditions for Smoothed Particle Hydrodynamics Simulations.* Publications of the Astronomical Society of Australia, 32, figure 1 (part), reproduced with permission.

how these various geometric arrays of microcrystals intersect on the spherical surface. With a large number of particles, the intersections of the various array geometries are quite smooth and the spherical shape is approximated by ever-smaller tiles.

Figure 5.4 shows models of cubic and hexagonal close packed spheres computed by a smooth particle hydrodynamics (SPH) algorithm by Diehl et al. (2015). The SPH approach is used to model a variety of astrophysical phenomena, including star formation, planet formation, cosmology, stellar collisions, stellar mergers, gas dynamics in the Galactic center, galaxy mergers, and supernovae. However, the basis of the method is to model individual particles that act according to hydrodynamic flow equations, and the results are equally applicable to framboids, where the individual spheres approximate microcrystals, as to galaxies. The spheres in Figure 5.4 contain 22,000 particles, equivalent to a 10 μm framboid containing 0.3 μm microcrystals. They are sectioned orthogonally to the c-axis of the sphere but at an angle to the layers of microcrystals. Even so, they display the cubic, hexagonal, and linear microcrystal arrays encountered in sections through framboids with single domain microarchitectures, as described earlier. With the larger number of particles, ideal sphericity, which is approached by successively decreasing the layer size, becomes even more closely approached, and the intersections between the various geometric arrays even more seamless.

Figure 5.5 shows a large part of the surface of a mainly cubic close packed framboid which shows that the various geometries associated with this packing are displayed at the surface of the framboid. Note also that the depth of penetration of the image is mainly not more than two microcrystal layers.

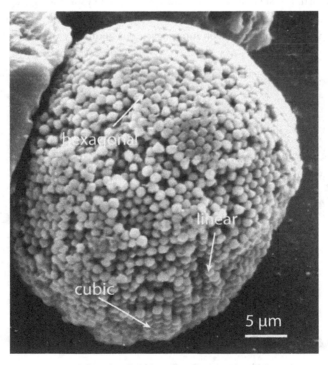

Figure 5.5. An organized framboid with cubic close packing showing areas of hexagonal, cubic, and linear geometries.
Scanning electron microscope image by Anders Elverhøi.

5.3. Multiple Domain Microarchitectures

Framboids displaying more than one geometry of microarchitecture were first described by Rickard (1970). He noted that framboids in which the microcrystals were apparently totally disordered often displayed areas with close packed geometries. The point was that any theory which explained the development of organized framboids needed to take into account these partially ordered forms. There is no strict separation between ordered and disordered framboids, but they grade into each other from the end-member organized and disorganized types.

The problem was underlined by reports of framboid sections showing a variety of spectacular geometric arrays of microcrystals (e.g., Figure 5.6). The patterns displayed by the microcrystals in these framboids included patterns that seemed to be due to twinning, as seen in the X-, Y-, and V-patterns in Figures 5.6(a), (c), and (d), where the observed pattern seems to be reflected on different hemispheres of the framboids. Closer examination of the radiating pattern (Figure 5.6(b)) also suggests a twinned structure. The importance of these observations was that the observed patterns seemed to suggest that framboids were odd forms of skeletal crystals and not simple crystalline aggregates. It was not until this century that the question was resolved by the Cardiff group, as described in Chapter 6. An even more egregious geometry was displayed by the

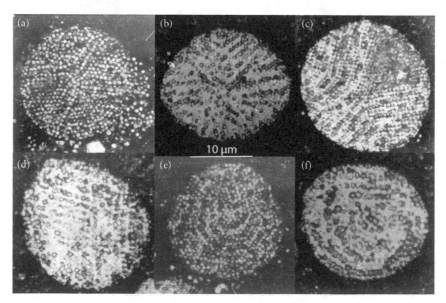

Figure 5.6. Sections of framboids in reflected light showing various complex geometries of their constituent microcrystals: (a) X-shaped, (b) star-shaped, (c) Y-shaped, (d) V-shaped, (e) pentagonal, (f) circular.

framboids with pentagonal ordering, as shown in Figure 5.6(e), which probably was the origin of the apparent circular organization in Figure 5.6(f).

The presence of pentagonal symmetry in the microcrystal ordering of pyrite framboids disproved the hypothesis that framboids were extreme single crystals since this was a forbidden crystal geometry in classical crystallography. The problem is that tiling of pentagons leaves gaps, which means that pentagonal symmetries cannot produce the long-range ordering required by conventional crystals.

5.3.1. Pseudo-icosahedral Domain Structure

The possibility of icosahedral packing was suggested by Large et al. (2001) as an explanation for the pentagonal packing geometries they observed in framboid sections. A complete explanation was provided by Ohfuji (Ohfuji, 2004; Ohfuji and Akai, 2002). Ohfuji realized that icosahedra displayed twofold, threefold, and fivefold symmetry axes (Figure 5.7), which could explain many of the various geometries exhibited by framboids in section. The icosahedral domain

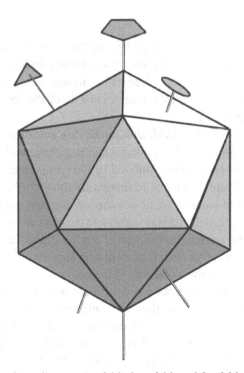

Figure 5.7. Icosahedron showing twofold, threefold, and fivefold axes of symmetry.

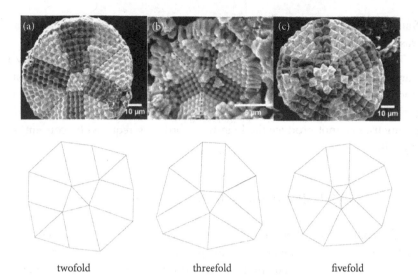

twofold   threefold   fivefold

Figure 5.8. Sections through icosahedra perpendicular to (a) twofold, (b) threefold, and (c) fivefold symmetry axes. The grayscales on the scanning electron micrographs are enhanced in order to display the symmetries.
Images from Ohfuji (2004).

structure in framboids consists of 20 tetrahedral sub-units, sharing a common vertex at the center, in which the microcrystals are regularly packed in ccp arrays. The three types of symmetrical microcrystal arrangements are produced from sections perpendicular to each symmetrical axis of the icosahedrally packed domains (Figure 5.8).

The relative sizes of the individual domains vary according to the position of the section with respect to the framboid center (Ohfuji and Akai, 2002). This leads to the various geometries exhibited by some organized framboids in section (Figure 5.6). More detailed SEM images are shown in Figure 5.8. The three geometries obtained from views of icosahedral domains perpendicular to the three icosahedral symmetry axes are similar to the ideal sections.

The microcrystal packing is consistently ccp, and the different hexagonal and square microcrystal arrays reflect different angular views of the constituent ccp tetrahedral sub-units of the icosahedra (Figure 5.9).

Detailed observations of the microcrystals in the tetrahedral sub-units of the icosahedra reveal that they are uniformly octahedral (Figure 5.9). However, the faces of the octahedral are not in contact; rather, the octahedra are linked by shared vertices and edges. The octahedra in the icosahedral framboids orient their {1 1 1} faces parallel to the hexagonal array (Ohfuji and Akai, 2002). This contrasts with the packings observed in single domain

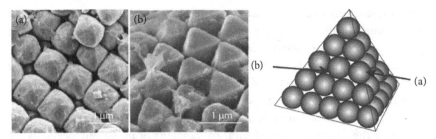

Figure 5.9. Detail of octahedral microcrystal packing in icosahedral framboids:
(a) square array viewed perpendicular to the edges of the tetrahedral subunits;
(b) hexagonal array viewed perpendicular to the faces of the tetrahedral subunits.
Modified from images in Ohfuji (2004).

framboids where the $\langle 1\ 1\ 0 \rangle$ is parallel to the hexagonal array and the $\{1\ 1\ 1\}$ faces are in contact.

Perfect icosahedral geometry of framboids would result in their surfaces being covered exclusively by $\{1\ 1\ 1\}$ crystal planes. In this arrangement, each microcrystal maximizes the number of its nearest neighbors, thereby minimizing its surface energy (Hendy and Doye, 2002). Obviously, the perfect icosahedron then only results from fixed number of microcrystals dependent on the icosahedron size. These numbers have been called magic numbers, by analogy with the lowest energy configurations of atomic clusters. However, whether the number of microcrystals in a framboid-nucleating system contains the exact number of microcrystals required to produce a perfect icosahedron is a matter of chance and seems statistically improbable. For framboids with microcrystal numbers between these ideal configurations, symmetries other than icosahedral and disorganized arrangements are competing (Doye and Calvo, 2001), and variously organized and disorganized domains result.

There is, however, a further problem with icosahedral packing of framboids. The tetrahedral sub-units that make up icosahedra are not regular tetrahedra but are necessarily distorted. The distance between the center and vertex of an icosahedron is about 5% shorter than the vertex-vertex distance (Mackay, 1962). But, if the tetrahedra are constituted by regular microcrystals in ccp, then they should be regular (Figure 5.9). The packing of 20 such regular tetrahedra into an icosahedron results in gaps between each unit, which is the fundamental reason why the icosahedron is not an allowed crystal symmetry. Icosahedral symmetry tends to minimize short-range attractive interactions and maximize entropy (de Nijs et al., 2015). However, fivefold symmetry is incommensurate with long-range positional order and, at greater length scales the icosahedral order becomes complex.

As discussed in Chapter 4, macroscopic pyrite crystals form a plethora of forms, some of which are similar to icosahedra without being ideal geometric equivalents. The pseudo-icosahedron, formed from a combination of the octahedron and pyritohedron, is a well-known—if rare—pyrite habit. In this habit, the tetrahedral sub-units are not all of the same shape, similar to the relationship between the pyritohedron and the pentagonal dodecahedron. Detailed inspection of the framboids displaying icosahedral symmetry in Figure 3.12 in Chapter 3 shows that the constituent tetrahedral subdomains of pyrite microcrystals are not of the same shape, and the form is therefore a pseudo-icosahedron.

Mackay (1962) revolutionized packing theory by looking at packing structures as a series of layers or concentric shells of particles, and Large et al. (2001) noted that some framboids show clear evidence of concentric shells of microcrystals (e.g., Figures 5.7, 5.6, and 5.12). In this view of framboid packings, repeated icosahedral shells produce increasing gaps in the structure due to the problems of tiling of pentagonal symmetries mentioned earlier.

That is, an initial ideal or Mackay icosahedral nucleus of 13 microcrystals develops necessarily complex ordering as it extends by the addition of further layers of microcrystals. There are two possible results: (1) that the gaps in the structure are compensated by structural flaws in the framboid structure, which was favored by Ohfuji (2004); or (2) that the final polygon is not an ideal icosahedron. In fact, as discussed in the following, both solutions occur in framboids.

The gaps produce strains in the structure which can ultimately be accommodated by intercalating anti-Mackay layers (Figure 5.10), areas with different particle geometries, or simply by introducing structural defects. The anti-Mackay layers are less dense than the Mackay layers and release stress on the surface plane and increase the sphericity of the aggregate (Kuo, 2002; Mackay, 1962; Wang et al., 2019a).

Hendy and Doye (2002) computed packing geometries for clusters of lead atoms and found that they resembled anti-Mackay icosahedra, which have an ideal icosahedral core but with hexagonal close-packed outer layers. Hendy and Doye remarked that the icosahedron illustrated in Figure 5.11, which is very similar to observed framboids (e.g., Figure 5.8(c)), had the highest binding energy of any of the clusters they had computed up to that date. This is consistent with Large et al's (2012) remark that the repulsions between the 12 neighboring particles in an icosahedral geometry are minimized.

De Nijs et al. (2015) showed that entropy maximization and spherical confinement alone are sufficient for the formation of icosahedral aggregates. This was an important conclusion since it paved the way for the synthesis of spheroidal aggregates of inert colloidal spherical particles, with the aggregating force being due to evaporation of the medium in which the spheroidal aggregates are

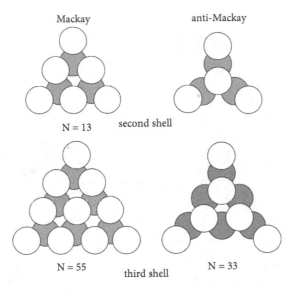

Figure 5.10. Mackay and anti-Mackay layers for the third shell of an icosahedral aggregate looking down a single icosahedron face.

developing. This contrasts with configurations of atoms—which were studied by Hendy and Doye (2002)—where the electronic interactions between the atoms are responsible for the aggregation. It also contrasts with framboids where surface forces on the pyrite microcrystals provide the impetus for aggregation (see section 13.1 of Chapter 13). Wang et al. (2018) used microfluidics to produce a library of hundreds of experimentally-produced icosahedral and related spheroidal aggregates of spherical polystyrene particles of colloidal size. The results have considerable implications for the origins of self-organization of pyrite framboids, and these are discussed in section 13.2.

The results of de Nijs et al. (2015) and Wang et al. (2018) show that the curved interface enforces deviation from the ccp structure. Icosahedral symmetry is favored by medium-sized aggregates between a few hundred and about 10 000 colloidal particles. Experiment shows that these aggregates can be well-ordered where the number of particles approaches the magic number required for ideal icosahedral symmetry (Wang et al., 2019a). In between these numbers, decahedral aggregates occur. Over 30 000 discrete 10-faced polyhedra, or decahedra, are known, and none is regular, although some have regular faces. The occurrence of decahedra in the development of icosahedral framboids is consistent with the infinite variety of microcrystal organizations observed in framboids. Above 10^5 particles, ccp packing reoccurs (Wang et al., 2019b), which is coincidental with the largest number of microcrystals in pyrite framboids (Chapter 4).

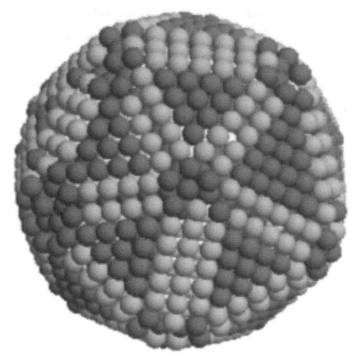

Figure 5.11. A simulation of a 2057-atom icosahedron with anti-Mackay surface structure and an extra {100}-facet. Note the remarkable similarity to the icosahedral framboid shown in Figure 5.8(c).

Reprinted figure with permission from Hendy, S. C., and Doye, J. P. K., *Physical Review B*, 66, Figure 4, 2002. Copyright (2002) by the American Physical Society.

5.3.2. Other Multiple Domain Microarchitectures

A number of other geometries of the microcrystals in framboids have been reported over the years which are not easily explained by ccp or pseudo-icosahedral packing of framboidal domains.

One of the earliest was by Morrissey (1972), which showed a pentagonal framboid with apparently hexagonally packed microcrystals showing sixfold symmetry (Figure 5.12). The pentagonal form of the framboids could result from a section through a cube or more likely a rhombic dodecahedron. However, the layered hexagonal geometry of the microcrystals is more difficult to relate to the packing microarchitectures discussed earlier. I suppose that it could possibly represent the cornice of a polyhedron with ccp viewed perpendicular to the intersection of ⟨1 1 1⟩ and ⟨0 0 1⟩ (Figure 5.2).

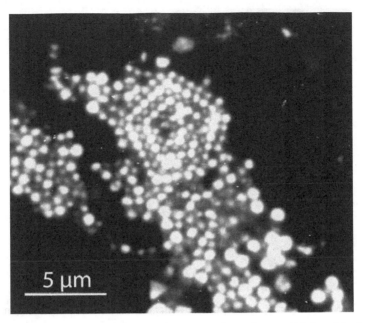

Figure 5.12. Microcrystal packing in the section through a ca. 10 Ma polyhedral framboid from Tynagh, Ireland.

Reflected light microphotograph by C. J. Morrissey.

A similar form reported by Skipchenko and Berber'yan (1976) was interpreted by Ohfuji (2004) to be the result of a duplex X-shaped pattern similar to that obtained through sections through pseudo-icosahedral framboids on twofold axes (Figure 5.8).

Rarely, microcrystal orderings made up of fewer ordered domains (e.g., two to four units), lacking distinctive symmetry, are also found. The framboid section in Figure 5.13 shows a square domain characterized by a cubic microcrystal arrangement, which is clearly separated from the rest of the framboid where the microcrystals are all hexagonally arranged. It might possess fourfold symmetry, although the detail of the domain distribution is uncertain. A similar pattern was described by Kalliokoski and Cathles (1969), suggesting such a domain structure might also be intrinsic to some pyrite framboids.

5.4. Disordered Framboids

It is easy to mistakenly assume that most framboids display ordered microcrystal domains because of their eye-catching nature. In fact, in the majority of

Figure 5.13. Square array of ccp pyrite microcrystals surrounded by hexagonal domains.

Scanning electron microscope image by Hiroaki Ohfuji.

framboids, normally microcrystals are mainly disordered and ordered framboids are less abundant. Ian Butler (personal communication, 2020) noted that he always got the impressions that ordered framboids were the more abundant in the 390 Ma Devonian Chattanooga Shale, a classic framboid *lagerstätte* popularized in the 1960s as a site especially rich in ordered framboids.

In disordered framboids, the microcrystals are apparently completely randomly arranged (Figure 5.14). In nanocrystal assemblies, these have been described as *amorphous superlattices* (Naik and Caruntu, 2017). The microcrystals themselves are jumbled together, with their crystallographic axes pointing in all directions. Even so, the microcrystals retain the same habits and are of uniform size. The uniformity of habit and size of framboidal microcrystals is therefore not the fundamental reason for the development of extreme self-organization displayed by some framboids, although, of course, it is a necessary framboid attribute. Wang et al. (2019) report that disorganized spheroidal aggregates are more common in their microfluidic syntheses.

At the same time, Rickard (1970) concluded that totally disordered framboids might also be less common. Detailed examination of many disordered framboids reveals local areas of ordering, even under the light microscope (Figure 5.15). Ohfuji (2004) reported similar features displayed under the SEM. This suggests

Figure 5.14. Disordered framboid from the ca. 2 Ma Uonuma formation, Japan, show apparently completely random arrangement of microcrystals: (a) general view of the disordered framboid; (b) detail of randomly assembled octahedral microcrystals.
Scanning electron microscope images by Hiroaki Ohfuji.

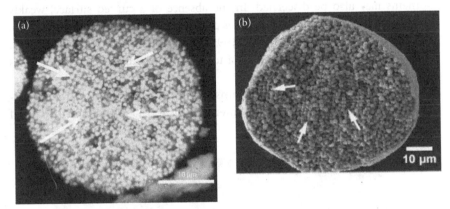

Figure 5.15. Disorganized framboids with some of the areas where the microcrystals appear to be organized marked by arrows: (a) from Portuguese soil in reflected light; (b) from the ca. 2Ma Pleistocene Uonuma Formation.
(b) Scanning electron microscope image by Hiroaki Ohfuji.

that self-organization evolves with time in framboids and that these partially organized framboids are essentially works in progress.

Ohfuji (2004) went further and differentiated entirely randomly organized framboids into a group in which the disorder is not obvious on external examination and a group where the exterior is obviously disorganized. In their library of spherical aggregate geometries, Wang et al. (2019) also describe spherical colloidal clusters with outer organized shells and a disordered interior. This provides a caveat to observations of whole framboids, as might be provided by SEM, for example. The observation of ordering on the outer surfaces of framboids does not necessarily imply that the framboid microcrystals are ordered internally. This group of disordered framboids grade into non-framboidal aggregates of pyrite crystals which frequently display interpenetrate habits.

5.5. The Microarchitecture of Microcrystalline and Nanocrystalline Pyrite

Organized domains in microcrystalline and even nanocrystalline pyrite are not limited to discrete bodies such as framboids. Amorphous masses of these materials can display organized domains. Figure 5.16 is interesting since this shows nanocrystalline pyrite from a recent sediment displaying distinct organized domains. The geometries of the domains are various and include arrays mainly associated with ccp, although patterns explained by pseudo-icosahedral domains may also be discerned. In the absence of a curved surface, weakly interacting spherical colloidal particles preferentially self-assemble into a cubic close packed structure (Woodcock, 1997). The imposition of a curved surface tends to facilitate the development of icosahedral packing and its variants, as described earlier.

These areas of organized domains in masses of microcrystalline pyrite are not limited to recent sediments. Figure 5.17 illustrates a mass of

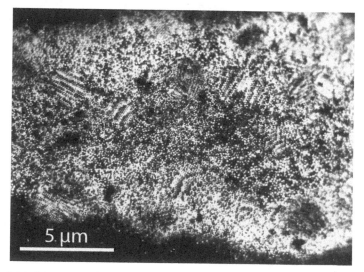

Figure 5.16. Nanocrystalline pyrite from a recent sediment showing organized domains.

Reflected light microphotograph.

microcrystalline pyrite from a 4 Ma Tertiary mudstone showing domains with hexagonal geometries as well as framboids. Similar microarchitectures have been reported by Jiang et al. (2001), and Ohfuji (2004) described a tubular array from the ca. 390 Ma old Chattanooga shale. These images show that the microcrystals in these masses are equidimensional and show similar habits to those in framboids.

Figure 5.16 is further interesting in that the individual pyrite crystals involved are not resolvable by light microscopy and appear as pinpoints of light. They are about 100 nm in size, or slightly less, and suggest that the processes involved in self-organization are not limited to colloid-sized microcrystals but also occur in nanocrystals. Pyrite nanoparticles contribute a significant component of the nanoparticulate budget of the oceans (Hochella et al., 2012) and may contribute up to 10% of the dissolved iron emanating from black smoker hydrothermal vents (Yücel et al., 2011). Gartman et al. (2014) further defined the sulfide nanoparticles from hydrothermal vents at Lau Basin, MAR and EPR 9°50′N.

The occurrence of organized domains in amorphous masses of pyrite crystals suggests that the process of organization is distinct from processes involved in the development of spheroidal, sub-spheroidal, or polyhedral bodies shapes of framboids *sensu stricto*.

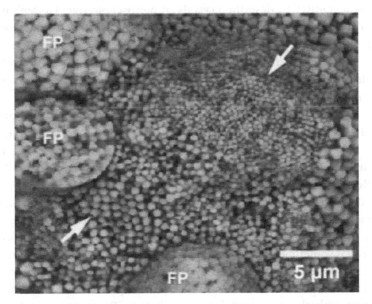

Figure 5.17. Organized areas of pyrite microcrystals (arrows) and framboidal pyrite (FP) from a 4 Ma Tertiary mudstone.
Scanning electron microscope image by Hiroaki Ohfuji.

6

The Crystallography of Pyrite Framboids

One of the fundamental problems about framboids was that their exact nature was not known until the early years of this century. That is to say, it was not known whether they were microcrystalline aggregates or some end-member form of skeletal crystals or something else entirely. Their internal structure, often revealing the remarkable ordering of tens of thousands of individual microcrystals of the same size and apparently oriented in the same direction (Figure 6.1(a)), seems to suggest some sort of governing crystalline force or preexisting template. Until their detailed crystallography could be resolved, there was no robust basis on which to base any theory of how they actually formed.

In 2005, Helmut Cölfen proposed the term *mesocrystal* to describe superstructures of small crystals with similar sizes and shapes with a common crystallographic orientation (Cölfen and Antonietti, 2005). Actually, this was a redefinition since the term *mesocrystal* had been used for over a century to describe crystals of mesoscopic size. Cölfen's use of the term derived from the idea that the small crystals became organized by self-assembly at the mesoscale and the term *mesocrystal* is short for mesoscopically structured crystalline materials or types of colloidal crystals. Since this time, hundreds of papers have been published using the term *mesocrystal* in this sense (Strum and Cölfen, 2017). Strum and Cölfen (2017) revisited the definition of mesocrystals in view of developments since the original study. They reported that the constituent crystals of mesocrystals are, by definition, nanocrystals since, if they are larger, the size range is no longer mesoscale. Nanoparticles are generally classified as particles <100 nm or 0.1 μm in size.

In 2017, Strum and Cölfen reported (p. 7) that framboids "can sometimes be classified as mesocrystals." I do not think this is the case, although mesocrystals may be useful analogies for framboids. As reported in Chapter 2, the mesocrystal size range is sometimes observed in framboids, but normally these are micron-sized and consist of micron-sized microcrystals. Framboid-like nanoclusters with nanosized microcrystals rarely display the superstructure that would be consistent with the mesocrystal characterization. By contrast, as discussed in section 12.2.3 of Chapter 12, the formation of individual pyrite framboid microcrystals might have involved a mesocrystal phase.

Framboids. David Rickard, Oxford University Press. © Oxford University Press 2021.
DOI: 10.1093/oso/9780190080112.003.0006

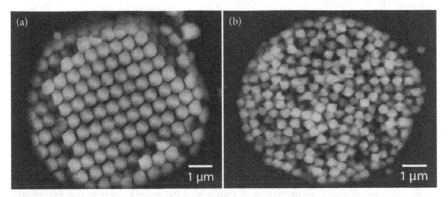

Figure 6.1. (a) well-ordered, cubic close packed, and (b) randomly packed framboids from the Devonian Chattanooga Shale, Tennessee.
Scanning electron micrograph images by Hiroaki Ohfuji.

6.1. Pyrite Structure

The crystal structure of pyrite determines its habit and form. This structure is well known and one of the earliest to be determined (Rickard, 2015). Barlow (1883) first developed the theory of close packing and then, with W. H. Pope, made predictions of real structures (Barlow and Pope, 1907). In 1891, the Russian mathematician E. S. Fedorov showed that there are only 230 possible spatial symmetries that crystals can adopt (Federov, 1891). The distribution of the Fe and S atoms in pyrite was particularly important here, since the space group Fedorov assigned to pyrite was considered by many contemporaries to be merely a mathematical construct, unrelated to the real material world. In fact, it turned out that the arrangements of atoms in the pyrite unit cell are exactly as predicted by the Fedorov system.

Pyrite was one of the first crystalline materials investigated by X-ray diffraction. However, its structure proved more difficult to unravel. W. H. Bragg (1913) published a famous paper on the atomic structure of common salt, sphalerite,

Table 6.1 Crystallographic Conventions Used in Text

[$h\,k\,l$]	A crystallographic direction
⟨$h\,k\,l$⟩	A set of directions equivalent to [$h\,k\,l$]
($h\,k\,l$)	A plane of the crystal structure
{$h\,k\,l$}	A set of planes equivalent to ($h\,k\,l$) also known as crystal form

fluorspar, and calcite in which the difficulties with the pyrite structure interpretation were mentioned. Finally, W. L. Bragg (1914) succeeded in solving the pyrite structure. He wrote (1961, p.148):

> *The structure of pyrite provided the greatest thrill. It seemed impossible to explain its queer succession of spectra, until I discovered, going through Barlow's geometrical assemblages, that it was possible for a cubic crystal to have non-intersecting trigonal axes. The moment of realization that this explained the iron pyrite result was an occasion I well remember.*

Bragg's results also demonstrated that separate atoms occupy the symmetry centers of Fedorov's space group and confirmed Fedorov's original theoretical results. The structure of pyrite was confirmed by Ewald and Friedrich (1914) and the space group for pyrite was given by Schoenflies (1915).

In the conventional stick-and-ball rendering of the pyrite structure (Figure 6.2), the S_2^{-2} groups of the FeS_2 moieties are situated at the cube center and the mid-points of the edges of an NaCl-type structure, with low spin Fe(II) atoms (d^6, t_{2g}^6) located at the corners and face centers. In this representation the disulfides are depicted as dumbbells and it is clear that the structure, although cubic, has a reduced symmetry compared with the simple NaCl-type structure.

In particular, pyrite displays only twofold symmetry along the $\langle 1\ 0\ 0 \rangle$ directions (see Table 6.1 for crystallographic conventions) compared with

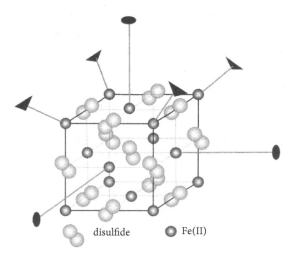

disulfide   Fe(II)

Figure 6.2. Conventional stick-and-ball rendering of the molecular structure of pyrite showing the Fe and S atoms and the 3 orthogonal twofold symmetry axes and 4 diagonal threefold symmetry axes.

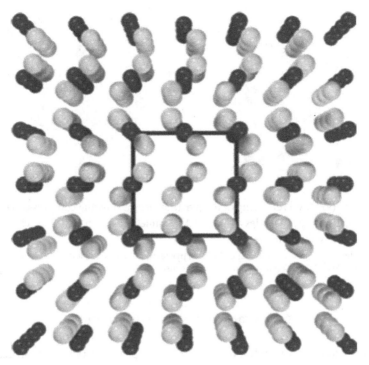

Figure 6.3. Structure of pyrite projected along $\langle 0\,0\,1\rangle$. The cubic unit cell is outlined and apparently provides fourfold rotational symmetry. However, the atomic distribution in projection only displays twofold symmetry.

Modified from image in *Ultramicroscopy* 160, Figure 2. Pattern matching approach to pseudosymmetry problems in electron backscatter diffraction. Nolze G., Winkelmann A., and Boyle A. P. Copyright (2016) with permission from Elsevier.

fourfold symmetry for the NaCl structure (Figure 6.3). The structure has three-fold axes along the $\langle 1\,1\,1\rangle$ directions and twofold axes along the $\langle 1\,0\,0\rangle$ directions (Figure 6.2). The twofold symmetry means that 90° rotations around the a, b, and c crystallographic axes (equivalent to the [1 0 0], [0 1 0], and [0 0 1] zone axes) do not return to similar geometries as occurs in the simple face-centered cubic structure. This is discussed in some detail in section 6.4.1 and has significant consequences for framboid crystallography.

A more realistic molecular model was computed by Birkholz (2014), which renders the disulfide molecules as ellipsoidal and reveals that these molecules themselves are asymmetric with the ellipsoids being flattened along the $\langle 111\rangle$ symmetry axes (Figure 6.4(a)). Birkholz's insight was that the deformed ellipsoids enhance the stability of the pyrite structure: the bonding coordination for the sulfur ellipsoids produces two interpenetrating tetrahedrons with the neighbor S on top and three Fe forming the base of an S_2Fe_6 cluster (Figure 6.4(b)).

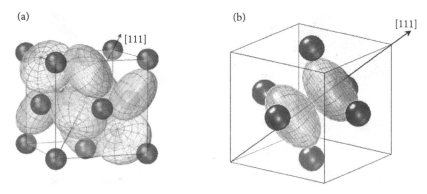

Figure 6.4. (a) Novel rendering of the pyrite structure showing compression of ellipsoidal disulfide molecules along [1 1 1] symmetry axes. (b) The S_2Fe_6 cluster units enhance stability to the pyrite structure.

From Birkholz, M., Modeling the shape of ions in pyrite-type crystals. *Crystals* 4, 2014, Figure 2, Published by MDPI AG. Creative Commons BY license.

6.2. Framboid X-ray Diffraction

Compared to the X-ray diffraction spectra of simple NaCl-type cubic structures, the pyrite X-ray diffraction pattern is relatively complex. Instead of reflection peaks marching across the spectrum at regular intervals, the pattern shows more lines reflecting the lowest cubic symmetry of the pyrite space group $Pa3$ and point group $2/m3$ resulting from the absence of a fourfold rotation axis due to orientations of the S_2 (-II) groups. Consequently, pyrite normally displays 19 reflections at resolution greater than 0.9 Å, compared with just 9 reflections for simple high symmetry face-centered NaCl-type compounds. This means that a single crystal XRD image of a pyrite framboid consisting of several thousand individual microcrystals could become extremely complex.

Rickard (1968a) adapted the classical single crystal XRD method for mounting individual framboids on a glass fiber tip for X-ray diffraction studies In this system, framboids were separated by lightly crushing a framboid concentrate between two glass slides and then placing an individual framboid on a tip of a glass fiber coated with a rubber cement (Cowgum™) under a microscope (Figure 6.5). We later substituted silicone grease for the rubber cement. The resulting mount could be centered in a Debye-Scherrer X-ray powder diffraction (XRPD) camera and the spectrum obtained. A similar setup was later used for individual framboid mounts in single crystal XRD cameras. The beauty of the system is that it is non-destructive and that the chemical and physical properties of the individual framboid can be probed before and after the XRD investigation.

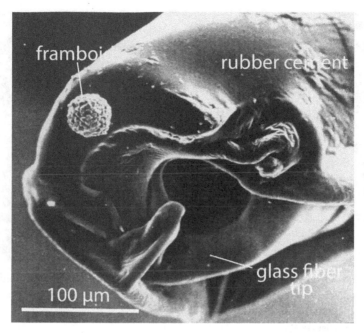

Figure 6.5. SEM of mount of single framboid on a glass fiber with rubber cement.
Scanning electron micrograph image by Ian Butler.

Butler (1994) used this technique to examine framboids in a single crystal diffractometer. He found that the single crystal XRD pattern was typical of an aggregate of microcrystals with myriads of individual reflections being recorded. Organized framboids (e.g., Figure 6.1(a)) did not display single crystal XRD patterns, which might be expected of a mesocrystals in which the constituent microcrystals have a common crystallographic orientation. Rather the framboids appeared to show both powder-diffraction style patterns indicative of random aggregates of microcrystals and brighter spots that occurred due to the superposition of some of these diffractions. These brighter spots suggested some morphologic or crystallographic ordering in the framboid. Butler (1994) observed that these bright spots appeared and disappeared as the framboid was rotated (Figure 6.6), which suggested a number of crystal domains in the framboid containing groups of microcrystals with similar relative orientations. However, the computing power needed to deconvolute these data were not available at that time.

Three types of XRD-spectra were obtained in both the Debye-Scherrer XRPD camera and 2D (nominally 1°) slices through the 3D Ewald sphere data set in the single crystal camera (Butler 1994; Ohfuji, 2004). These are illustrated in Figure 6.7 from the single crystal XRD study.

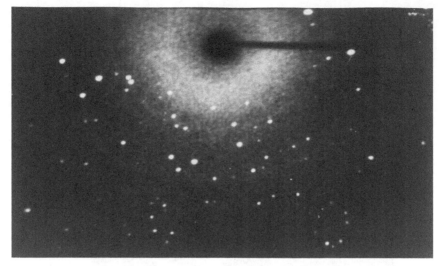

Figure 6.6. Original single crystal X-ray diffraction image of a pyrite framboid showing myriads of reflections and a number of brighter spots which appear and disappear as the framboid was rotated.

Image from Butler (1994).

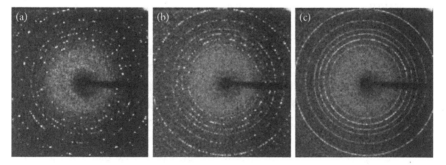

Figure 6.7. Framboid X-ray diffraction patterns: (a) ring pattern with complete population; (b) coarse ring pattern with partial population; (c) multicrystalline patterns.

Modified from image in Ohfuji (2004).

(1) Ring patterns with complete populations of diffraction spots, similar to X-ray powder diffraction patterns (Figure 6.7(a)). These result from framboids consisting of random aggregates of microcrystals.

(2) Coarse ring patterns with incomplete populations (Figure 6.7(b)). These result from a superposition of several diffraction spots, suggesting partial crystallographic and/or morphologic ordering of the microcrystals in the framboid.

(3) Multicrystalline patterns resulting from the superposition of tens of single crystal diffraction spots (Figure 6.7(c)), suggesting some greater degree of crystallographic and/or morphologic ordering of the framboid.

Each XRD pattern in Figure 6.7 consists of six rings corresponding to a lattice plane in a particular ($h\,k\,l$) direction in pyrite and their d-values correspond to $d_{111}, d_{200}, d_{210}, d_{211}, d_{220},$ and d_{311}.

Ohfuji (2004) computed the numbers of microcrystals in 25 large ca. 2Ma framboids (>80 µm diameter) from Japan and showed the type of XRD pattern obtained was proportional to the number of microcrystals in the framboid (Figure 6.8). The larger the number of microcrystals, the more probable that the framboid displayed typical powder pattern ring patterns. The numbers of microcrystals in framboids displaying ring patterns ranged from 2742 to 114 468. Framboids displaying coarse ring patterns had 1803–4600 microcrystals, whereas microcrystalline framboids had 880–2277 microcrystals. The ring-patterned framboids included both disordered and ordered framboids, whereas the framboids with coarse ring and multicrystalline patterns displayed ordered microstructures. The results suggest that, as expected, the numbers of XRD diffraction spots increase with the numbers of microcrystals.

However, more detailed probes of the data reveal a more complex relationship. In particular, the degree of ordering of the microcrystals within a framboid also affects the XRD pattern displayed: icosahedrally packed framboids mostly display coarse ring structures and many well-organized cubic close packed framboids show multicrystalline patterns. Framboids with icosahedrally packed

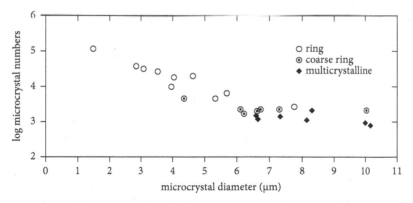

Figure 6.8. Logarithm of the number of microcrystals in a framboid versus the microcrystal diameter for framboids displaying ring, coarse ring, and multicrystalline XRD patterns.

From image in Ohfuji (2004).

structures consist of 20 tetrahedral subunits in which the microcrystals are arranged in similar orientations, but the orientations are different in each domain. Thus, the number of crystallographic domains in icosahedral framboids is greater than that in cubic close packed framboids, and their XRD patterns display coarse ring structures, indicative of both aggregates of microcrystals and similar microcrystallographic orientations.

The 2 0 0 diffractions for pyrite are perpendicular to the unit cell and have the strongest intensity. Ohfuji (2004) collected the 2 0 0 diffractions from a complete Ewald sphere data set for an organized framboid (Figure 6.9). Each microcrystal within the framboid produces six vectors mutually at right angles for the 2 0 0 reflection: [2 0 0], [0 2 0], [0 0 2], [$\bar{2}$ 0 0], [0 $\bar{2}$ 0] and [0 0 $\bar{2}$]. This framboid contains 1228 microcrystals so that 7368 2 0 0 diffraction spots would ideally be produced. The number of diffraction spots collected was 1421, much smaller than the total number of discrete diffraction spots produced. The geometrical

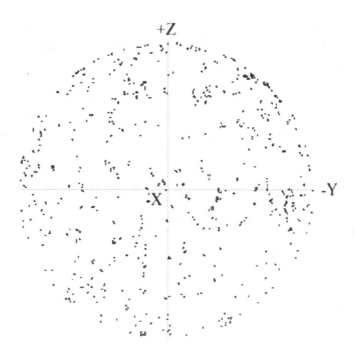

Figure 6.9. Pyrite 2 0 0 reflections plotted in reciprocal orthogonal space from a section through the frontal hemisphere of an Ewald sphere for an organized framboid. The reflections are clearly not uniformly distributed, but instead form clusters of spots.

From image in Ohfuji (2004).

limitations of the goniometer and the attachment point of the framboid limits the amount of reciprocal space through which the framboid can be rotated. Correcting for these blind regions suggests that the total number of diffraction spots collected was around 1500. This suggests that each measured diffraction spot results from the superposition of several individual diffractions. That is, the framboid is composed of small groups of microcrystals with similar crystallographic orientations. Ohfuji (2004) estimated that the average number of microcrystals in each group is around 5 microcrystals. This is a surprising result since the framboid appeared to display a well-organized cubic close packed microstructure which involved most of its 1228 microcrystals. It suggests that there is a disconnect between the morphologic and crystallographic orientations of microcrystals in framboids. That is, although the microcrystals look as if they are crystallographically ordered, they display variable crystallographic orientations for similarly morphologically orientated microcrystals. Common crystallographic orientations are restricted to small groups of, on average, 4–5 microcrystals.

The result explains the often-conflicting XRD patterns collected from framboids in the past. The three end-member patterns (ring, coarse ring, and multicrystalline) result from framboids with disorganized to organized microstructures. The increase of the frequency of ring patterns with the number of microcrystals results not only from disordered microcrystals but also from the larger numbers of differentially oriented groups in these framboids. For example, a framboid with 100 000 microcrystals might display 20 000 diffraction spots, even if apparently containing a perfectly morphologically organized microstructure, resulting from 20 000 domains of 5 crystals. These 20 000 diffraction spots would be spread over the 6 rings, providing, on average, around 4000 diffraction spots per ring, which would not be readily discriminated from the spectrum obtained from a random aggregate of microcrystals. Large framboids with large numbers of microcrystals will therefore display powder-like ring patterns independently of whether or not they have an organized microstructure.

6.3. Electron Backscatter Diffraction

The XRD results of framboids show that they are not much different from powders, with most of the particles being apparently randomly oriented, and are therefore aggregates, not mesocrystals. This is in contrast to observations which show that these same framboids that give XRD images closely comparable with powder patterns contain microcrystals that show similar habits and sizes and which appear to be perfectly regularly arranged (Figure 6.1(a)). The final piece of the jigsaw describing exactly what framboids are crystallographically was to reconcile the apparently incompatible results of the crystallographic arrangement of microcrystals in framboids with the observed organized arrays of pyrite microcrystals. During the 1990s, electron backscatter diffraction (EBSD) became available for examining the crystallography and microstructure of polycrystalline material, and we used this technique to resolve the framboid problem (Nolze et al., 2016; Ohfuji, 2004; Ohfuji et al., 2005).

Electron backscatter diffraction (ESBD) is a technique used to characterize the microstructure and crystallography of polycrystalline materials. Here, instead of the XRD spots, the target is the Kikuchi bands which are generated by the interaction of electrons with the structure (Figure 6.10). Although ideally suited to framboid studies, since the crystal orientations of multiple microcrystals can be obtained, the original problem was the treatment of the amount of data generated. We used eight Kikuchi bands per microcrystal in the analysis of framboids. In our original investigation into the framboid microstructure, ESBD patterns were collected on a rectangular grid at 50 nm spacing and the Kikuchi bands at each pixel were analyzed. In order to increase resolution, at least three camera frames are averaged for each individual Kikuchi pattern. We can see that if this is averaged over a 10 μm diameter framboid section, then a total orientation analysis involves solving the solutions to over 10^5 Kikuchi bands. Presently commercial systems can acquire up to 1200 patterns per second that enables highly detailed orientation maps of large sample areas.

It was obvious from the beginnings of modern EBSD in 1993 that the technique would be well suited to resolving the pyrite framboid microstructure.

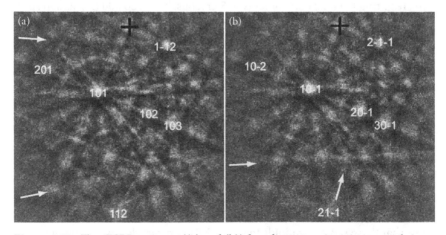

Figure 6.10. Two ESBD patterns ((a) and (b)) for adjacent pyrite microcrystals in an organized framboid showing 90° misorientation. The arrows indicate different Kikuchi bands in otherwise similar patterns.
From images in Ohfuji (2004).

However, our early attempts foundered over the data treatment problem, although they did provide proof of concept. The first successful EBSD analysis of a pyrite framboid was reported by Ohfuji (2004).

6.3.1. Accuracy of Pyrite EBSD

There are particular problems with EBSD analyses of crystallographic orientation in pyrite. One problem is due to mis-indexing of the Kikuchi bands in pyrite due to pseudosymmetry. This leads to mistaken relative orientation results of individual pyrite microcrystals since different crystal faces are being compared (Nolze et al., 2016).

  The mis-indexing is summarized in Table 6.2, where it can be seen that the $m\bar{3}$ point group of pyrite displays different multiplicities with respect to only 9 of the 24 possible face forms, allowing discrimination from the other cubic point groups. All Kikuchi bands representing $\{0\ 0\ 1\}$, $\{1\ 1\ 0\}$, $\{1\ 1\ 1\}$, and $\{h\ h\ l\}$ in pyrite will reflect apparently fourfold rotational symmetry (Figure 6.3), whereas—except for the general plane $\{\underline{h}\ k\ l\}$—only planes of the type $\{h\ k\ 0\}$ will enable differentiation between $m\bar{3}$ and $m\bar{3}m$. This is a significant restriction because $\{0\ 0\ 1\}$, $\{1\ 1\ 0\}$, $\{1\ 1\ 1\}$, and $\{h\ h\ l\}$ are represented by strong Kikuchi bands. It means that two pyrite microcrystals in a framboid oriented at 90° to each other have the majority of Kikuchi bands in common (Figure 6.10).

Table 6.2 Multiplicity of Face Forms for All Cubic Crystal Classes: Only 9 of the 24 Face Forms in the Cubic System (Underlined) Enable a Differentiation from Pyrite ($m\bar{3}$)

Face Form	Point Groups				
	$m\bar{3}$ (Pyrite)	23	432	$\bar{4}3m$	$m\bar{3}m$
$\{1\ 0\ 0\}$	6	6	6	6	6
$\{1\ 1\ 1\}$	8	$\underline{4}$	8	$\underline{4}$	8
$\{1\ 1\ 0\}$	12	12	12	12	12
$\{h\ k\ 0\}$	12	12	$\underline{24}$	$\underline{24}$	$\underline{24}$
$\{h\ h\ l\}$	24	$\underline{12}$	24	$\underline{12}$	24
$\{h\ k\ l\}$	24	$\underline{12}$	24	24	$\underline{48}$

Modified from Nolze et al. (2016).

Ohfuji et al. (2005) overcame this problem by manually examining EBSD patterns collected from adjacent pairs of microcrystals showing 90° misorientations to identify key differences in the Kikuchi patterns (Figure 6.10). Nolze et al. (2016) subsequently developed a pattern-matching technique to resolve the problem computationally.

The ability to resolve issues of 90° misorientation between framboid microcrystals is obviously fundamental to understanding framboid microstructure. Pyrite framboids produced relatively poor-quality Kikuchi patterns, and this in turn affects the accuracy of band detection. The estimated precision of orientation measurements in framboids by Ohfuji et al. (2005) is 0.5°, although this can be increased to 0.25° with pattern-matching (Nolze et al., 2016).

6.3.2. Framboid Microcrystals Are Single Crystals

Band contrast images are the result of summing all the EBSD patterns from the analyzed areas of the framboid. Figure 6.11 shows two such images from a randomly organized (Figure 6.11(a)) and a well-organized framboid (Figure 6.11(b)) from the 390 Ma Devonian Chattanooga shale, Tennessee, displaying ccp. These are not conventional SEM images, such as shown in Figure 6.1: the microcrystals are outlined in darker gray because of the poor pattern quality in the gaps between microcrystals. Within these boundaries, the microcrystals are evenly shaded, suggesting that they represent individual crystal domains. This means that the microcrystals are single crystals of pyrite. This was confirmed by Ohfuji et al. (2005), who showed, using EBSD, that the crystallographic

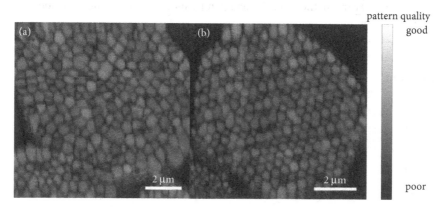

Figure 6.11. Band contrast images of (a) a disorganized framboid, and (b) an organized framboid. The grayscale indicates pattern quality.
Modified from images in Ohfuji (2004).

orientations of individual pixels across single microcrystals within framboids were constant.

6.3.3. Microcrystal Orientations

The microcrystals in a random section through a framboid usually will not, of course, all be aligned with their axes normal to the plane of the section. However, if the microcrystals are crystallographically aligned, as in a mesocrystal, their major crystal planes will be similarly oriented with respect to the plane of the section.

Microcrystal orientation maps (Figure 6.12) are produced by color-coding the crystallographic orientation of the individual microcrystals based on the Kikuchi band computations from the ESBD observations. Surprisingly, both randomly packed and organized framboids show adjacent microcrystals with orientations rotated 90°. The effect is clearly shown in the color-coded images where even in the ccp organized framboids the microcrystals are randomly colored depending on their orientation.

As expected, the pole figures for all EBSD patterns collected from framboids displaying random packing of their constituent microcrystals show no preferred orientation (Figure 6.13). By contrast, the pole figures for organized framboids show that the preferred orientations for the ⟨1 0 0⟩ and ⟨1 1 0⟩ directions are clustered. The ⟨ 1 0 0 ⟩ pole figure shows three clusters around poles defined by [1 0 0], [0 1 0], and [0 0 1]. The ⟨1 1 0⟩ pole figure shows six clusters defined by [1 1 0], [1 0 1], [0 1 1], [1 1̄ 0], [1̄ 1 0], and [1̄1 0].

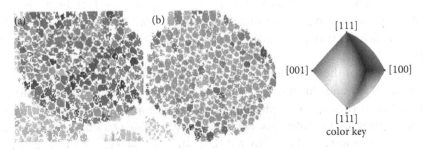

Figure 6.12. Original (unprocessed) EBSD orientation maps of the framboids in Figure 6.11: (a) disorganized framboid; (b) organized framboid and orientation color key. Basically, darker microcrystals are oriented at 90° to lighter microcrystals.

Compiled from images in Ohfuji (2004) and from *Ultramicroscopy* 160, Figure 5, Pattern matching approach to pseudosymmetry problems in electron backscatter diffraction. Nolze G., Winkelmann A., and Boyle A. P. Copyright (2016) with permission from Elsevier.

random packing

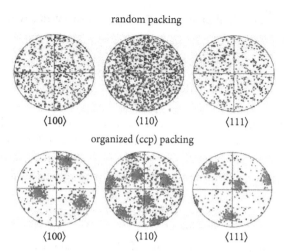

⟨100⟩   ⟨110⟩   ⟨111⟩

organized (ccp) packing

⟨100⟩   ⟨110⟩   ⟨111⟩

Figure 6.13. Upper hemisphere pole figures of all EBSD patterns collected from the randomly organized framboid (Figure 6.11(a)) and the organized (ccp) framboid (Figure 6.11(b)).
From images in Ohfuji (2004).

The surprising result is that the ⟨1 1 1⟩ pole figure, which shows clusters defined by four directions ([1 1 1], [1 1̄ 1], [1̄ 1 1], and [1̄ 1̄ 1]) shows a 90° rotation for one crystallographically preferred orientation cluster (Figure 6.13). Such 90° misorientations are not distinguishable in the ⟨1 0 0⟩ and ⟨1 1 0⟩ directions, as discussed earlier and shown in Table 6.2. However, their detection in the ⟨1̄ 1 1⟩ direction shows that some of the microcrystals in the organized framboids are crystallographically rotated at 90° to each other. This proves that framboids are not crystallographically homogenous and that therefore their microstructure is not determined by framboid-wide crystal growth processes. Framboids are not mesocrystals, and they do not represent some extreme variety of single crystals, nor are their microstructures determined by epitaxial processes or a preexisting mineral template.

The poles of the crystallographic planes cluster around fixed points and are not all exactly coincidental. This suggests that, apart from the major 90° rotation of the pyrite microcrystals, there are also minor misorientations of adjacent microcrystals of a few degrees. Distributions of misorientation angles between adjacent microcrystals in disordered (i.e., randomly packed) and ordered (i.e., ccp) framboids are shown in Figure 6.14. The detection limit of misorientations was 2°. Adjacent microcrystals in disordered framboids display misorientations which are similar to that expected from random packing of microcrystals (Figure 6.14(a)). By contrast, adjacent microcrystals in ordered framboids where the microcrystals are cubic close packed show two distinct modes in the

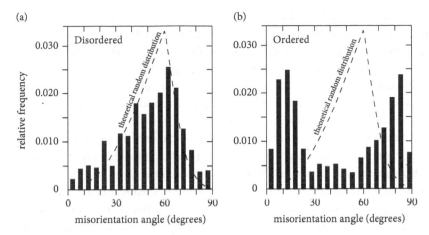

Figure 6.14. Misorientation angle distributions between adjacent microcrystal pairs for (a) disordered framboids and (b) ordered framboids. The expected distribution for random orientation distribution is indicated.
Compiled from data in Ohfuji (2004).

misorientation angles (Figure 6.14(b)). One mode is around 10° and reflects low-angle misorientations, and the other is around 85° and reflects the high-angle, 90° rotations between adjacent microcrystals.

6.4. Framboid Crystallography and Self-Organization

As discussed in previous chapters, self-organization in framboids, where thousands of pyrite microcrystals become stacked in ordered arrays, has fascinated and puzzled researchers. The results of single crystal XRD and framboid EBSD studies have clearly shown that the microcrystals are assembled by an aggregation process, rather than being the result of a crystallographic template or preexisting structural control.

The surprising result is that pyrite microcrystals in organized framboids show two distinct modes of misorientation between adjacent microcrystals. One mode reflects low-angle misorientation, as might be expected through the slight imperfections in packing of natural microcrystals. These low-angle misorientations clustering around a modal value of about 10° are small, but have been observed in detailed visual observations of framboid microcrystals, as described earlier. They are also observed in the large diffraction spots in the single crystal XRD patterns of ordered framboids. These large spots reflect groups of microcrystals with similar—but not exactly coincident—orientations.

The high-angle modal value of around 85° reflects adjacent microcrystals which have been rotated at 90° with respect to each other, together with the natural variation of small displacements described by the low-angle misorientations. Note that 270° rotations would also be indistinguishable from 90° rotations, and 180° rotations indistinguishable from no relative rotation. Also, note that the rotations may occur in any crystallographic direction (see Figure 6.15) and are not limited to just one rotational axis.

These 90° rotations can occur because of the shapes of the pyrite microcrystals. Commonly observed habits in organized framboids include the truncated octahedron and the simple octahedron. An example is given in Figure 6.15. The rotation of the truncated octahedron produces an identical shape, but the crystallographic orientation is rotated 90° as shown by the ⟨1 0 0⟩ faces. The cause of this dichotomy between morphology and crystallographic orientation is the lower symmetry of pyrite as shown in the atomic structure projection in Figure

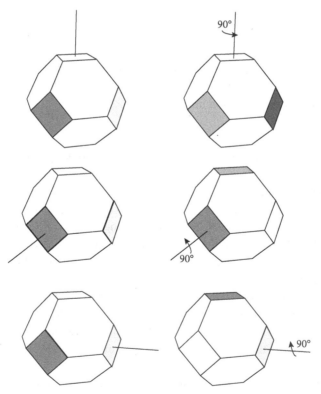

Figure 6.15. Rotation of a truncated pyrite octahedron through 90° in any crystallographic direction presents identical crystal habits, although the crystallographic orientation varies, as indicated by the shaded <100> faces.

6.3. Of course, pyrite octahedra, which are also common framboid microcrystal habits, behave in a similar fashion, as do less common forms such as the cube. The property of maintaining crystal habit during rotation is perhaps fundamental to limiting the minerals that can form framboids with organized microstructures.

Ohfuji (2004) found similar results for random pairs of microcrystals in organized framboids as with adjacent microcrystals. We can assume that all those microcrystals with relative misorientations >~80° represent microcrystals which have been rotated 90°, and that those with misorientations <~10° are microcrystals which have a relative crystallographic orientation of 0°. The variations from these exact numbers are caused by natural variations in the packing, as noted earlier. The quantitative data (e.g., Figure 6.14) suggest that around half of the microcrystals in an organized framboid have rotated crystallographic orientations. This would be expected if the rotation was a purely random event: that is, there is a 50% chance of a microcrystal being rotated.

6.4.1. Origins of Different Crystallographic Orientations of Microcrystals in Framboids

The question, then, is whether the pyrite microcrystals grow originally with these crystal orientations, or if they develop subsequent to their formation. The crystallographic data clearly show that crystal formation was well developed before ordering occurred. Framboid microcrystals result from simultaneous nucleation and crystal growth (see Chapters 11 and 12) which produce microcrystals with homogeneous sizes and habits. Obviously, then, a microcrystal cannot be formed in a void space defined by preexisting microcrystals, since all are formed at the same time.

The crystallographic results suggest that the microcrystals are formed originally with random orientations. Microcrystal packing in the majority of framboids is dominantly random, and the organized and partially organized microstructures are subsets. In organized domains, the microcrystals need to be equal sized and display similar habits. As discussed in section 12.1 of Chapter 12, the crystal habit development in pyrite is a reflection of the environment in which crystal growth takes place. The environment within the confined microscopic space of a framboid is likely to be homogeneous, and physical and chemical changes occur simultaneously throughout the volume. The result is that the pyrite microcrystals within the framboid are similar both in size and shape.

In some framboids with highly equant microcrystals with common habits, random original crystallographic distributions rotate into similar crystallographic directions. These rotations occur in three dimensions: that is, the rotation is not limited to just one crystallographic axis (Figure 6.15). Half of the pyrite microcrystals end up with orientations normal to the other half. It appears that microcrystals with original orientations between ca. 15°–80° rotate to 0° or 90° with respect to their neighbors, depending on which direction of rotation is the closest. Note that all odd numbered multiples of 90° rotation are indistinguishable crystallographically, as are 0° and 180° rotations and their multiples. However, these rotations >90° seem unlikely since there is no advantage to the microcrystal.

The pre-formed framboid microcrystals which are initially randomly organized throughout the framboid volume then, in some cases, begin to wholly or partly self-order. This is effected by rotation of the microcrystals until an ordered array is produced. Framboids are mostly formed in an aqueous medium and the consequence of this rotation must be that the microcrystals are initially packed loosely enough for rotation to occur. The processes involved in the rotation could include forces intrinsic to the microcrystals themselves, such as surface forces, or forces imposed from outside the framboid, such as Brownian motion. The fundamental driving force for microcrystal rotation and the development of organized microcrystal arrays in framboids is entropy maximization. This is described in more detail in Chapter 13.

It is interesting that clusters of colloidal and nano-particles where the particles are organized into ccp, hcp, and their variants, are currently described as *supracrystals* by some authors (e.g., Naik and Caruntu, 2017). This seems to bring the discussion of the exact nature of framboids full circle.

7

Organic Matter in Framboids

The association between organisms, organic matter, and pyrite framboids has been the subject of much confusion. Obviously, the sulfide in framboidal pyrite in sediments is bacterial in origin. There is thus a genetic relationship between framboids and microorganisms. Likewise, these sulfate-reducing prokaryotes are mainly heterotrophic, so there is also a close relationship between organic matter and pyrite framboids in sediments. However, many framboids are formed in hydrothermal environments and have been produced experimentally in the absence of both microorganisms and organic matter. Therefore, organic matter and microorganisms are not necessary for the formation of framboids. However, it is possible that there are several routes for the framboid formation, and the inorganic route does not preclude, a priori, the involvement of either organisms or organic matter.

The close association of framboids with organisms has been recognized and discussed since Schneiderhölm (1923) first suggested that framboids represented pyritized bacterial colonies. In fact, the relationship goes back to the first reports of framboids: Rhumbler (1892) originally described framboids from within foraminifera. The relationship between sulfate-reducing bacteria and pyrite was described by Issatshenko (1912).

The association of framboids with organic matter goes back even further. Framboids were originally described from peat deposits by Früh (1885), and Papunen (1966), from his studies of framboids in Finnish peats, proposed that framboids formed within colloidal humus drops.

The micropaleontologist Leonard Love originally extracted organic matter from within framboids, and his identification of these extracts as microfossils resulted in the idea of an apparent direct relationship between framboids and organic matter (Love, 1957). This mistaken thesis proposed (1) that framboids are the fossilized remains of microorganisms, and (2) that organic matter is responsible for the development of the framboid form and the organization of the constituent microcrystals. These fallacious ideas dominated discussions about the relationship between framboids, organic matter, and organisms during much of the latter half of the 20th century, and continue to surface occasionally in current research literature.

Framboids. David Rickard, Oxford University Press. © Oxford University Press 2021.
DOI: 10.1093/oso/9780190080112.003.0007

7.1. Framboids as Fossilized Microorganisms

The earliest bacteria identified were the large sulfur bacteria since these were big enough to be identified with the naked eye. These were described by the micropaleontologist Christian Gottfried Ehrenberg in 1838. In 1888 Winogradsky used sulfur bacteria in the discovery of chemolithotrophic autotrophy, the process whereby microorganisms obtain their energy from the metabolism of inorganic compounds like the oxidation of sulfide and their carbon from CO_2. In 1895, Beijerinck discovered sulfate-reducing bacteria, which reduce sulfate to sulfide, thereby completing the microbiological sulfur cycle.

7.1.1. Schneiderhöhn's Fossil Bacteria

The major impetus for the idea that framboids were fossilized bacteria derived from Schneiderhöhn's (1923) classic study of the 255 Ma old Kupferschiefer at Mansfeld in the Harz mountains of Germany. Schneiderhöhn identified framboids as fossilized *Monas* (i.e., *Chromatium*), *Thiophysa*, *Beggiatoa*, and *Thiotrix* (sic *Thiothrix*) species. Schneiderhöhn thought that the source of the framboid microcrystals were the large numbers of sulfur globules within their large spherical cells. Schneiderhöhn did not actually identify any fossil organic remains with the framboids. He was particularly taken by the coincidence of the size ranges of these large sulfur bacteria and the framboids he observed in the Mansfeld shales.

Schneiderhöhn's major sources for his thesis were the standard bacteriological handbooks by Migula (1900), Bruno Doss's work on sedimentary iron sulfides (Doss, 1912a, 1912b) and Hunt's (1915) work on the role of sulfur bacteria in the formation of Sicilian sulfur deposits. He missed Thiessen's (1920) paper which reported organic spheres 30–40 um in diameter which were often associated with pyrite in coal. Thiessen thought these were microorganisms. Schneiderhöhn's idea was rapidly taken up by Bergh (1928), who interpreted framboids from the ca. 500 Ma Alum Shale in Sweden as fossil sulfur bacteria, citing Schneiderhöhn's paper as his only source.

Little further process was made on the idea that framboids were fossilized microorganisms until Schouten (1946) debunked the whole idea. Schouten had been working on this in pre-war Holland and was not able to widely publish his ideas until after the end of the Second World War. Schouten (1946) thought that apart from some overlap in size range, there was little or no morphological similarity between modern sulfur bacteria and framboids. He also pointed out that framboids occurred abundantly in hydrothermal systems where bacteria were unlikely to be present.

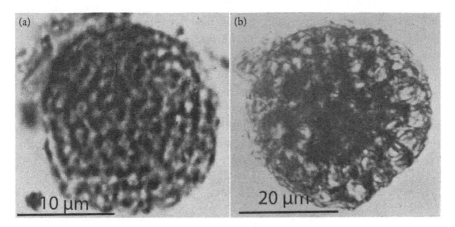

Figure 7.1. (a) *Pyritosphaera barbaria* (Love, 1957) and (b) *Pyritella polygonalis* (Love, 1957). Transmission microphotographs of type specimens, extracted from 300 Ma Carboniferous pyrite framboids.

Reprinted from Love, L., Micro-organisms and the presence of syngenetic pyrite. *Quarterly Journal of the Geological Society, London*, 113: 428–440. 1957, courtesy of the Geological Society of London.

7.1.2. Love's Microfossils

Love (1957) used standard maceration techniques on rocks containing framboids and isolated a number of organic structures from the pyrite framboids. He showed that the individual framboid microcrystals were each contained in a cell-like compartment within a spherical organic structure (Figure 7.1).

In contrast to Schneiderhöhn, Love thought these were the remains of multicellular organisms rather than single-celled bacteria, but noted that their biological affinities were unclear. He identified two species, which he named *Pyritisphaera barbaria* and *Pyritella polygonalis*, both within a new genus of what he supposed were sulfate-reducing eukaryotes he called *Pyritosphaera*.

Vallentyne (1962) disproved Love's theory when he showed that the framboids separated from lake sediments did not contain any organic residues. That meant that framboids could form without the intervention of any biological intermediary. The inorganic syntheses of framboids (e.g., Berner, 1969; Sweeney and Kaplan, 1973) further repudiated Love's theory, and Love himself helped correct the record with descriptions of framboids formed in andesitic lavas (Love and Amstutz, 1969).

With hindsight, it is apparent that Love was not observing organisms but rather the impressions left by pyrite crystals in organic residues. In particular, *Pyritella polygonalis* was characterized by polygonal interior cell shapes, which are obviously the impressions of microcrystals.

The impressions left in organic residues after pyrite removal can be highly variable (Figure 7.2). Figure 7.2(d) shows how clearly impressions of preexisting pyrite crystals can be formed.

The idea that framboids represent bacterial colonies or multicellular organisms continues to reappear in the literature in spite of the overwhelming negative evidence. Folk (2005) thought that framboids were colonies of "nannobacteria" [*sic*]. Gong et al. (2007) described the hollow microcrystals described in section 4.5.6 of Chapter 4 as "mouth-like openings" and thought that the framboids represented bacterial colonies.

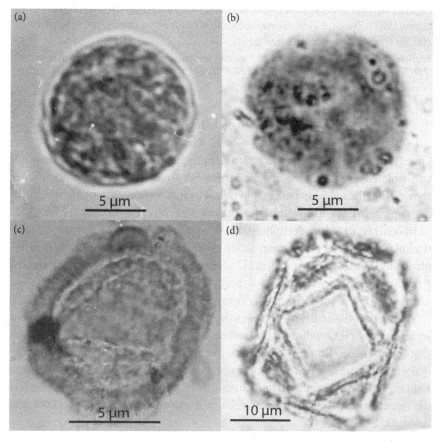

Figure 7.2. Photomicrographs of organic residues released by maceration of pyrite in the 170 Ma Jurassic Dogger formation. Figures 7.2. (a) and (b) might be classified as *Pyritosphaera barbaria* (Love, 1957), (c) is a different form, and (d) shows the impression left by a pyrite crystal in an organic residue.

7.1.3. Pyrite Precipitation within Prokaryotic Cells

One of the simple tests to see if a culture of sulfate-reducing bacteria is growing is to observe the blackening of the culture as FeS is precipitated on the cells. Precipitation of FeS on and within bacterial cells has been studied in some detail (see Rickard, 2012a, for review) since Issatshenko (1912) originally reported pyrite within the cells of cultured sulfate-reducers. FeS penetration and envelopment appear to occur when the culture is declining. The work of Hatton and Rickard (2008) emphasized the requirement for healthy organisms to protect themselves against FeS coating and penetration when they showed that FeS had a deleterious effect on DNA.

The intimate spatial relationship between iron sulfides, including pyrite, and prokaryotes is not questioned. The problem is then to extend this into prokaryotic cells or colonies being templates for framboids. Obviously, they are not *necessary* templates for framboids since framboids can be produced inorganically. However, this does not rule out a priori the possibility that they might be, in some cases, fossil microorganisms.

7.1.4. Framboids as Fossil Microbial Colonies

Apart from the fact that pyrite framboids form inorganically and do not require preexisting organic templates, the problem with the microbial colony hypothesis is that prokaryote cell walls are unlikely to provide a template. The cell envelopes of bacteria and archaea are complex structures in which the cell wall *sensu stricto* forms only a part. The cell wall, in the sense of a rigid structural element, is represented in bacteria by a peptidoglycan layer, a polymer of sugars and amino acids which determines, to a large extent, bacterial shape (Beveridge, 2001). All sulfur bacteria possess a plasma membrane, a fluid phospholipid bilayer about 7 nm thick, which encloses the interior cytoplasm. At cell death, the integrity of the cytoplasmic membrane is not retained. Archaea have variable cell envelopes which are differentiated from bacterial cell walls by the absence of peptidoglycan. Many bacteria and archaea possess a pseudocrystalline surface layer of proteins or glycoproteins. In the case of archaea, this S-layer appears to be the only component of the cellular envelope which may provide some rigidity to the cell form (Engelhardt, 2007), but even here it is not universally present.

The shapes of microorganisms are maintained by an internal turgor or fluid overpressure. The internal overpressure is commonly of the order of two bars, and the cell walls are constructed such that these overpressures can be contained. Prokaryote cell walls are thus relatively thin structures with little chance of maintaining their shape after cell death and therefore being preserved as fossils.

Experiments with modern bacteria show that desiccation is a process that can preserve at least an approximation of the original bacterial form (Edwards et al., 2006). In aqueous sedimentary environments, cells are not easily preserved. They autolyze rapidly and their contents are utilized by other organisms. The conditions for preservation of prokaryotes in sulfides which might involve initial desiccation are theoretically possible, but appear to be uncommon.

Rickard (2012a) estimated that the devolatilization of organic matter in an average microbial cell during fossilization would result in a carbon-rich layer less than 100 nm thick. It would therefore seem improbable that colonies of prokaryotes might explain the organic remains reported by Love (1957) from framboids.

7.1.5. Framboids as Fossil Giant Sulfur Bacteria

Schneiderhöhn's idea that framboids are fossilized giant sulfur bacteria is also untenable, although the cell envelopes of the large sulfur bacterium *Beggiatoa* are among the thickest known among bacteria (Maier and Murray, 1965) and are more likely to be preserved as fossils. However, like *Thioploca*, where the trichomes are enclosed in a thick sheath, these are filamentous forms and cannot act as templates for spherical framboids. The shape of the spherical sulfur bacteria is unlikely to be maintained after cell death, for the reasons discussed earlier, and cannot act as the form for sub-spherical framboids.

As originally pointed out by Schouten (1946), the idea that each framboid microcrystal was coincident with a sulfur globule in a large spherical sulfur bacterium is untenable as this would result in a majority of partially spherical aggregates since most sulfur bacteria are not completely packed with sulfur globules (e.g., Figure 7.3).

7.1.6. Framboids as Fossil Eukaryotic Microorganisms

Pyrite microspheres and microcrystals were described encased in organic matter from 1.6 Ga Proterozoic pyritic rocks associated with the giant Mt. Isa orebodies, Queensland, Australia (Love, 1965; Love and Zimmermann, 1961) (Figure 7.4). In polished section, the organic material appears as a rim of moderately reflective mineral surrounding the pyrite grains. The optical properties of this material suggest that it is graphitic. This is consistent with the high C fraction analyzed by Love and Amstutz (1966), as described later in this chapter. The organic bodies extracted from the pyritic shale often show both biologic and crystallographic affinities. The idea that they are fossil organisms is supported by the double (and

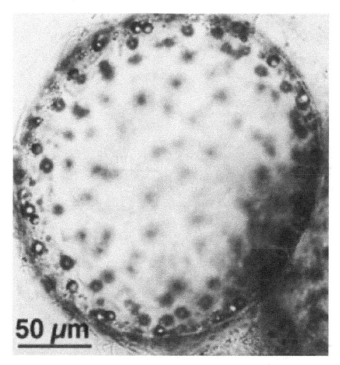

Figure 7.3. The giant sulfur bacterium *Thiomargarita namibiensi*. Like most of these colorless sulfur bacteria, the cells consist of a large central vacuole surrounded by a rim of sulfur globules.
Microphotograph by Heidi Schulz-Vogt.

rarely) triple wall structures, as well as the appearance of groups. The crystallographic affinities are suggested by the common polygonal internal cavity and the frequent polygonal external form. This dichotomy was resolved by Muir (1981), who commented that the original organic remains have been distorted by internal crystal growth. This is important since this suggests that these forms cannot be dismissed as abiologic because of their crystalline internal forms.

It is significant that the pyritic grains within these organic sacs do not show any framboidal textures. Etching reveals often relatively complex pyrite crystals with a history of overgrowths. However, relict framboid textures are not revealed on etching, as would be expected if they were originally present (e.g., Rickard and Zweifel, 1975). The implications of the absence of framboids in these Proterozoic shales is discussed further in section 1.3.2 of Chapter 1.

These microfossils are common in many Proterozoic and Phanerozoic sediments, where they are often described as acritarchs. This is a bit of a circular definition since acritarchs were originally defined by Evitt (1963, p. 200) as

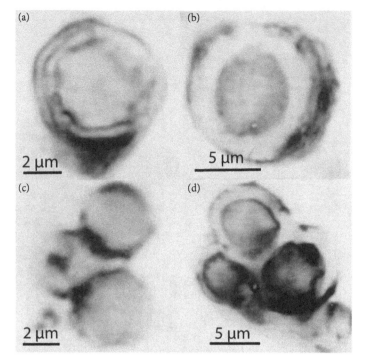

Figure 7.4. Organic residues from pyrite in 300 Ma Carboniferous ((a), (b)) and 1650 Ma Proterozoic ((c), (d)) shales.

Reprinted from Love, L., Micro-organic material with diagenetic pyrite from the Lower Proterozoic Mount Isa Shale and a Carboniferous shale. *Proc. Yorks. Geol. Soc*, 35: 187–202. 1965, courtesy of the Geological Society of London.

small microfossils of unknown and probably varied biological affinities consisting of a central cavity enclosed by a wall of single or multiple layers and of chiefly organic composition.

They are common in the Urquhart Shale, a host rock of the Mt. Isa orebodies, and non-pyritized varieties have been described by Muir (1981). Acritarchs in the Phanerozoic often have dinoflagellate affinities, and the infilling of dinoflagellate cysts by pyrite has been described by Szczepanik et al. (2017). In some cases, clusters of small pyrite framboids were observed in these dinoflagellate cysts. That is, the cysts did not define the shapes of individual framboids (see section 7.1.7).

The possibility that these Proterozoic pyrite forms were replacement of algal cell walls by pyrite was investigated experimentally by Rickard et al. (2007). They showed that pyrite does not replace the cell walls, and this was confirmed in the natural samples by Szczepanik et al. (2017). Rather, pyrite infills cells and the cell walls are lysed by enzymes released on cell death. The cell wall is mostly fluid, and the small amount of particulate carbonaceous material it contains is compressed and deformed during pyrite crystal growth, creating a boundary between the pyrite grains. The overall effect is to form, quite commonly, a good mold of the original cell. Rickard et al. (2007) demonstrated experimentally that infilling of algal cells by pyrite was a plausible explanation for the small, sub-spherical pyrite grains enclosed in carbon sacs observed in the Australian Proterozoic stratiform sulfide deposits in the Mt. Isa area.

Single-celled eukaryotes obviously do not provide a template for the formation of the individual microcrystals that constitute framboids. However, the lack of any common relationship between eukaryotic single cells and framboids is notable. It would seem that intuitively, the thick resistant cell walls of algae and their relatives would provide a ready explanation for the origin of the organic bodies extracted from some framboids. However, this does not seem to be the case, and algal infillings tend, at best, to result in pyrite spherules rather than framboids. This in turn suggests that the processes that form framboids are inhibited within these organic bodies with relatively thick cell walls.

7.1.7. Polyframboids as Steinkerns

As described in section 3.4 of Chapter 3, polyframboids in this book specifically refer to spheroidal aggregates of sub-spheroidal, similarly sized, framboids. The central problem, then, is how the framboids aggregate to a spherical form. The weight of a cluster of relatively dense pyrite framboids in a fluid medium necessitates some template to maintain sphericity. As shown in section 3.2.3 of Chapter 3, gravitational effects may even cause deformation of individual framboids by processes akin to soft-sediment deformation. Then the formation and maintenance of sphericity in polyframboids suggest a special environment of formation. Indeed, the various forms of irregular clusters of framboids support the conclusion that the development and maintenance of sphericity in polyframboids require particular conditions.

In fact, framboids commonly occur in groups in sediments and, when they are extracted, they may appear as stuck together in clusters. There is obviously a gradation between framboid clusters and polyframboids as illustrated in Figure 7.5. Figure 7.5(a) shows an irregular cluster of framboids similar to those commonly observed in sediments and within fossils, often aligned along bedding planes. Close inspection shows some evidence for soft-sediment deformation. Figure 7.5(b) shows what, at first sight, appears to be a spheroidal cluster of framboids. However, closer inspection shows the constituent framboids to be deformed, varying in size and with individual framboids having noticeably different sized

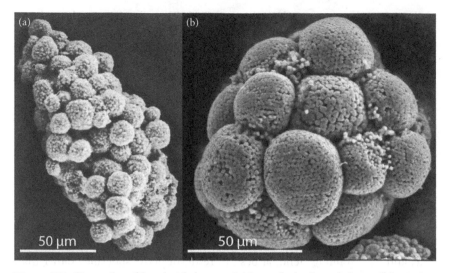

Figure 7.5. Examples of framboid clusters: (a) irregularly shaped cluster (b) quasi-spheroidal cluster of framboids with varying microcrystal sizes. Both examples show evidence of soft-sediment deformation.

Scanning electron micrographs from ca 150 Ma, late Jurassic, northern France, courtesy of Nicolas Tribovillard.

microcrystals. Comparison with the polyframboid in Figure 3.14 (Chapter 3) underlines the differences in these textures. The framboidal clusters in Figure 7.5 have been suggested to have been formed within fecal pellets (Tribovillard et al., 2008). This is consistent with the widespread idea that these framboidal clusters are formed within organic matter including the humic-fulvic acid gels, discussed later (section 7.3) with respect to the origin of framboidal organic matter (cf. Sawlowicz, 2000).

It would appear that the spheroidal clusters characterizing polyframboids evidence a more rigid template than organic gels. The first use of the old German word *steinkern* in English to describe mineral casts of fossils was probably by the US paleontologist J. Brookes Knight in 1931. He remarked as a footnote (1931, p. 222): "I much prefer the simple and expressive German *steinkern* to our vague and in-exact internal mold (or is it internal cast?)." Since that time, there have been many reports of pyrite steinkerns, often including framboids. In fact, pyritized fossil molds have proven an important source of information about the nature and evolution of life on Earth (e.g., Rickard, 2012b).

Schallreuter (1984) found a close similarity between framboids packing a diatom test and polyframboids (Figure 7.6). He noted that the framboids in the foraminifer-nannofossil-diatom ooze formed polyframboids, whereas in black shales they were dispersed. He related the formation of polyframboids to the

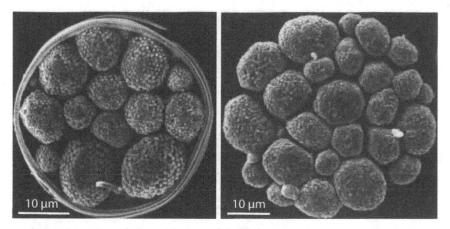

Figure 7.6. Framboids filling a diatom and similar association of framboids where the diatom skeleton is not present.

Scanning electron microscope images by Schallreuter, R., 1984. Framboidal pyrite in deep-sea sediments. *Initial Reports of the Deep Sea Drilling Project*, 75: 875–891 published courtesy of the IODP.

presence of diatom skeletons in the ooze. However, the diatoms are disk-shaped and would not be consistent with the 3D sphericity displayed by polyframboids.

McNeil (1990) showed clearly that the outermost framboid layer of the polyframboid in 40 Ma Eocene diatoms from the Beaufort-Mackenzie Basin, Canada, formed against the diatom skeletal wall, retained their sphericity. The result was confirmed by De Jonghe et al. (2011) for 45 Ma Eocene diatoms from the Isle of Wight, United Kingdom (Figure 7.7).

This observation is significant since other protists, such as radiolarians, do display spherical skeletons with the same size range as that of polyframboids. Schallreuter's observation suggests that these spherical skeletons might provide the stable environment over a period of weeks or months necessary for the formation of polyframboids such as the one displayed in Figure 3.14 (Chapter 3).

Szczepanik et al. (2017) described framboids filling dinoflagellate cysts. Their images show elliptical cysts about 50 μm in length, filled with framboids around 5 μm in diameter (Figure 7.8). Nicolas Tribovillard (personal communication, 2019) reports similar framboidal structures lacking organic membranes from 160 Ma claystones of southeastern France.

Fabricius (1961) noted that these polyframboids had similar dimensions to some species of the giant sulfur bacteria. As discussed previously, the identification of framboids as fossil sulfur bacteria has a long history. Since Fabricius's time, further giant sulfur bacteria have been discovered, including the largest *Thiomargarita nambiensi*, by Schulz et al. (1999), which ranges up to 750 μm in

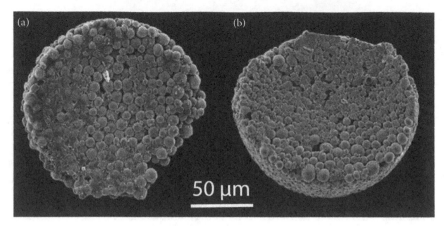

Figure 7.7. Steinkerns of framboid-filled diatoms showing plane view (a) and tilted view (b).

Scanning electron micrographs reprinted from *Proceedings of the Geologists Association*, 122(3): 472–483. Middle Eocene diatoms from Whitecliff Bay, Isle of Wight, England: Stratigraphy and preservation. De Jonghe, A., Hart, M. B., Grimes, S. T., Mitlehner, A. G., Price, G. D., and Smart, C. W. Copyright (2011) with permission from Elsevier.

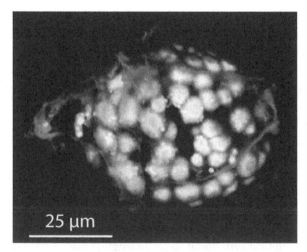

Figure 7.8. Framboids filling a dinoflagellate cyst from 167 Ma old Bathonian mudstones of the Polish basin, showing flattened framboids against the cyst cell wall.

Scanning electron back scattered electron micrograph reprinted from *Review of Palaeobotany and Palynology*, 247: plate 2E. Pyritization of dinoflagellate cysts: A case study from the Polish Middle Jurassic (Bathonian). Szczepanik, P., Gize, A., and Sawlowicz, Z. Copyright (2017) with permission from Elsevier.

diameter. The consonance between these sulfur bacteria, their obvious relation-ship with pyrite-forming environments, and potential polyframboids is striking. However, as discussed earlier, it is unlikely that they are involved in polyframboid formation. The problem is that on death, turgor is not maintained and the cell rapidly collapses. In other words, there is not enough time for the formation of the polyframboid before the enclosing spheroidal casing collapses. The ta-phonomy of *Thiomargarita* has been studied experimentally by Cunningham et al. (2012) and their results confirm this conclusion.

There is a body of evidence that is consistent with the suggestion that polyframboids are steinkerns. This includes (a) the coincidence of sizes be-tween polyframboids and protists, (b) the apparent requirement for the de-velopment of close packed ordering in a spherically constrained system, and (c) the common distribution of polyframboids in specific sedimentary layers or lenses (e.g., Love 1962). If this is the case, the organisms involved must have a robust spherical exoskeleton which would remain intact for months after death for the polyframboids to form. This could be satisfied by carbonate or silica-based protists such as foraminifera or radiolarians, or organic-walled bodies, such as dinoflagellate cysts.

7.2. Framboidal Organic Matter

The fact that framboids can form within organic matter has been known since the original description of framboids in peat by Früh (1885). Since that time there have been many studies of framboids in peat and, because of the economic and environmental impact of pyrite in coal (e.g., Finkelman et al.,2019). Framboids can constitute a significant part of the pyrite-loading of coals and often occur as part of a complex paragenesis involving pyrite for-mation at all stages of coal development (e.g., Chou, 2012; Querol et al., 1989; Wiese and Fyfe, 1986). However, the syngenetic stage of framboid formation in coals relates to their formation in the precursor peat deposits (Chou, 2012; Früh, 1885; Papunen, 1966). In this environment, framboids form within the hydrated organic macromolecular polymers that develop within plant and algal remains during early diagenesis (Papunen, 1966; Rickard, 1970, 2012a).

Investigating the organic matter enclosed within framboids, rather than the bulk organic matter enclosing framboids, is especially challenging and requires advanced microscopic and analytic methods. Analytical high-resolution elec-tron microscopy, in particular, has enabled the visualization and identification of organic matter in framboids in situ rather than secondarily after treatment to remove the pyrite.

7.2.1. Framboids and Biofilms

One potential source of the organic matter in framboids was originally resolved by Large et al. (2001) when they identified framboids forming with biofilms in canal sediments. Maclean et al. (2008) confirmed and extended Large et al.'s work with a high-resolution study of framboids in a biofilm scraped from the inner surface of a 9 m, 30-year-old borehole drilled into an exploration tunnel within a Witwatersrand gold mine (Figure 7.9).

Charaklis and Marshall (1990) defined a biofilm as a group of cells immobilized at a substratum surface and frequently embedded in an organic polymer matrix of microbial origin. Biofilms are characterized by high cell densities ($\leq 10^{10}$ cells mL^{-1} hydrated biofilm) in a matrix of EPS. EPS has been variously defined as extracellular polysaccharides, extracellular polymer slime, and extracellular polymeric substances. The amount of EPS produced by a bacterial population is remarkable. I cultivated a halotolerant strain of the sulfate-reducing bacterium *Desulfovibrio vulgaris*, for example, which could rapidly fill the test tube with EPS (Rickard, 1968b). Biofilms are ubiquitous in natural aqueous environments and include flocs which are known as marine snow in the marine environment and feature particularly in areas of high productivity, such as outside deep sea hydrothermal vents.

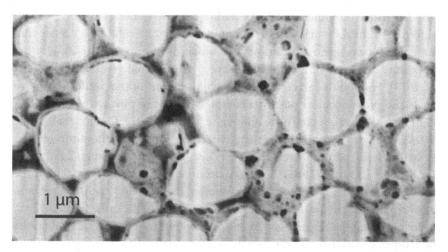

Figure 7.9. Organic matter filling the interstices of an ion-milled <30-year-old framboid.

Transmission electron micrograph reproduced from *Geobiology*, 6(5): 471–480. A high-resolution chemical and structural study of framboidal pyrite formed within a low-temperature bacterial biofilm. Maclean, L. C. W., Tyliszczak, T., Gilbert, P., Zhou, D., Pray, T. J., Onstott, T. C,. and Southam, G., courtesy of John Wiley and Sons.

The composition of bacterial EPS is highly dependent on bacterial species and the environmental conditions under which it is formed (e.g., Guo et al., 2018). EPS may be composed of polysaccharides, proteins (both structural proteins and enzymes), extracellular DNA, lipids, and surfactants (Jahn and Nielsen, 1995). However, the major component is water, which can constitute 97% of the mass of the biofilm (e.g., Nazir et al., 2019).

A detailed electron micrograph of organic matter in an ion-milled Holocene framboid is shown in Figure 7.9 from Maclean et al. (2008). The borehole from which the framboids were collected was 30 years old, so this gives a maximum limit to the age of the biofilm. The organic matter has already developed small holes, which I suppose may be due to dehydration, although condensation and devolatilization could also play a part. Ian Butler (personal communication, 2020) notes that his team had observed many, often large, gas bubbles in fresh biofilm, representing evolved gases from microbial respiration. This means that the small holes in the biofilm shown in Figure 7.9 may be primary. Note that, although the organic matter is concentrated around the pyrite microcrystals themselves, close inspection reveals that it is already separating from the surfaces of some microcrystals, leaving only strand-like connections.

Simple air drying of biofilms can reduce the water content to less than 10% (Tamaru et al., 2005), which means a substantial decrease in the volume of the biofilm. So we might not expect ancient biofilms in virtually water-free framboids to be completely preserved. Continuation of the dehydration process illustrated in the <30-year-old biofilm over ca. 50 million years is shown in Figure 7.10.

Figure 7.10. Detail of organic matter coating and linking pyrite microcrystals within organized framboids from within fossilized wood from the 50 Ma Eocene London clay.

Scanning electron micrograph image by Ian Butler.

The organic matter within framboids does not completely enclose the pyrite microcrystals. Indeed, relatively little organic matter is retained in the framboids and, without detailed studies of individual framboids, this may be difficult to spot. Certainly, dissolving the pyrite from the framboid shown in Figure 7.10 will not produce a contiguous Love-style organic body. There is thus a continuum between limited organic matter in framboids and complete enwrapping of the pyrite by organic matter.

Biofilm has been proposed to be the source of the organic matter encasing individual microcrystals in 560 Ma framboids which were enclosed within pyrite crystals (Wacey et al., 2015). Figure 7.11 shows details of the framboids. Both the framboid and its constituent microcrystals are encased in organic matter-rich nanoporous pyrite. The organic matter distribution was traced by ion mapping with nano-scale secondary ion mass spectrometry (NanoSIMS) of $^{12}C^-$ and $^{26}CN^-$ and was coincident with the nanoporous pyrite zones. Wacey et al. (2015) concluded that the organic matter represents ancient biofilms in which the framboids originally developed. It is interesting to speculate on what would be the result of Love-style chemical extraction of the organic residues in these framboids. The observations suggest that the result would be a thicker organic wall encasing a matrix of organic material. Wacey et al.'s (2015) observations provide a possible route for the preservation of biofilm in ancient framboids, and organic matter–rich nano-porous pyrite has been observed in 3500 Ma old stromatolites (Baumgartner et al., 2019), underlining the potential of this process for preserving ancient biofilms.

A similar relationship between organic matter and framboids was described by Hu et al. (2018), who also used NanoSIMS to trace the distribution of organic

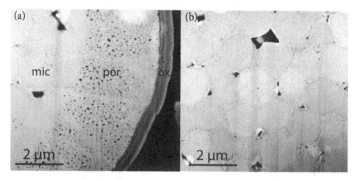

Figure 7.11. Dark-field scanning transmission electron micrographs through ultrathin sections of 560 Ma framboids: (a) framboid showing inner microcrystals (mic), nanoporous OM-rich zone (por) and outer oxidized zone (ox) section; (b) microcrystals showing porous rims.
Images by David Wacey.

matter, as $^{26}CN^-$, in mixed ZnS-pyrite framboids from metamorphosed pelites of uncertain geologic age from the Otago Schist in New Zealand. Again, the organic matter coated the individual framboid microcrystals as well as the framboids themselves. In this case, the authors were not able to demonstrate the presence of nanoporous pyrite, but the NanoSIMS trace showed that the organic matter was accompanied by pyrite, suggesting a similar relationship.

Biofilms have been observed on framboids in dinosaur bones (Kaye et al., 2008), where they were originally mistaken to represent fossilized dinosaurian soft tissue. This raises a caveat for the identification of framboidal organic matter as biofilm. Tribillovard (personal communication, 2020) identified biofilm growing on framboids after just three weeks of routine storage. Framboid mounts were observed in SEM and were shown to be free of biofilm. They were then routinely stored in a small, sealed plastic box and re-examined three weeks later. The framboids were found to be covered in biofilm. That is, biofilms associated with framboids may represent modern material. Perhaps equally significantly, it does suggest that biofilms develop readily on framboids, and their formation may post-date framboid formation. Thus, they may not represent a sample of syngenetic framboidal organic matter.

The conclusion is that bacterial biofilm may constitute a major component of the organic matter observed in ancient framboids. However, it could be modern, and care needs to be taken to exclude the development of modern biofilms during framboid storage, using cold storage for example. If present at the time of framboid growth, the biofilm must have been in a gel-like form, which left an even coating on all the microcrystals in the framboid. The organic matter of the biofilm was excluded from the pyrite microcrystals as they developed, resulting in an organic matrix to the framboid. Preservation of ancient biofilm was enhanced by reprecipitation of pyrite around the microcrystal margins.

7.2.2. Geopolymers and Framboid Organic Matter

The organic material extracted from ancient framboids is a tough, refractory geopolymer which would be generally classed as kerogen. Kerogen is formed from the condensation of the products of microbial degradation that escape further degradation by the microbial community. The organic geopolymers in framboids developed in a sulfide-rich environment and the original organic moieties reacted with sulfide, polysulfide, and sulfonate species to produce organic sulfur polymers (see Rickard, 2012a, for review). Sulfide builds large three-dimensional network of altered biomolecules connected by mono-, di-, and tri-sulfide bridges. These large organic geopolymers with extensive sulfur cross-linking are protected from enzymatic decomposition and microbial attack and this may contribute to organic matter preservation. The result is an increase in the content of macromolecular organic geopolymers during diagenesis (Eglinton et al., 1994).

The evidence from peat deposits suggests that it is probable that this framboidal organic matter is a product of early diagenesis. The dominant organic component in the dysoxic to anoxic environments where framboids form is amorphous organic matter (AOM): that is, all the organic matter that appears structureless at the light microscopic scale. Pacton et al. (2011) originally showed that AOM was not only the degradation products of terrestrial or marine components but also derived from biofilms, which may include prokaryotic cells, EPS, and algae, especially in stagnant and oxygen-depleted environments. The fraction of the insoluble organic matter in recent sediments that is resistant to acid hydrolysis is called protokerogen. It seems probable that this is the organic matter originally incorporated into framboids.

Protokerogen is mainly formed from phytoplankton and initially evolves through microbial degradation and condensation. The microbial degradation products are mostly derived from proteins, polysaccharides, and lipids. The random condensation of these degradation products leads to the formation of complex, cross-linked geopolymers exhibiting a high resistance to subsequent diagenetic degradation. In sulfur-rich environments, protokerogen formation involves sulfurization which results in high molecular weight structures cross-linked by sulfur and resistant to acid hydrolysis (Sinnighe Damsté and de Leeuw, 1990). In their study of euxinic Holocene sediments in the Cariaco Trench, Aycard et al. (2003) found that protokerogen formation through carbohydrate sulfurization was faster than lipid sulfurization and only sulfurized carbohydrates were found in the youngest (<900 a old) samples. This time frame envelopes the time required for framboid formation (Rickard, 2019b) and means that it is probable that the bulk of the protokerogen incorporated into framboids is likely to have been sulfurized carbohydrates. The sulfurization is important since it renders the carbohydrates unavailable as metabolites for sulfate-reducing bacteria.

Although the incorporation of polysulfide linkages into biomolecules results in polymerization and initial preservation of organic matter, the polysulfide linkages are thermally labile (Aizenshtat et al., 1995). Lewan (1998) showed that it is the formation of sulfur radicals, rather than the weakness of the S–S bonds, that causes early maturation of sulfur-rich petroleum precursors. This is interesting from the point of view of framboid organic matters since it may suggest that the incorporated organic material becomes more plastic during late diagenesis, due to polymerization, resulting in the even redistribution of the material throughout the framboid.

7.2.3. The Form of Protokerogen

Aycard et al. (2003) found that most of the AOM in recent sediments in the euxinic Cariaco Trench consisted of brown, heterogeneous flocs around 15 nm in size which were frequently associated with pyrite (Figure 7.12). Orange gel-like drops 2–10 µm in size were dispersed within these flocs and their abundance increased with depth. These orange gel-like drops are interesting from the point of view of framboid formation since they have both similar sizes and shapes to framboids and would provide an ideal medium for pyrite nucleation and growth. The orange color is usually taken as a proxy for S-rich compounds (Tribovillard et al., 2001).

Similar orange spheres were observed in AOM extracted from 350 Ma Mississippian black shales by Emmings et al. (2019). Emmings et al. noted the similarity of these spheres to Love-type organic matrices extracted from some pyrite framboids and showed that the size distribution of the organic spheres was similar to that of coexisting pyrite framboids. It appears that there is strong evidence that these orange drops are the source of at least part of the organic matrices observed in pyrite framboids. These organic forms were predicted by Rickard (1970) and were described at that time as "coacervates." They are possibly similar to the spherical, yellow to brown "organisms" reported by Thiessen (1920) from a peat bog near Hayton, Wisconsin.

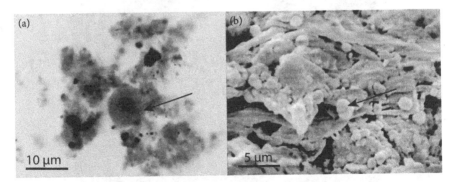

Figure 7.12. Spherical organic forms (examples arrowed) extracted from Holocene sediments from the Cariaco Trench: (a) transmitted light microphotograph and (b) scanning electron micrograph.

Reprinted from *Organic Geochemistry*, 34(6): 705. Figure 2. Formation pathways of proto-kerogens in Holocene sediments of the upwelling influenced Cariaco Trench, Venezuela. Aycard, M., Derenne, S., Largeau, C., Mongenot, T., Tribovillard, N., and Baudin, F. Copyright (2003) with permission from Elsevier.

7.2.4. Extrusion of Framboidal Organic Matter

Another attribute of the Love-type organic bodies (Figure 7.1) associated with pyrite framboids is that they have the same size distribution as framboids. The organic material often forms a continuous membrane around the framboid (Figure 7.13). How this membrane developed is a key issue: did the framboid initially fill a sac-like body now defined as the membrane, or did the membrane develop during or after framboid formation? In other words, did the membrane form the original template for the final framboid form, or did the membrane result from pyrite formation?

One problem with the organic drop theory of framboidal organic matter is the apparent correlation in size between the pyrite framboid and the organic body. Correlation is not causation, and there is no a priori reason for the size of any individual drop to be the same size as the framboid.

The original framboid organic material was also coherent and not a fluid-filled cell like an algal cell, since the organic matter permeates the framboid structure. More detailed analyses show that the organic form was not necessarily sub-spheroidal since framboids with faceted exterior forms are also enclosed in organic membranes (e.g., Large et al., 2001). The conclusion is that the form of the organic matter was inherited from the framboid, and not vice versa. That

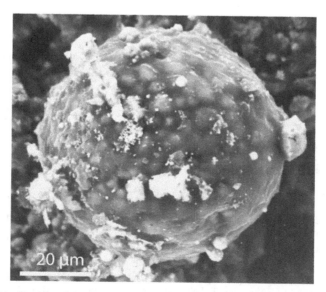

Figure 7.13. Framboid from the 380 Ma Gowanda shale from New York, encased in organic matter.

Scanning electron micrograph image by Ian Butler.

is, the sub-spherical form of the organic protokerogen gel is not a template for framboid formation.

The forms of the Love-type organic bodies extracted from framboids clearly retain the shapes of the individual pyrite microcrystals. That is, the final distribution of organic matter in framboids occurred after, and probably as a result of, pyrite microcrystal growth. The size of the original drop of organic matter can be estimated. In the limiting case, ccp results in a maximum packing coefficient of 0.74 (section 4.2 of Chapter 4) for spheres, and the unoccupied volume is then just 26% of the framboid volume. Assuming this space is occupied by organic matter, the volume of organic matter in the framboid is 26% of the framboid volume. Since the volume of a sphere is proportional to the cube of the radius, the radius of a sphere of 26% of the framboid volume is equal to the cube root of 0.26, or ~0.64. That is all the organic matter that could be contained in a sphere with radius equivalent to 64% of the framboid radius. In fact, the packing efficiency is likely to be less than 0.74, and the organic matter in the framboid would be contained in a sphere nearer the diameter of the final framboid. This means that if the framboid was originally formed in a volume of organic matter similar in size to the volume of the framboid, then an excess of organic matter would be extruded to form the observed membrane enclosing the framboid. If all this extruded organic matter were to form the enclosing membrane, then this would produce a membrane with a maximum thickness of approximately 10% of the framboid radius. For example, for a 10 μm diameter framboid, the maximum membrane thickness would be about 0.5 μm. This is surprisingly similar to the estimated thicknesses of the membranes of organic matter observed surrounding pyrite framboids and the sacs or "cell-walls" enclosing the Love-like organic bodies, since it does not take into account changes in the organic matter during diagenesis. In particular, evolution of the original organic materials will include condensation, devolatilization, and dehydration, all of which act to shrink its volume.

7.2.5. Epigenetic Organic Matter

Love and Amstutz (1969) were especially influenced by Vallentyne's (1963) report of the absence of organic matter in modern lacustrine framboids. Since organic matter had been widely extracted from geologically ancient framboids, they suggested that framboidal organic matter was a later infilling. In fact, as discussed earlier, recently formed framboids do contain organic matter, although, if it is biofilm, it may be relatively sensitive to more robust extraction methods. Love and Amstutz (1966) discovered that framboids had substantial interstices and these would permit organic matter to penetrate into the framboid after formation. Unfortunately, one might expect intermediate stages of this penetration to be displayed by this process and these types of organic infillings have not been reported.

7.2.6. Composition of Infilling Materials within Framboids

Apart from the biofilms described in section 7.2.1, the actual nature of the interstitial material in framboids has not been widely studied. The observed material often appears as translucent and amorphous, is resistant to acid dissolution, and is usually identified as organic in origin. Love and Amstutz (1966) analyzed the major element composition of the organic residues left after treatment of the specimens with HCl, HF, and HNO_3. The results showed carbon fractions varying from 35% in the 390 Ma Rammelsberg framboids to over 70% in the 1650 Ma spherules in the Mt. Isa shale. The results confirmed that the residues were indeed organic and suggested that the organic matter in framboids was subjected to the usual coalification processes in the diagenesis of organic matter in sediments.

One feature of the Love and Amstutz (1966) analyses of framboidal residues was the highly variable fraction of residual material left after pyrolysis of the organic matter. Since the framboids were initially collected as heavy concentrates, this material was assumed to represent refractory heavy minerals which were resistant to acid attack.

Scheihing et al. (1978) reported the clay mineral kaolinite, $Al_2Si_2O_5(OH)_4$, from within framboids (Figure 7.14). They considered this a rare occurrence since they found no kaolinites in samples from another 50 coals. However, Love et al. (1984) reported similar forms of kaolinite-quartz matrices, associated with organic matter in this case, from both marine and non-marine horizons in 300 Ma (Pennsylvanian) mudrocks. Zheng et al. (2017) described framboids in a kaolinite matrix in Chinese coals. Scheihing et al. (1978) interpreted the kaolinite as penecontemporaneous with the pyrite. They noted that the kaolinite was evenly distributed around the framboid microcrystals and reported that framboids without kaolinite showed no interstices between the microcrystals. They concluded that the framboids formed within a kaolinitic gel.

The importance of these observations is that they give a clue to how the organic matrices may have formed within framboids. The distribution of kaolinite within framboids is very similar to the distribution of organic matter: the kaolinite and organic matter evenly coat individual pyrite microcrystals. It thus seems that the organic matter was in an analogous state to the kaolinite gel: an immiscible organic polymer within which the framboid nucleated crystallized and grew. In contrast to the rare kaolinite matrices, organic matrices are relatively

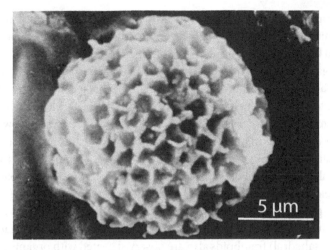

Figure 7.14. Kaolinite matrix from framboid in 300 Ma coal, from Ohio.

Scanning electron micrograph from *Journal of Sedimentary Petrology*, 48, Figure 2c, p. 727, Interstitial networks of kaolinite within pyrite framboids in Meigs Creek Coal of Ohio by Scheihing, M. H., Gluskoter, H. J., and Finkelman, R. B., reproduced courtesy of US Government.

common, which might be expected if the pyrite-S was sourced from microbial sulfate-reduction using the organic matter itself as a metabolite.

7.3. Organic Matter in Framboids

As to be expected in one of Nature's most widespread mineral textures, the organic matter recovered from or observed in framboids has several origins. However, it is obvious that organic matter is not necessary for framboid formation since (a) most framboids do not have any organic matter associated with them, (b) framboids can be synthesized entirely inorganically, and (c) framboids are formed naturally in environments where organic matter is not especially prevalent, such as hydrothermal systems. Thus, although organic matter may contribute to the formation of an individual framboid (e.g., by enhancing local Fe(II) and S(-II) concentrations or acting as a template), it is not generally responsible for framboid formation.

The organic matter associated with framboids can be broadly divided into two broad classes: (1) biofilm and (2) kerogen. The two groups overlap, so that the protokerogen precursor to kerogen commonly incorporates biofilm as well as phytoplanktonic and plant material. End-member biofilms appear to be fragile and subject to extreme shrinkage with age due to dehydration. Kerogen is derived from protokerogens and is often orange in color, suggesting that it is sulfurized.

This is important since sulfurization of organic matter renders it resistant to further microbial degradation. However, not all the kerogenous organic matter recovered from framboids displays this orange or yellow color.

The organic matter in framboids is contemporaneous with framboid growth, and framboids grew within the organic matter. This is consistent with the organic matter originally being in the form of gel-like material which could coat all the framboidal microcrystals evenly. Subsequent microcrystal growth caused extrusion of organic matter from the framboid, contributing to the outer organic matter membrane observed on many of the organic bodies extracted from framboids—as well on similarly sized pyrite microspheres in, particularly, Proterozoic shales. Biofilms also left a membrane on the framboids. In all cases, the pyrite growth determined the form of the organic matter bodies associated with framboids. Thus, the internal matrices are often polygonal, and faceted (including icosahedral) framboids also are observed coated with organic matter.

The observations on organic matter associated with framboids does not, a priori, exclude the possibility of framboid templates provided by microorganisms. The microspheres in Proterozoic shales, for example, appear to be acritarch relicts with probable algal affiliations, and it has been demonstrated experimentally that it is possible to fill algal cells with iron sulfides. However, detailed studies of these pyrite textures have not revealed framboids. If organisms are involved as templates for framboid formation, they have not, as yet, been defined. Generally, known prokaryotic cells, such as giant sulfur bacteria, are too delicate to provide a scaffold for pyrite framboid formation. Eukaryotic organisms which might provide the internal template necessary for framboid formation have not, as yet, been identified.

8

Framboid Mineralogy

8.1. Non-pyritic Framboids

In discussing framboids constituted of minerals other than pyrite, it is important to recollect how framboids are defined. The key attributes are (1) a sub-spheroidal form and (2) constituted by equant and equidimensional microcrystals. As discussed in section 11.5 of Chapter 11, the basic process giving rise to these attributes is simultaneous nucleation, which implies that the minerals need to have relatively low solubilities. Organization of the microcrystals is not a fundamental attribute of framboids; it arises when there is sufficient freedom of movement for the microcrystals to move into a maximum entropy situation by Brownian motion (section 13.3.2 of Chapter 13).

There is a further possible process for the formation of framboids: the replacement of pyrite framboids such that the original pyrite microcrystals are pseudomorphed by the new phase. This is defined as a secondary process, in contrast to the primary processes giving rise to the original pyrite framboids. In this secondary process, it is not necessary for the normal habit of the mineral to be equant, nor is burst nucleation a requirement: the mineral habit and size are inherited from the original pyrite.

There has been some controversy regarding whether minerals other than pyrite can form framboids directly since Schneiderhöhn (1923) first described chalcopyrite ($CuFeS_2$) framboids. Schouten (1946) originally argued that earlier reports of framboids formed by minerals other than pyrite were actually descriptions of replacement of pyrite framboids. In this interpretation, pyrite framboids reacted with later, usually hydrothermal solutions to produce framboids of other minerals. Even the framboids which Rust (1935) originally defined had chalcopyrite matrices.

The introduction of machine-based microanalysis, such as electron probe microanalysis (EPMA), helped resolve the sites of exotic minerals in pyrite. One of the early studies detailing the replacement process was the report describing pyrite framboids with a matrix of cobaltite, $CoAsS$, in the Polish Kupferschiefer by Large et al. (1999). These authors used energy dispersive X-ray analyses to locate the distribution of not only Co and As in the framboids, but also Ni, Cu, and Ag (Figure 8.1). They tracked the cobaltite replacement process (Figures 8.1(a) and (b)) and showed that other elements associated with the later replacement

Framboids. David Rickard, Oxford University Press. © Oxford University Press 2021.
DOI: 10.1093/oso/9780190080112.003.0008

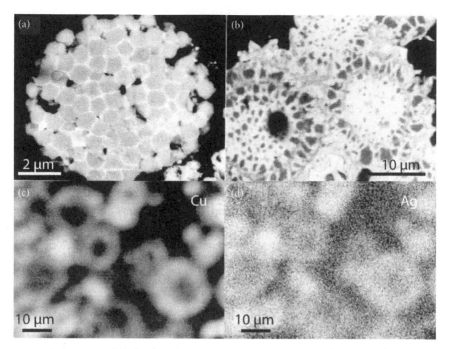

Figure 8.1. Cobaltite, CoAsS, cemented framboids from the ca. 250 Ma Kupferschiefer in Poland: (a) microphotograph of cobaltite matrix (white) cementing pyrite microcrystals (gray); (b) remnant pyrite framboid (gray) almost entirely replaced by cobaltite (white); (c) and (d) energy dispersive X-ray maps showing distribution of Cu and Ag around framboid rims

Images from Large, D.J., Sawlowicz, Z. and Spratt, J., A cobaltite-framboidal pyrite association from the Kupferschiefer: Possible implications for trace element behaviour during the earliest stages of diagenesis. *Mineralogical Magazine*, 63(3): 353–361. 1999. Reproduced with permission.

reaction, such as Cu and Ag, were concentrated in the edges of the framboids. Similar results were found by Zhao et al. (2018) for framboids from Carlin-type gold ores in southwestern China. They found that Cu and Pb were homogenously distributed within the pyrite microcrystals, whereas As and Se were concentrated around the microcrystal margins.

A step-change in the definition of the sites of exotic minerals in pyrite occurred when Deditius et al. (2011) reported an investigation into pyrite nanomineralogy. They used high-resolution transmission electron microscopy (HRTEM) and high-angle annular dark-field scanning transmission electron microscopy (HAADF-STEM) with beam diameters less than 1 nm to discriminate nanoparticles of discrete minerals in pyrite. They reported four groups of nanoparticles in pyrite: native metals (e.g., Au, Ag), sulfides (e.g., PbS, HgS, Cu-Fe-S), sulfosalts (e.g., Pb-Sb-S, Ag-Pb-S), and complex Fe sulfide phases (e.g., Fe-As-Ag-Ni-S). They concluded that nanoparticles <10 nm in size resulted from exsolution from the pyrite matrix, and that nanoparticles >20 nm in size were formed directly. Deditius et al. (2011) suggested that relatively high concentrations of trace elements in pyrite are mostly due to the presence of nanoparticles of discrete mineral phases. Investigations of nanoparticle distributions in framboids are limited, but it would seem probable that the nanomineralogy of framboids is far more varied than that of crystalline pyrite because of the potential mineral contents of the ≥26% matrix volume in framboids.

It is obvious that framboids constituted of minerals other than pyrite occur; the key question, then, is whether these are primary textures or inherited after original pyrite. It is significant since other materials forming framboids cast further light on the framboid-forming processes themselves. Experimental framboid formation with polystyrene microspheres discussed in section 13.3 of Chapter 13 is a clear example. This discrimination is important in determining the conditions in which the framboid formed and the geochemical evolution of the local environment. Thus, the significance of other minerals forming natural framboids has a more general consequence: the geochemistry of framboids has become an important probe into natural processes, including processes involved in meteorite formation and the early history of the solar systems, ore formation, and the secular geochemical history of the Earth's surface environment.

8.2. Sulfide Minerals, Other Than Pyrite, in Framboids

8.2.1. Marcasite Framboids

Marcasite, the orthorhombic dimorph of pyrite, is a classic case of a mineral missing one of the attributes that are required from framboid formation. Marcasite crystals are commonly tabular and do not usually display equant habits. Zhang et al. (2014), for example, in a detailed HRTEM study of pyrite and marcasite in sediments, noted that the marcasite crystals are elongated compared with the more equant pyrite. The result is that marcasite forms crystal aggregates which, although they may have a sub-spheroidal form, are obviously not framboids. The result is clearly shown in Figure 8.2. These marcasite aggregates were produced by Butler (1994) in one of many attempts in the Cardiff laboratory to reproduce the pyrite framboid syntheses of Sweeney and Kaplan (1973).

Few natural occurrences of purportedly framboidal marcasite have been reported (e.g., Ixer and Vaughan, 1993; Youngson, 1995), but no details of the form

Figure 8.2. Synthetic marcasite aggregates.
Scanning electron micrograph image by Ian Butler.

of these marcasite textures were presented. By contrast, marcasite overgrowths on pyrite framboids are common (section 1.3.4 of Chapter 1).

8.2.2. Greigite, Fe$_3$S$_4$, Framboids

Framboids constituted by the ferrimagnetic, metastable, iron sulfide mineral greigite have been widely reported. These framboids were of particular interest to those researchers who originally believed that greigite had to be formed for pyrite to be produced. In fact, as discussed in Chapter 10, this is not the case in pyrite framboid formation. Greigite-containing framboids have been widely suggested by the identification of pyrite framboids appearing in the magnetic fractions of sediment preparations (e.g., Roberts and Turner, 1993).

Jiang et al. (2001) reported microcrystalline greigite closely associated with framboidal pyrite (Figure 8.3) from a ≲2 Ma Pleistocene mudstone from Japan. They reported that the greigite post-dated pyrite framboid formation. Using magnetic force microscopy, Ebert et al. (2018) showed that apparent ferrimagnetic pyrite framboids from the Dead Sea contained greigite nano-inclusions. It seems highly probable that pyrite framboids which regularly appear in the

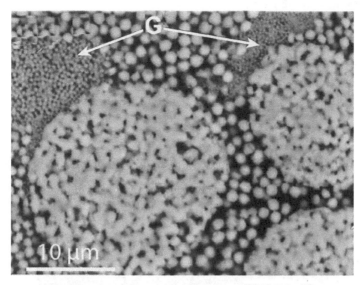

Figure 8.3. Greigite (G) associated with framboids (modified from Jiang et al., 2001).

Scanning electron back scattered electron micrograph reprinted from *Earth and Planetary Science Letters*, 193(1–2): Figure 8d. Contradictory magnetic polarities in sediments and variable timing of neoformation of authigenic greigite. Jiang, W. T., Horng, C. S., Roberts, A .P., and Peacor, D. R. Copyright (2001) with permission from Elsevier.

magnetic extracts from, especially, lacustrine sediments should contain greigite inclusions or closely associated greigite particles.

Framboids wholly composed of greigite microcrystals have been reported very rarely. Perhaps the most frequently cited is a sub-spheroidal greigite aggregate from lacustrine sediments shown in Figure 8.4, reported by Nuhfer (1979) and Nuhfer and Pavlovic (1979). The greigite was identified by X-ray diffraction analyses of single spheroidal aggregates similar to the one illustrated. Interestingly, the X-ray analysis did not show any pyrite, consistent with the idea that greigite is a dead end in pyrite formation (Rickard and Luther, 2007). The greigite aggregate is associated with kaolinite, suggesting that it grew in a kaolinite gel in a similar manner to the pyrite framboids associated with organic matter described in section 7.2.6 of Chapter 7. Detailed examination of the greigite aggregate reveals varying sizes and forms of the constituent microcrystals, suggesting that it is not strictly framboidal but is better described as an aggregate.

Herndon et al. (2018) described greigite-containing framboids from Fayetteville Green Lake, New York. They identified greigite on the basis of Fe:S ratios obtained from energy dispersive X-ray spectrometry of thin sections. The authors did not report the results of optical microscopic examination of these

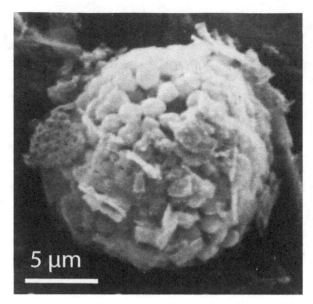

Figure 8.4. A sub-spheroidal aggregate of greigite, Fe_3S_{4g} microcrystals of various shapes and sizes associated with kaolinite and a diatom frustule. The texture was described as a greigite framboid by Nuhfer and Pavlovic (1979).
Scanning electron micrograph image by Edward Nuhfer.

framboids which would have readily confirmed the presence of greigite. The "proto-framboids" described by Large et al. (2001) showed indeterminate Fe:S ratios which were interpreted by the authors as greigite. However, as discussed earlier (section 2.1 in Chapter 2 and section 4.2 in Chapter 4), these textures are better described as nanoparticle clusters.

The properties of greigite, including the isometric crystal class and its low solubility, suggest that it should form framboids. However, the process of self-ordering resolved from experimental studies (section 13.3) is a secondary process following assembly of the microcrystals. The primary driving force for this process is entropy maximization, and the microcrystals must retain sufficient freedom of movement such that Brownian motion can move the microcrystals into ordered arrays. The problem with greigite microcrystals is that Brownian motion is unlikely to be a strong enough force to overcome their intrinsic intense ferrimagnetism. This means that greigite microcrystals are likely to remain in the same geometry as they aggregated. Large et al. (2001) originally noted this problem. This might explain the more frequent irregular aggregates of greigite microcrystals illustrated in Figure 8.3, for example.

8.2.3. Copper Sulfide Framboids

Secondary enrichment processes, including supergene enrichment, are important in sulfide mineral deposits. They involve reaction of primary sulfides with later groundwater, late-stage low temperature hydrothermal solutions, or tectonically driven regional fluid flow. The products can be economically important, especially in copper deposits, where the reaction between copper-rich solutions and primary sulfides leads to the formation of copper-rich minerals such as chalcopyrite ($CuFeS_2$), bornite (Cu_9FeS_4), digenite (Cu_9S_5), chalcocite (Cu_2S), and covellite (CuS). Since pyrite framboids are often formed diagenetically in sediments, they are primary materials for these later reactions.

As mentioned earlier, chalcopyrite framboids were originally described by Schneiderhöhn (1923) from the ca. 250 Ma old Mansfeld Kupferschiefer. He interpreted them as fossilized bacterial colonies. Schouten (1946) re-examined them and concluded that they were secondary replacements after pyrite. The formation of chalcopyrite framboids through the oxidative alteration of primary pyrite framboids by later cupriferous fluids was subsequently reported by Li et al. (1998) from the 445 Ma Bendigo goldfield, by Oszczepalski (1999) from the 255 Ma Kupferschiefer, by Merinero et al. (2019) from ca. 100 Ma manto-type ore bodies in Chile, and by Hu et al. (2015) from ca. 100 Ma New Zealand gold deposits (Figure 8.5).

As early as 1916, Zies et al. showed experimentally that the continued reaction of copper (II) solutions with chalcopyrite produced bornite (Cu_5FeS_4) as a primary product. Bornite framboids have been described by Oszczepalski (1999) from the Kupferschiefer, and Wilson and Zentilli (1999) and Merinero et al. (2019) from the Chilean manto deposits. Figure 8.5(a) shows a pyrite framboid which has been clearly successively replaced by chalcopyrite and bornite.

The endpoint of the copper enrichment process is the formation of minerals of the digenite-chalcocite group with formulae varying between Cu_9S_5 and Cu_2S, and framboids constituted by this mineral group have been described by Sawlowicz (1990) and Oszczepalski (1999) from the 250 Ma Permian Kupferschiefer, Poland (Figure 8.6). This mineral group includes djurleite, roxbyite, and yarrowite, but only chalcocite and digenite have been identified in framboids. The minerals have similar optical and chemical properties and are normally discriminated by X-ray diffraction. This means that the minerals identified in framboids as chalcocite or digenite may in fact include any of the five members of the group. Schneiderhöhn (1923) originally described chalcocite framboids from the Mansfeld Kupferschiefer. Schouten (1946) described these framboids as replacements of original pyrite framboids by chalcocite, and this is the generally accepted origin of these framboids today (e.g., Merinero et al., 2019; Oszczepalski, 1999; Wilson and Zentilli, 1999).

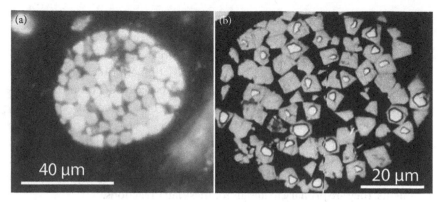

Figure 8.5. Chalcopyrite (CuFeS$_2$, gray) and pyrite (white) framboids from the ca. 250 Ma Kupferschiefer, Poland: (a) whole pyrite microcrystals selectively replaced by chalcopyrite with bornite (Cu$_5$FeS$_4$) in the matrix; (b) detail of pyrite microcrystals partially replaced by chalcopyrite.

Reflected light photo micrographs reprinted by permission from Springer. *Mineralium Deposita* 34(5–6): 599–613. Origin of the Kupferschiefer polymetallic mineralization in Poland. Oszczepalski, S. Copyright (1999).

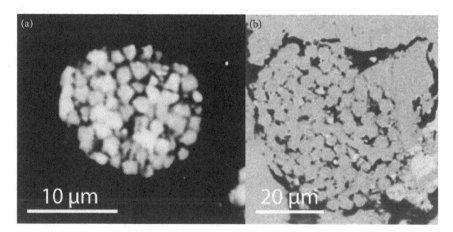

Figure 8.6. Reflected light microphotographs of digenite (Cu$_9$S$_5$) and chalcocite (Cu$_2$S) framboids from the ca. 250 Ma Permian Kupferschiefer: (a) digenite (gray) and chalcocite (white) microcrystals; (b) digenite framboid in massive digenite.

Reflected light photo micrographs reprinted by permission of (a) Zbigniew Sawlowicz and (b) from Springer, *Mineralium Deposita* 34(5–6): 599–613. Origin of the Kupferschiefer polymetallic mineralization in Poland. Oszczepalski, S. Copyright (1999).

However, using scanning electron microscopy, Alyanak and Vogel (1974) showed that chalcocite, Cu_2S, "framboids" in the 1100 Ma old White Pine deposit, Michigan, were ellipsoidal aggregates of mixtures of euhedral crystals and platelets 0.1–0.5 mm in size (Figure 8.7). Alyanak and Vogel suggested that these chalcocite aggregates were primary features and were not formed by replacement of original pyrite framboids. Sawlowicz (1990) agreed and described primary digenite and chalcocite framboids from the Kupferschiefer.

As described in section 10.4.1 of Chapter 10, the syntheses of copper sulfide "framboids" reported by Farrand (1970) were actually sulfide sols consisting of spherical nanoclusters <1 μm in diameter and not true framboids. Chalcocite is monoclinic below 105°C and generally forms tabular or prismatic crystals. The crystal form is not normally equant and thus the mineral does not normally provide one of the basic attributes necessary for framboids. The result is the formation of microcrystalline aggregates, rather than framboids, as illustrated in Figure 8.7.

8.2.4. Sphalerite Framboids

Sphalerite, cubic ZnS, possesses one of the basic attributes required for framboid formation: it forms equant crystals. However, the solubility of ZnS is highly sensitive to the chemistry of the environment and it is usually many magnitudes greater than that of pyrite. This means, as discussed with respect to pyrite in

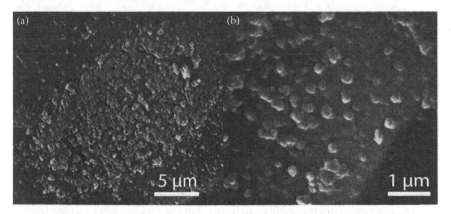

Figure 8.7. Framboid-like chalcocite aggregates from 1080 Ma White Pine copper ore, Michigan: (a) general view of ellipsoidal aggregates; (b) detailed view of irregular chalcocite, Cu_2S, microcrystals.

Early SEM microphotographs by Alyanak and Vogel (1974) reprinted by courtesy of the Society of Economic Geologists.

section 11.3.1 of Chapter 11, that the supersaturations required to initiate sphalerite nucleation are not extreme under most natural conditions, and the precipitation of sphalerite from aqueous solutions is facile. These properties are not ideally suited to the process of burst nucleation which is necessary for framboid formation (section 11.1).

Nanoclusters of ZnS, sometimes constituted by 1–5 nm nanoparticles, have been widely synthesized (e.g., Farrand, 1970; Labrenz and Banfield, 2004; Moreau et al., 2004) and have been described from natural environments, including peatlands (Yoon et al., 2012), Lake Kivu (Degens et al., 1972) and from the ca. 200 Ma old Bleiberg Zn-Pb ore deposit (Kucha et al., 2010). Framboidal sphalerite was also reported by Kucha et al. (2010) from the Bleiberg ore deposit. However, the images indicated ZnS nanoclusters rather than framboids, and the exact relationship with the pyrite framboids is unclear.

Hu et al. (2018) presented a detailed study of ca 100 Ma Zn-rich framboids from the Otago schist of New Zealand. They tracked Zn enrichment in pyrite framboids and confirmed earlier work of Labrenz et al. (2000) and Moreau et al. (2004) that ZnS concentrations in framboids were collocated with the organic matrices. They found framboids with trace Zn, framboids with Zn concentrated in the microcrystal boundaries and rimming pyrite microcrystals, and framboids with mixed ZnS and pyrite microcrystals (Figure 8.8). Pure ZnS framboids were not reported, and I note that the apparent octahedral forms of the ZnS and pyrite microcrystals are similar (Figure 8.8(b)). This is analogous to the situation with

Figure 8.8. General view (a) and detail (b) of ZnS (darker gray) and pyrite (lighter gray) microcrystals in a framboid from the ca. 220 Ma, Otago schists. Octahedral ZnS and pyrite microcrystals are clearly defined in (b).

NanoSIMS maps of $^{65}Zn^{32}S$ and $^{56}Fe^{32}S$ distributions reprinted from *Geochimica et Cosmochimica Acta*, 241: 95–107. Sequestration of Zn into mixed pyrite-zinc sulfide framboids: A key to Zn cycling in the ocean? Hu, S. Y., Evans, K., Rempel, K., Guagliardo, P., Kilburn, M., Craw, D., Grice, K. and Dick, J. Copyright (2018) with permission from Elsevier.

chalcopyrite framboids discussed earlier, suggesting that ZnS formation in these framboids is a secondary process. The difference from the Hu et al. (2018) results is that the Zn is interpreted by these authors as having been introduced together with the organic matter.

8.3. Oxide Framboids

8.3.1. Magnetite Framboids

Micrometeorites are interplanetary dust particles that mainly represent samples of asteroids and comets (Genge et al., 2008). The magnetite micrometeorites display varying textures, mainly related to the degree of melting they experienced on entering the Earth's atmosphere. Some of these micrometeorites display framboid-like textures. Modern samples are often collected from Antarctic ice, where tens of thousands may be recovered during a single sampling expedition. The oldest samples may be the micrometeorites extracted from 2.7 Ga limestones in the Pilbara area, Australia (Figure 8.9).

Magnetite framboids are also found in carbonaceous chondrites (Jedwab, 1971). Miyake et al. (2012) described magnetite framboids from the Murchison meteorite as fossil bacteria and suggested that these were evidence that biotic

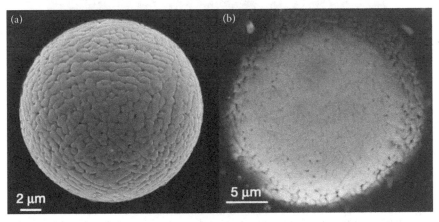

Figure 8.9. Magnetite micrometeorites from 2.7 Ga limestones: (a) scanning electron micrograph; (b) micrograph of polish section.

Reprinted by permission of Springer Nature, *Nature*, 533(7602): 235–238. Ancient micrometeorites suggestive of an oxygen-rich Archaean upper atmosphere., Tomkins, A. G., Bowlt, L., Genge, M., Wilson, S. A., Brand, H. E. A., and Wykes, J. L. Copyright (2016).

processes occurred at the beginning of the solar system. Xu et al. (2019) reported that the magnetite microcrystal habits were similar to those of pyrite and concluded that the magnetite was an oxidation product of earlier pyrite framboids. Kimura et al. (2013) described magnetite framboids from the Tagish Lake carbonaceous chondrite that fell in 2000 (Figure 8.10). They interpreted the magnetite as primary and the irregular edges to the magnetite microcrystals (Figure 8.10(c)) as due to magnetic field effects during crystal growth. Kimura et al.

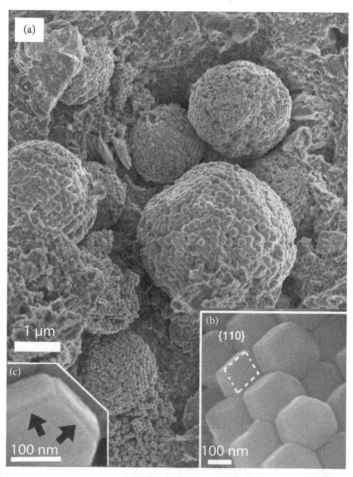

Figure 8.10. Magnetite framboids from the Tagish Lake carbonaceous chondrite: (a) SEM image; (b) cubic close packed microcrystals; (c) rhombic dodecahedral microcrystals with irregular edges (arrowed).

Images reprinted by permission of Springer Nature, *Nature Communications*, 4. Vortex magnetic structure in framboidal magnetite reveals existence of water droplets in an ancient asteroid. Kimura, Y., Sato, T., Nakamura, N., Nozawa, J., Nakamura, T., Tsukamoto, K., and Yamamoto, K. Copyright (2013).

showed that the magnetite microcrystals in the framboids were ordered and displayed ccp (Figure 8.10(b)). If these were primary magnetite framboids, they are the only ones described to date to display regular packing. As noted in section 13.2.4 of Chapter 13, experimental evidence suggests that enhanced interparticle magnetic attraction would inhibit the secondary readjustment of magnetite microcrystals into ordered domains. The rhombic dodecahedral form of the magnetite microcrystals (Figure 8.10(c)) is relatively rare in terrestrial magnetites. It is, however, a relatively common pyrite crystal habit (see Table 4.3 in Chapter 4).

The question of whether the magnetite framboids found in carbonaceous chondrites are primary or result from pyrite oxidation may have profound implications. The original framboids formed in a fluid—and possibly aqueous—phase and the question of whether the framboids were originally pyrite or not has implications for the original chemistry of that phase (cf. Kimura et al., 2013; Xu et al., 2019).

Magnetite framboids in ancient rocks have been demonstrated to form as a result of the oxidation of primary pyrite framboids. For example, Suk et al. (1990) described magnetite framboids in 390 Ma limestones which had been formed by the oxidation of primary pyrite framboids by later regional fluid flow related to tectonism. Magnetite framboids are also a common constituent of fly ash. The coal feed to power plants often contains abundant pyrite framboids, and the combustion process produces magnetite framboids with identical size distributions to the pyrite framboids in the feed (Lauf et al., 1982). Apart from the implications of these observations for the origin of magnetite framboids in meteorites, they have further implications for the magnetite framboids in sediments and sedimentary rocks. The particles produced by coal-fired power stations can travel long distances before deposition, and thus it is possible that some magnetite framboids found in recent sediments may be anthropogenic.

8.3.2. Iron and Mn (Oxyhydr)oxide Framboids

The term *(oxyhydr)oxide* describes undefined mixtures of oxides, hydrated oxides, hydroxides, and oxyhydroxides. Hematite and goethite framboids formed by the oxidation of pyrite framboids were described from the 65 Ma old Campanian-Maastrichian ironstones of Nigeria by Mucke et al. (1999). These observations are interesting since the iron (oxyhydr)oxide framboids are associated with kaolinite, which is reminiscent of the kaolinite matrices associated with pyrite framboids described in section 7.2.6 of Chapter 7. Unspecified iron (oxyhydr)-oxide framboids were reported as pyrite framboid pseudomorphs from dinosaur bones of various ages by Kaye et al. (2008). Sgavetti et al. (2009) described iron (oxyhydr)oxide framboids from the 6 Ma old Messinian evaporites. Gao et al. (2017) described unspecified iron (oxyhydr)oxide framboids from 360 Ma stratiform barite deposits from China which pseudomorphed original pyrite.

Mn-oxide, Mn-Fe oxide, and Fe-(oxyhydr)oxide framboids were reported by Soliman and El Goresy (2012) from 65 Ma Maastrichian organic-rich shales. The Mn- oxide and Mn-Fe oxide minerals were not identified, but the Fe-(oxyhydr)-oxides were reported to include goethite.

Goethite spheres and Mn-oxide framboids derived from pyrite framboid oxidation were reported by Soliman and El Goresy (2012). Oxidative dissolution of pyrite microcrystals produced goethite spheres with a honeycomb structure reflecting the original microcrystal locations (Figure 8.11(a)). The formation of a goethite sphere from a pyrite framboid results in a theoretical increase in volume of about 7%, assuming that (1) the original pyrite microcrystals in a randomly packed framboid occupied ca. 60% of the framboid volume (section 4.2, Chapter 4); (2) that the holes approximate this volume; (3) that goethite has a molar volume of 12.75% greater than pyrite; and (4) that there was no net loss of Fe. The relatively small volume change is consistent with the preservation of the integrity of the pyrite framboid. The key to the formation of this texture is obviously the local retention of Fe. This suggests that the oxidizing system was not entirely open to water.

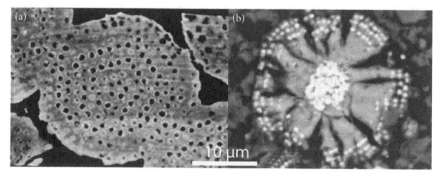

Figure 8.11. Framboid oxidation: (a) honeycomb structure resulting from the oxidative dissolution of pyrite microcrystals and the reprecipitation of goethite; (b) rosette structure resulting from the oxidation of pyrite microcrystals and the formation of ferrous sulfate.

(a) Reprinted from *Geochimica et Cosmochimica Acta*, 90: 195–220. Framboidal and idiomorphic pyrite in the upper Maastrichtian sedimentary rocks at Gabal Oweina, Nile Valley, Egypt: Formation processes, oxidation products and genetic implications to the origin of framboidal pyrite.? Soliman, M. F., and El Goresy, A. Copyright (2012) with permission from Elsevier.

(b) Reprinted by permission of Springer. *Bulletin of Engineering Geology and the Environment*, 71: 113–117. Sulphate heave: A model to explain the rapid rise of ground-bearing floor slabs. Hawkins, A. B. Copyright (2012).

The Mn-oxide framboids are pseudomorphic after pyrite framboids and often occur in an Mn-oxide groundmass. The Mn-Fe oxide framboids were formed during weakly oxidizing conditions coincident with a period of negative sedimentation and the formation of glauconite and phosphates. The pseudomorphing of pyrite by Mn-oxide was probably a two-stage process since they have no elements in common. It is likely that the oxidation first produced iron (oxyhydr)oxide framboids which then reacted with dissolved Mn to produce Mn-Fe and ultimately Mn oxide framboids.

Soliman and El Goresy's (2012) observations are interesting since they emphasize how delicate these replacement reactions can be under the right circumstances, a point originally made by Schouten (1946). In Soliman and El Goresy's case, the system was open enough to permit water flow and removal of the dissolved sulfate and Fe products, but not sufficiently open to permit entire dissolution. The individual pyrite microcrystals retain their form through exchange of pyrite-S and O_2 and then exchange Fe with dissolved Mn_{soln} (Equations 8.1 and 8.2). The products are dissolved sulfate (SO_{4soln}) and iron (Fe_{soln}).

$$FeS_2 + O_2 + H_2O \rightarrow FeO(OH) + SO_{4soln} \qquad 8.1$$

$$FeO(OH) + Mn_{soln} \rightarrow MnO(OH) + Fe_{soln} \qquad 8.2$$

Hawkins (2012) showed that the oxidation of pyrite framboids was a major factor in the phenomenon known as pyrite heave, where the ground floors of buildings may rise up to 48 mm within five years of construction. The problem, which occurs worldwide and causes major social disruptions and vast economic costs, happens when pyritiferous shales are used as backfill in the foundations or when the building is built upon pyritiferous shale bedrock. Although it had long been known that pyrite oxidation was the cause, it was Hawkins's work which showed that the oxidation of pyrite framboids was the trigger for rapid heave. The effect is illustrated in Figure 8.11(b). The initial effects of exposing crushed pyritiferous shale to the atmosphere is the oxidation of pyrite to ferrous sulfate. This results in a volume increase of over 350% depending on the exact hydration state of the ferrous sulfate. Subsequently the sulfate dissolves to and reacts with calcium carbonates in the shale to produce gypsum ($CaSO_4 \cdot 2H_2O$) which has a volume of approximately double that of the original components of the shale. In the initial phase the crystallization of ferrous sulfate within the framboids causes the development of large crystalline rosettes some five times larger than the original framboids. In these rosettes remnant pyrite microcrystals are carried out toward the edges of the rosette as the ferrous sulfate crystal grows.

Again, the formation of these ferrous sulfate rosettes from framboid oxidation requires specific environmental conditions. Ferrous sulfate itself is very soluble, which means that during this initial phase, the system cannot be entirely open to the influx of groundwater. In fact, the initial oxidation of the framboids initiates that cracking of the rocks allows increased water ingress.

9

Geochemistry of Framboids

The geochemistry of pyrite has been widely investigated for over a century, and each technical advance has resulted in more precise probes into its composition. Fleischer (1955) published the benchmark paper on trace and minor elements in pyrite, which summed up knowledge of pyrite geochemistry in the period before microscopic instrumental analysis methods became available. Fleischer concluded that exotic elements in pyrite are either in solid solution or exist as discrete mineral phases. The same conclusion can be applied to framboids except that these bodies additionally contain a minimum 26 vol % interstitial space even at the tightest microcrystal close packing, but more usually this approaches 40 vol %. This interstitial space is normally filled with either material that is syngenetic with the original framboid formation or material that is added later. Material that is syngenetic with original framboid formation includes organic matter, material such as silica in which the framboid grew, and particles of the host rock entrapped during framboid formation. Material which is epigenetic to framboid growth includes organic matter that is forced into pore spaces during diagenesis, material which is exsolved from the pyrite structure through equilibration during diagenesis and metamorphism, and minerals that replace the pyrite through secondary fluid reactions and oxidation.

As mentioned in section 8.1 of Chapter 8, Deditius et al.'s (2011) investigation of the nanomineralogy of framboids suggested that trace elements that appear to be in solid solution in pyrite may actually be discretely-sited mineral nanoparticles, and thus bulk analyses may give widely varying results. It is also probable that the concentration of nanoparticles in framboidal pyrite is far greater than in crystalline pyrite because of their inclusion in the framboid matrices.

The problem is to distinguish between primary framboids and secondary geochemistry. Both report some aspects of ambient geochemical conditions, but discrimination between the events recorded can be difficult. For example, if pyrite framboids have been replaced or cemented by ZnS, as described in section 8.2.4 of Chapter 8, framboid Zn concentrations may reflect the Zn concentrations in the sediment when the framboid is being formed (as suggested by Hu et al., 2018) or some later hydrothermal event (e.g., Schouten, 1946). The problem can become difficult when it appears that just a few microcrystals are replaced by another sulfide, as in the case of galena in the 2.9 Ga Witwatersrand framboid

Framboids. David Rickard, Oxford University Press. © Oxford University Press 2021.
DOI: 10.1093/oso/9780190080112.003.0009

shown in Figure 1.11 (Chapter 1). On the other hand, the difficulty of discriminating between the trace element load of the matrix and the pyrite microcrystals may not constitute a barrier to paleoenvironmental reconstructions. For example, framboid matrix organic matter may concentrate trace elements such as Zn, Cu, Mo, and U relative to the pyrite, but still record variations in the geochemistry of the ambient environment.

9.1. Pyrite Stoichiometry in Framboids

Various aspects of framboidal pyrite composition have been reported, such as the trace element composition and isotopic constitution. However, the fundamental question of the stoichiometry of pyrite in framboids remains a mystery. The reason for the uncertainty is analytical. The problem is that whole framboids have masses typically of the order of nanograms and a substantial proportion of this can be non-pyritic matrix material such as clays or organic matter, depending on packing geometries and environments. So accurate framboid pyrite analysis requires microbeam techniques to analyze individual microcrystals.

The chemical composition of stoichiometric pyrite is 46.55% Fe and 53.45% S. Nano- and micro-crystal pyrite analyses have commonly been performed by wavelength or energy dispersive X-ray techniques and the results derived from ratios of the major elements. The analytical problem mainly concerns the accuracy of the pyrite-S totals since S has a lower mass than Fe but is more abundant. For example, non-stoichiometric pyrite with a formula $FeS_{1.9}$ contains 47.83% Fe and 52.17% S. Each 0.1 apfu (atoms per formula unit) S deficiency involves a decrease in S % of just 0.12 wt %. This is far beyond the precision of microbeam techniques at present. The effect is shown in Table 9.1, where high precision electron microprobe (EPMA) analyses show formulae with apfu values for S between 1.93 and 1.97 and between 1.97 and 1.99 if all the error is assumed to be due to uncertainties in the S analyses. Electrical measurements of pyrite can be used as a measurement of pyrite composition by assuming that variations in resistivity are contributed by S vacancies. Using these techniques, Zhang (2015) showed that all the resistivity measurements of synthetic pyrites could be explained by compositional variations between $FeS_{1.9}$ and $FeS_{2.0}$.

The data in Table 9.1 are consistent with Zhang's results and Kullerud and Yoder's (1959) original proposition that natural pyrite is essentially stoichiometric FeS_2. The analyzed framboids show S:Fe ratios all within 1.9 and 2.0 and, if the total error is assumed to reside mainly in the S analyses, between 1.97 and 2.00. Kolker's results are further interesting since the minor elements analyzed make little difference to the S:Fe ratios: all show apfu values of less than the second decimal place. These results are consistent with EPMA measurement of the Fe

Table 9.1 EPMA Analyses of 3 Framboids (Numbered 1, 2, 3) from Coal Showing wt %, the Totals, the Pyrite Formulae, and the Pyrite Formulae Assuming the Total Error is in the S Analyses

	1 wt %	2 wt %	3 wt %
Ni	0	0.01	0.02
Cu	0.06	0.03	0.01
Zn	0	0.06	0.01
As	0	0	0
Se	0	0	0
Pb	0	0	0
S	52.88	51.94	52.07
Fe	46.78	46.63	46.95
Total	99.72	98.67	99.06
Formula	$FeS_{1.97}$	$FeS_{1.94}$	$FeS_{1.93}$
Formula S error	$FeS_{1.98}$	$FeS_{1.99}$	$FeS_{1.97}$

Computed from data in Kolker (2012).

and S contents of 20 framboids from Carlin-style gold deposits in SW China reported by Zhao et al. (2018), which give a geometric mean S:Fe ratio of 1.9 for analyses which totaled between 90.7 and 99.0 wt %. However, in this recalculation I have distributed the error equally between the Fe and S analyses, whereas it is more likely to reside mostly in the S analyses. This means that the computed S:Fe ratio is a minimum value. Kohn et al. (1998) analyzed pyrite framboids from recent sediments from Monterey Bay, southern California, with EPMA. They reported mean S:Fe ratios of 1.99 for mean analytical totals of 97.9%.

There has been considerable recent interest in synthesizing nanoparticulate pyrite, and there are often reports of S:Fe ratios widely deviating from stoichiometry in this literature. The problem here of course is that accurate totals are often not available by microbeam techniques, so there is no way of knowing the accuracy of the results. Sophisticated algorithms normalize the counts for the individual elements to 100 wt %. Estimates of the analytic uncertainties are usually around 0.1 on the S:Fe ratio even after multiple measurements on relatively pure synthetic pyrite crystals (e.g., Voigt et al., 2019).

Inductively coupled mass spectrometric analyses of the products of laser ablation (LA-ICPMS) of framboids have also been widely employed in framboid analyses. This method is well suited to trace element compositions of framboidal pyrite but, again, accurate pyrite-S analyses remain a problem because of the mass interferences of Ar and O_2 on the main S isotope, combined with the relatively low ionization efficiency of S in the plasma. Pyrite-S analyses are subject to potentially significant errors and this appears to be consistent with the widely reported extreme sulfur-deficient compositions, which are S:Fe ratios in reality (e.g., as low as $FeS_{1.4}$ according to Kar and Chaudhuri, 2004). It seems that developments in ICPMS technology, including increased precision in mass discrimination (Martinez-Sierra et al., 2015), may soon enable accurate analyses of pyrite-S in framboids.

Early work on pyrite stoichiometry was reviewed by Rickard (2012b) and Rickard and Luther (2007), who concluded that there were no robust analytical reports proving that pyrite was a non-stoichiometric mineral and that significant deviations from stoichiometry have not been reported for natural pyrites. They concluded that pyrite has a very narrow homogeneity range (<1‰). Voigt et al. (2019) confirmed this conclusion. They showed that sulfur vacancies gave rise to n-type pyrite, where the conductivity is due to free electrons, and that these occurred throughout the pyrite crystal and not just at the surface. However, the amount of sulfur deficiency was well within the analytical uncertainty of 0.1 apfu and probably ranges up to $FeS_{1.97}$.

The importance of the accurate analyses of microcrystalline pyrite is obvious in materials science since pyrite is a semi-conductor which displays both p-type, where the conductivity is due to positive holes, and n-type semi-conductivity, which is due to the free electrons (Agaev and Emujazov, 1963). There are no reports on measurements of the conduction in framboidal pyrite. Certainly, the ability to synthesize framboid-like arrays of pyrite microcrystals with specified conduction types would be interesting to the electronics industry. However, the determination of n- or p-type conduction in pyrite is less straightforward than may appear at first sight from the literature. In particular, the discrimination between n- and p- type conduction is only really robust when the electron mobility is high. For this reason, many reports of, particularly, p-type behavior in pyrite are less secure. There seems little debate over the observation that sulfur vacancies in pyrite give rise to n-type conduction; that is, n-type of conductivity occurs in sulfur-deficient pyrite (Abraitis et al., 2004; Favorov et al., 1974; Khalid et al., 2018). Khalid et al. (2018) showed that p-type conductivity may occur in pyrite with S:Fe ratios >2. However, Zhang et al. (2017) showed that reports of p-type conduction in pyrite are much more uncertain.

While n- and p-type conductivities have been reported from trace amounts of other elements in the pyrite structure, such as As, Se, Te, Co, and Ni (Li et al., 2011), the data are often controversial. It has been suggested that P, As, and Sb produce p-types and Ni, Co, Au, Cl, and Br, n-type. However, robust data are only available for Co, which is undoubtedly n-type, whereas P, Sb, and As give rise to p-type conduction and the status of Ni is uncertain (Voigt et al., 2019).

Natural pyrite therefore exhibits both n-type and p-type semi-conduction, sometimes apparently within the same crystal (Favorov et al., 1974). It has been suggested that low-temperature pyrite, such as may occur in many framboids, tends to be Fe-deficient and display p-type semiconduction, whereas high-temperature pyrite is mainly S-deficient and is an n-type semiconductor (Pearce et al., 2006). However, this is misleading. The idea stems from studies of arsenian pyrites, and high As contents are likely to result in p-type semiconduction. Arsenian pyrites are commonly associated with arsenopyrite in hydrothermal deposits, whereas this association is less common at lower temperatures. Abraitis et al. (2004) note that natural pyrites are typically n-type, even when formed at high temperature when arsenopyrite is not present in the mineral assemblage.

9.2. Pyrite Composition

9.2.1. Trace and Minor Elements in Sedimentary Pyrite

Huerta-Diaz and Morse (1992) pioneered the investigation of the relationship between trace elements and pyrite in sulfidic sediments. They used classical wet chemical methods and atomic adsorption spectroscopy on leached samples from the Gulf of Mexico. They listed their results as the ratio of moles of trace elements in pyrite to the moles of Fe in pyrite. Since that time there have been several reports of trace element loads on sedimentary pyrite of different geologic ages (see Berner et al., 2013, and references therein). As remarked by Berner et al. (2013), the results of these studies have been inconclusive due to the large variabilities of the trace element concentrations within any particular environment. These results of these studies were also complicated by contributions of various forms of pyrite of various ages included in any bulk sample. Framboids were not generally analyzed as a discrete fraction.

Large et al. (2014) used LA-ICPMS on samples of pyrite from the euxinic sediments from the Cariaco basin. The advantage of LA-ICPMS is that it is an in situ method, so that individual grains of pyrite can be analyzed. The results of Large et al. (2014) are compared with those of Huerta-Diaz and Morse (1992) in Table 9.2. As can be seen, there is considerable variation in trace element contents of sedimentary pyrite between the two reports. The variation is at least partly due to the different methods employed, and the geometric means of the 6500 metal analyses by Huerta-Diaz and Morse data mainly fall outside the maximum and minimum values of the 690 metal analyses listed by Large et al. (2014). The

Table 9.2 Recalculated Average Geometric Mean ($\bar{x}^*$) Concentrations of Selected Trace Elements (ppm) in Pyrite from Marine Sediments from the Cariaco Basin (Large et al. 2014) and Gulf of Mexico (Huerta-Diaz and Morse, 1992)

	Cariaco Basin			Gulf of Mexico
	$\bar{x}^*$	max	min	$\bar{x}^*$
Mn	1973	5261	116	1335
Co	48	291	5	549
Ni	415	1132	75	1947
Cu	458	1574	77	3539
Zn	612	1556	132	1303
As	267	1954	71	1016
Se	162	847	21	
Mo	818	4064	169	68
Ag	7	33	1	
Cd	14	51	2	112
Sb	39	186	5	
Te	5	12	1	
Tl	23	75	7	
Pb	49	192	3	438
Bi	2	34	0	

Note: The Gulf of Mexico concentrations are mainly outside the ranges (maximum and minimum reported values) of the Cariaco basin pyrites.

Huerta-Diaz and Morse data include sediments which are described as anoxic, non-sulfidic, even in sampling sites coincident with the Cariaco basin. Even so, the data suggest considerable variations in the trace element contents of pyrites of modern marine sediments.

There is some independent evidence for potential geographical variations in the trace element contents of pyrite in modern sediments. Huerta-Diaz and Morse (1992) analyzed pyrites from a variety of locations around the Gulf of Mexico, including two sites on the northern continental shelf and slopes and four on the southern slopes over 600 km away. They reported ranges of trace element contents varied over two orders of magnitude both geographically and with depth. The concentrations of all the trace elements in pyrite decreased with depth, although concentration of the pyritic trace metals per gram dry sediment did not change significantly with depth. That is, the amount of pyrite increased with depth so that the trace element concentration in the pyrite decreased. This is an interesting suggestion since it shows that the more obvious cause of these trends, that the concentration of the trace element is generally more abundant in the surface sediments, is not supported by these results. The conclusion from these results is that the trace element content of sedimentary pyrites is not directly related to thermodynamic or kinetic factors, but simply reflects the amount of pyrite formed and the concentration of available trace elements at any point in the sediment. However, as discussed later, my calculations reveal that this is not the case in framboids.

So does the trace element composition of pyrite reflect the trace element composition of the contemporaneous ocean, as suggested by Berner et al. (2013) and Large et al. (2014)? Table 9.3 shows an estimate of the concentration factors for Cariaco basin pyrites. The ocean concentrations of trace elements are quite variable, and the figures listed in Table 9.3 are the best estimates for the average composition of normal ocean waters by Bruland et al. (2013). The data show that the concentration factors vary over four orders of magnitude. This variation is to be expected on the basis of the variable chemistry of these elements in the pyrite system.

Huerta-Diaz and Morse (1992) showed that the trace metals they studied could be divided into three groups whose behavior was largely independent of the sedimentary environment. Hg, Mo, and As were more or less completely removed from the sediments by pyrite and were enriched in the pyrite relative to the bulk sediment concentration. Co, Cu, Mn, and Ni were equally partitioned between pyrite and the reactive fraction, suggesting that the concentration of these elements in pyrite reflects the chemistry of the Fe-Co-Cu-Ni-Mn-S system at SATP. Cr, Cd, Pb, and Zn were depleted in pyrite relative to the bulk sediment, and Huerta-Diaz and Morse (1992) suggested that they formed stable metal sulfide complexes in sulfidic sediments. Berner et al. (2013) found that the relative proportion of trace elements in their 180 Ma authigenic pyrites increased in the order Ni<Cu<Mo = As<Tl in diagenetic pyrite formed in oxic-dysoxic environments, whereas As, Mo, and Sb were enriched in pyrite formed in euxinic environments. They postulated that the variation in the trace element contents of these syngenetic pyrite grains reflected changes in contemporary seawater composition.

9.2.2. Trace Elements in Pyrite in Coal

The trace element contents of pyrite in coal have been the subject of extensive investigations because of their health impacts, and environmental and economic

Table 9.3 Concentrations Factors for Trace Elements in Pyrite from Sediments in the Cariaco Basin

	Pyrite ppm	Ocean ppt	Factor $\times 10^4$
Mn	1973	16	11971
Co	48	2	2036
Ni	415	470	88
Cu	458	191	240
Zn	612	392	156
As	267	1723	15
Se	162	142	114
Mo	818	9978	8
Ag	7	2	324
Cd	14	67	21
Sb	39	195	20
Te	5	0	6531
Tl	23	14	161
Pb	49	2	2365
Bi	2	0	9570

The pyrite composition is from Large et al. (2014) and the ocean water concentrations are the best estimates from Bruland et al. (2013).

importance. Elements such as Hg can lead to hazardous air pollutants during combustion of coal feed, and As is an important potential hazard in acid mine drainage; both elements are potentially hazardous to health. From the scientific point of view, coal is derived mainly from freshwater peat deposited in a dominantly terrestrial environment and thus these data could complement the data from dominantly marine sediments. However, sulfur-rich coals, which contain more abundant pyrite, are characteristically affected by marine incursions (Chou, 2012) and therefore their pyrite trace element contents should be similar to those of other marine sediments. A further problem is that pyrite in coal is formed during its whole geologic history and not just during the early peat stage. For example, coal basins are widely subject to regional fluid migration (Gayer and Rickard, 1994) leading especially to the deposition of vein- and cleat-filling pyrite, often enriched in trace elements such as As, Se, Mo, Sb, Tl, Cu, and Hg (e.g., Diehl et al., 2004; Diehl et al., 2012). Even the pyrite in the original peat may also show several generations, although the earliest generation is often framboid-rich. Of course, not all framboids are from this early generation, and framboids also act as centers for further deposition of later pyrites, as in marine sediments.

The reports of the trace and minor elements in pyrite in coals show extreme variations in the concentrations of these elements (e.g., Kolker, 2012), even exceeding those of pyrite in marine sediments. There is a general thesis that the earliest generations of pyrite in coals have lower trace element loads than later generations. However, a contemporary study of the trace element concentrations of pyrite in peat, which would help determine the contribution of early generations of pyrite, has not been reported.

9.2.3. Trace and Minor Elements in Hydrothermal Pyrite

Pyrite is an abundant mineral in hydrothermal ore deposits and its trace and minor element compositions have been the subject of numerous geochemical studies, stretching back to the pioneering days of Goldschmidt (Goldschmidt and Strock, 1935) and Carstens (Carstens, 1941).

Analyses of pyrite in active ocean floor and terrestrial hydrothermal systems show consistent elevated values of Ag, As, Au, Cd, Pb, Sb, and Zn above the boiling horizon and increasing constancy in the Co/Ni ratio with depth and/or temperature (e.g., Hannington, 2014; Hannington et al., 1998; Libbey and Williams-Jones, 2016). Libbey and Williams-Jones (2016) found that Se enrichment was restricted to deeper levels in pyrite in the system. This may help to explain the reports of highly variable Se compositions of hydrothermal pyrite, which extend back to Goldschmidt and Strock's (1935) original analyses. By contrast, Hg, Sb, and Tl are relatively volatile elements and migrate to the upper parts of the hydrothermal system before being removed (Libbey and Williams-Jones, 2016). The problem with even these generalizations from a single active system is that hydrothermal fluid flow is extremely dynamic and various parts of the contemporary system are superimposed on earlier deposits. This is often evidenced by growth zones in individual pyrite crystals showing extreme variations in trace element loads (Tardani et al., 2017). The interpretation of results is further complicated, as noted in section 8.1 of Chapter 8, by the fact that high concentrations in pyrite are probably due to the presence of nanoparticles of discrete exotic phases within the pyrite grains (Deditius et al., 2011). This means that individual point analyses of hydrothermal pyrites are likely to be even more erratic and non-reproducible.

The result is that pyrite compositions in ancient hydrothermal deposits have shown highly variable trace and minor element loads in hydrothermal pyrite, even within individual deposits (see Tardani et al., 2017, for an extensive listing). In order to dampen some of the more extreme variations in trace element loads of hydrothermal pyrites, Gregory et al. (2017) used trace element ratios. They found that hydrothermal pyrites Co/Ni, Zn/Ni, Cu/Ni, As/Ni, As/Au, Ag/Au, Sb/Au, and Bi/Au were consistently higher in hydrothermal ore deposit pyrites than in sedimentary pyrite. However, comparisons with other published data sets showed that their Cu/Ni, As/Ni, Zn/Ni, and Tl/Co ratios were two magnitudes lower than those identified by Mukherjee and Large (2017) and Gregory et al. (2016).

A recent approach to solving the problem of inconsistencies in reported trace element contents of hydrothermal pyrites has been the application of sophisticated multivariate statistical analyses to pyrite trace element concentrations. This has been made possible by the large amount of relatively high-quality analytical data presently generated by LA-ICPMS analyses (Dmitrijeva et al., 2020; Gregory et al., 2019). The caveat here is the relatively large volume sampled by LA-ICPMS compared with EPMA or HRTEM analyses, the variable coupling of the laser beam with different materials (see section 9.3.1), and the problem of the concentrations below detection limits in the analyses of analytical distributions (Dmitrijeva et al., 2020).

Using these techniques, Dmitrijeva et al. (2020) suggested that Ag, Bi, and Pb were probably present as discrete telluride nanoparticles (although galena nanoparticles might give the same result), and the variation in Co/Ni was related to crystal zonation. Gregory et al. (2019) compared the results of cluster analyses with those of conventional X-Y scatter plots. They found that pyrite in low- to medium-temperature hydrothermal deposits tended to contain higher concentrations of most trace elements than high-temperature pyrites. They found that the trace element loads on pyrites in the low- to medium-temperature hydrothermal ores, found in sedimentary (SEDEX) and volcanic (VHMS) host rocks, were similar to those of sedimentary pyrites in the same vicinity. In contrast, they found that multivariate analyses of the Co, Ni, Cu, Zn, As, Mo, Ag, Sb, Te, Tl, and Pb contents of pyrite showed a high degree of consistency and could be used to discriminate different pyrite-forming environments with a high degree of confidence.

However, the reported trace element compositions of hydrothermal pyrites are extremely variable, even within the same grain. This may be partly affected by the presence of nanoparticles of discrete minerals. It seems that general conclusions about the behaviors of trace elements in hydrothermal pyrite cannot be presented in a simple manner. The problem could be resolved by unsupervised cluster analyses based on the machine-learning approach pioneered in this context by Gregory et al. (2019).

9.3. Framboid Composition

9.3.1. Trace and Minor Element Compositions of Framboids

There are limited data available on the trace and minor element compositions of framboids in coal, compared with framboids in other sediments and sedimentary rocks. Kolker (2012) reported that, in coals that have not been subjected to later mineralization, Ni is often relatively concentrated in framboids compared with other pyrite (Figure 9.1). He related this to Ni availability in the original peat.

Certainly, there are many accounts of heightened Ni concentrations in wetland sediments, especially those overlying nickeliferous regoliths (e.g., Rinklebe and Shaheen, 2017; Shotyk, 1988). By contrast, Kortenski and Kostova (1996) did not detect Ni, Co, or As in framboids from Bulgarian coals of varying ranks, although they did detect it in euhedral pyrite. They also noted Be and Zn in framboids and not in euhedral pyrite. Kortenski and Kostova used EDAX analyses on an SEM and their results are consequently indicative, but their comparative results between framboids and other forms should be robust. Kolker's (2102) framboid analyses (Table 9.1) also show significant Cu and Zn concentrations. Chalcopyrite and covellite were also found coating pyrite framboids in a copper-rich bog in British Columbia, although the framboids themselves were not analyzed (Lett and Fletcher, 1980). Diehl et al. (2012) reported that As and Se were generally depleted in framboids in coal

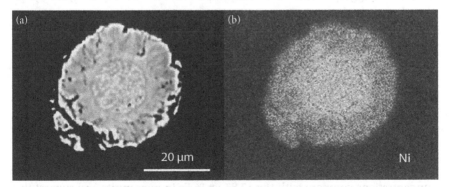

Figure 9.1 Electron probe microanalyzer images of Ni concentration in framboids in coal: (a) back-scattered electron image of a framboid with secondary overgrowth; (b) Ni distribution showing Ni concentration within the framboid and a depleted rim.

Modified from *International Journal of Coal Geology*, 94: 32–43. Minor element distribution in iron disulfides in coal: A geochemical review. Kolker, A. Copyright (2012) with permission from Elsevier.

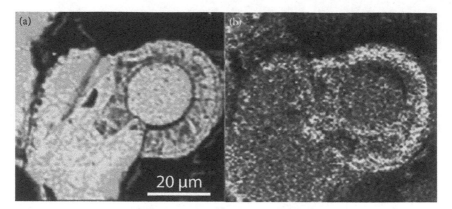

Figure 9.2 Arsenic distribution in a framboid in ca. 310 Ma Appalachian coal: (a) backscatter electron image of the framboid with rim of radiating FeS_2; (b) As distribution depleted in the framboid and enriched in the rim. Selenium shows a similar distribution.

Modified from *International Journal of Coal Geology*, 94: 238–249. Distribution of arsenic, selenium, and other trace elements in high pyrite Appalachian coals: Evidence for multiple episodes of pyrite formation. Diehl et al. Copyright (2012) with permission from Elsevier.

(Figure 9.2) and they concluded that reported enrichments were probably due to later reactions of the framboids with hydrothermal fluids (cf. sections 8.2.3 and 8.2.4, Chapter 8).

Trace element concentrations of framboids were reported by Tribovillard et al. (2008) from 155 Ma, late Jurassic shales from northern France. The framboids occurred as aggregates and were hand-separated before being analyzed by wet chemical methods.

The trace element contents of the framboids were compared with those of contemporary pyrite nodules (Table 9.4). I have recalculated the reported data, assuming that all Fe in the analyses represents pyrite. The framboids showed particularly high Mo contents relative to the nodules, which the authors explain as being either due to the framboids sampling different (younger) fluids relative to the nodules or being formed by a different mechanism. More recent data on framboid formation (Rickard, 2019) might suggest that the framboids formed more rapidly than the nodules. The analyzed Fe content is interesting, giving an average 33.66 wt % Fe for the framboids versus 32.66% Fe for the nodules. These concentrations compare with the 47.83 wt % Fe in pure pyrite and suggest that the analyzed samples included around 30% shale.

The team at the University of Tasmania under Ross Large pioneered the use of LA-ICPMS for the minor and trace element compositions of especially sedimentary pyrites (Large et al., 2014; Large et al., 2015). The program on the geochemistry of framboids has been led by Gregory (Gregory, 2013; Gregory et al.,

Table 9.4 Trace Element Contents (ppm) of Framboids and Pyrite Nodules from 155 Ma Jurassic Black Shales in Northern France

	Mo	Pb	U	V	Ni	Co	Cu	Cd	Zn	As
Framboids	97	36	8	56	45	6	39	2	313	60
Nodules	56	32	8	46	55	6	28	2	203	47

Recalculated from Tribovillard et al. (2008).

2014; Gregory et al., 2015; Gregory et al., 2017). The laser beams used varied between 10 and 40 microns in diameter, probably probing a depth into the sample of up to 50 microns (Gilbert, 2015). This is a far larger volume than the average single framboid, and the correction method used was to analyze nearby pyrite-free areas and subtract those totals from the framboid analysis to provide an estimate of the trace element composition of the framboid itself (Large et al., 2014). This is both computationally and technically challenging, not least because the laser couples in different ways with different materials. Framboids have an average diameter of around 6 µm (section 2.3 in Chapter 2) and thus the amount of material ablated is very small, leading to both higher errors and detection limits. This is compounded by the dearth of secondary standards for pyrite analyses. The laser drills into the framboid and therefore sudden discontinuities in composition can be related to exotic inclusions in the sample. Unfortunately, at present, only relatively large inclusions can be discriminated by this method, and even then, the results are partly dependent on how the laser couples with different materials. Ideally, of course, complete ablation with no melting would contribute to solving both problems. For framboids, then, the LA-ICPMS currently provides an approximate bulk analysis, including not only trace elements in the pyrite, but also suggesting discrete sulfide nanoparticles as well as possible matrix materials.

The pyrite content of framboids varies between ca. 60 vol % for randomly packed framboids to 74 vol % for cubic close packed framboids. This in turn means that framboids contain between 26% and 40% exotic material in their matrices. Since organized framboids constitute only a minor fraction of the total framboid population, the average framboid is more likely to contain up to 40% by volume matrix material.

As is discussed earlier and in sections 7.2.6 (Chapter 7) and 8.2 (Chapter 8), the composition of the matrix material may include organic matter and other non-pyritic minerals, each of which may carry their own trace element loads. If this material is syngenetic with framboid formation, then this may also sample the ambient environment. However, if this material is epigenetic with respect to framboid formation, then it is likely to sample a completely different environment.

There are few reports of the trace element contents of framboids in contemporary sediments. Gregory (2013) analyzed the trace element contents of 47 framboids from the Cariaco basin with LA-ICPMS. The results (Table 9.5) show variable concentrations of most elements over at least one order of magnitude.

The trace element contents of 193 framboids from black shales ranging in age from modern to Proterozoic are summarized in Table 9.6. Samples which were probably mistakenly described as framboids were excluded from the database. The data are presented as geometric means, $\bar{x}^*$, and geometric (or multiplicative) standard deviations, σ^*. Since it is obvious in a log normal distribution that the minimum concentrations mostly tend to 0 or undetectable and the maximum concentrations include outliers, the ranges are presented for 95% of the variance,

Table 9.5 Geometric Mean ($\bar{x}^*$), Maximum (max), and Minimum (min) Concentrations of 47 Framboids from Sediments in the Cariaco Basin

	$\bar{x}^*$ ppm	max ppm	min ppm
Mn	1995	5261	116
Co	48	291	5
Ni	419	1132	75
Cu	451	1574	77
Zn	608	1556	132
As	268	1954	74
Se	153	847	9
Mo	808	4064	169
Ag	8	33	2
Cd	19	51	4
Sb	40	186	5
Te	5	12	1
Tl	0	1	0
Pb	60	213	15
Bi	2	34	0

Recalculated from data in Gregory (2013).

Table 9.6 Trace Element Contents of Framboids (ppm) and Total Pyrite from Black Shales Deposited during the Last 2500 Myr in Terms of Geometric Means ($\bar{x}$ *), Geometric Standard Deviations (σ*), Number of Samples Analyzed (n), and the Maximum (max) and Minimum (min) Values for the 95% Range (^{x}/ 1σ*)

	Framboid				Total Pyrite			
	$\bar{x}$*	σ*	min	max	$\bar{x}$*	σ*	min	max
Mo	50	10	5	500	14	7.4	2	104
Co	181	4.5	40	8145	78	8.3	9	647
Ni	481	4.6	105	2212	354	4.2	84	1487
Cu	226	2.6	87	588	149	5.6	27	834
Se	27	4.8	6	130	20	5.2	4	104
As	696	7.9	9	5498	412	5	82	2060
n	193				1407			

Computed from data from LA-ICPM analyses by Gregory et al. (2015).

$\bar{x}$*/ σ* and $\bar{x}$* x σ*. The results reveal extremely broad ranges for trace element contents ranging up to over two magnitudes for Mo.

In terms of average contents, trace elements occur in framboids in the order As>Ni>Co>Cu>Mo>Se. The framboid data are compared with the total sedimentary pyrite variations in Table 9.6. The geometric mean values of the concentrations of these trace elements in the total sedimentary pyrite are consistently lower than in the contemporary framboid fractions and the ranges are more limited. This may be a mere artifact of the larger number of analyses (1407 to 193) or it may be real, reflecting either actual variation in the pyrite or the effect of up to 40 vol % interstitial material in the framboids. In their original study, Gregory et al. (2015) concluded that there was no significant variation in trace element loads in sedimentary pyrite displaying various textures. This conclusion is important since it suggested that the framboids were not sampling different (e.g., earlier diagenetic) fluids than the bulk sedimentary pyrite. The data listed earlier suggest that this might be worth revisiting.

Large et al. (2014) reported an investigation of the relationship between trace element content and framboid size (Table 9.7). The idea here was that smaller framboids were more likely to have been formed in the water column, whereas large framboids were probably formed in the sediments (see section 2.5.1 of Chapter 2). No significant trends were observed, except perhaps for Ni, which seems to decrease with framboid size. The basic parameter controlling framboid size is the framboid formation time (Rickard, 2019a): the 5 μm diameter

Table 9.7 Variation in Geometric Mean Trace Metal Contents (ppm) with Framboid Size (μm) from Late Permian (ca 250 Ma) Sediments of the Perth Basin, Australia

	Diameter (μm)					
	5	6	8	10	22	35
As	210	358	218	223	329	302
Se	12	50	3	27	10	16
Ni	260	209	196	217	164	161
Mo	13	31	22	22	24	12

Data extracted from figure 2 of Large et al. (2014).

framboids took perhaps 3–4 days to form, whereas the 35 μm framboids may have taken up to a year to form. The medium sampled by the framboids was then different. In particular, the trace element composition of the larger framboids averaged seasonal variations in dissolved trace element loads, whereas the smaller framboids sampled snapshots of just a few days in a single season.

Gregory et al. (2014), however, reported that the trace element contents of pyrite framboids varied with depth in the sediment (Table 9.8). The variation could be due to equilibration of pyrite during diagenesis, variations in the concentrations of pyrite in the sediment, or secular variations in the trace element contents of the ambient fluids. The results suggest that, if framboids are indeed sampling contemporary fluids, there is a variation between younger and older framboids in modern sediments.

Table 9.8 Variations in Trace Element Contents of Framboids with Depth from the Huon Estuary Tasmania

Depth (cm)	Co	Ni	Cu	Pb	Zn	As	Mo
10–15	14	18	11	2	37	181	14
20–25	3	3	17	0	26	354	12
30–35	4	3	19	2	27	590	5
40–45	7	9	31	1	65	209	14
50–55	1	4	68	1	59	194	30
60–65	26	19	82	3	68	122	40

From Gregory et al. (2014).

In order to test the hypothesis that the trace element content of framboids was proportional to the local pyrite concentration, which was originally suggested by Huerta Diaz and Morse (1992), I computed the ratio of total sulfur (wt %) to trace metal concentration in framboids (ppm) from the data for framboids from the Huon estuary, Tasmania, in Gregory (2013), assuming that the total sulfur concentration was an approximate proxy for the pyrite concentration. The results (Table 9.9) show that ratios do not show any consistent trends, which suggests that the trace element loads of pyrite framboids are not simply a function of the amount of pyrite formed. That is, the trace element concentrations reflect real variations in their local availability during pyrite framboid growth. Since this occurs over a relatively short time period (Rickard 2019), this result supports the hypothesis that the trace element contents of pyrite framboids represent a snapshot of the local trace element concentrations in the fluids at the time.

Tribovillard et al. (2008) and Gregory et al. (2017) asked the interesting question as to what extent the trace element concentrations of sedimentary pyrite differs from the trace element contents of the enclosing sediment. I have extended this question to framboids by extracting the framboid data from the respective reports. Tribovillard et al. (2008) found that Mo was mostly concentrated in pyrite framboids and framboids contributed the bulk of the total Mo in the contemporary sediment. Gregory et al. (2017), by contrast, concluded that pyrite constitutes a small percentage of the total rock mass so that the concentration of any element in pyrite is unlikely to represent the trace element concentration of the sediment as a whole.

Table 9.9 Variations in Framboid-Trace Metal (ppm)/Total Sulfur (wt %) Ratios with Depth in Sediments from the Huon Estuary, Tasmania

Depth	S	Co/S	Ni/S	Cu/S	Pb/S	Zn/S	As/S	Mo/S
cm	wt %	$\times 10^4$	$\times 10^4$	$\times 10^4$	$\times 10^4$	$\times 10^4$	$\times 10^4$	$\times 10^4$
10–15	1.06	13	17	10	2	35	171	14
20–25	2.02	1	1	9	0	13	175	6
30–35	1.96	2	1	10	1	14	301	3
40–45	2.25	3	4	14	0	29	93	6
50–55	2.6	0	2	26	1	23	75	12
60–65	2.45	10	8	33	1	25	50	16

Computed from data listed in Gregory (2013).

Table 9.10 Concentrations of Selected Trace Elements (ppm) in 9 Framboids from the 580 Ma Neoproterozoic Doushanto Formation, China, and Enclosing Shale

	Mo	Ni	Co	Mn	As	Pb	Sb	Zn	Cu
Shale	2	48	72	437	36	29	4	55	68
Framboid	41	784	47	3990	738	261	68	100	873
Framboid/shale	20	16	1	9	21		18	2	13

From data in Gregory et al. (2017).

Table 9.11 Key Trace Element Ratios for Framboids and Enclosing Shale from the 580 Ma Neoproterozoic Doushanto Formation, China, Compared with the Characteristic Ratios for Sedimentary Pyrite

	Co/Ni	Zn/Ni	Cu/Ni	As/Ni	As/Au	Ag/Au	Sb/Au	Bi/Au
Pyrite	<2	<10	<2	<10	>200	>2	>100	>1
Framboid	0.1	0.1	1.1	0.9	3500	16.9	323	3.2
Shale	1.5	1	0.2	1	—	—	—	—

From data in Gregory (2013); Gregory et al. (2015); Gregory et al. (2017).

I have extended Gregory et al.'s analysis to the nine framboids they analyzed and compared these results with the reported whole rock trace element concentration of the splits of samples in which the framboids occurred (Table 9.10). The number of analyses is obviously too small to have any statistical robustness, but the results for the framboids are generally consistent with Gregory et al.'s conclusion that these trace elements are generally preferentially concentrated in pyrite. The exception is Co, which is more enriched in the shale.

The ratios of trace elements that characterize sedimentary pyrite (Gregory, 2013; Gregory et al., 2015) are compared with the ratios in framboids and host shales in Table 9.11. The ratios fall well within the ratios characteristic of sedimentary pyrite. There are variations in the ratios which may suggest that framboids are better samples of contemporaneous fluid geochemistry than the shales. For example, the Co/Ni and Zn/Ni ratios in the framboids are at the extreme end of the sedimentary range. By contrast, As/Ni rations in both framboids and shale are similar and it appears that the shale does not concentrate Cu as efficiently as the framboids.

9.3.2. Trace Element Compositions
of Hydrothermal Framboids

The trace element compositions of hydrothermal framboids are unknown. The difficulties described previously in the chemical analyses of framboids are commonly multiplied in hydrothermal framboids because of (1) common sulfide overgrowths (see section 1.3.4 of Chapter 1), which makes discrimination of the trace element load of the framboid difficult, and (2) replacement and infilling of framboids by other sulfide minerals (see section 8.2 of Chapter 8).

Scott et al. (2009) analyzed hydrothermal framboids formed at 175°–220°C in late-stage ore veins in the Carlin Trend, Nevada, and the shear zone veins at Fosterville, Victoria, Australia. The difficulty in interpreting the analytical results is illustrated by the large contents of Al commonly reported, which suggests that the laser beam probed silicate country rock as well as framboids. The accuracy of the analyses in the Carlin trend vein framboids was further restricted by marcasite overgrowths. However, detailed factor analyses of the results suggest that the framboids are enriched in As, Sb, and Tl. The authors also reported analyses of hydrothermal framboids associated with the shear zone Au deposits at Fosterville. They found that framboids formed synchronously with or later than greenschist to amphibolite facies metamorphism were significantly enriched in As, Sb, Ni, and Co. Again, the analytical constraints on the framboid association constrained the accuracy of the reported trace element concentrations as representative of framboid compositions. Scott et al. (2009) also measured the Pb isotope composition of the framboids and found that this provided a broad confirmation that these framboids were formed during a metamorphic event up to 70 Myr after sedimentation.

9.3.3. Sulfur Isotope Compositions of Framboids

The sulfur isotopic compositions of sedimentary and hydrothermal pyrite have been widely reported. The data are generally listed in terms of the deviation, $\delta^{34}S$, of the ratio of the two most abundant isotopes ^{34}S and ^{32}S from a standard meteoritic value. This formalization then discriminates samples relatively enriched in ^{32}S ($\delta^{34}S$ <0) from samples enriched in ^{34}S ($\delta^{34}S$ >0). This representation is useful since microbial sulfate reduction results in a relative concentration of ^{32}S, and $\delta^{34}S$ for magmatic sulfur tends to 0. Then $\delta^{34}S$ values for sedimentary pyrites have been generally reported with negative values, but more detailed analyses show that they are actually characterized by highly variable values as depletion in one isotope leads to enrichment in the other, and vice versa. Large et al. (2020) reported 1500 ion microprobe sulfur isotope analyses of pyrite in black shales through time. They found that Archean pyrites have a mean $\delta^{34}S$ = + 3.7‰, suggesting a juvenile origin for the sulfide, whereas Proterozoic and Phanerozoic black shale pyrites have an average value of + 25‰, not significantly different from the seawater sulfate composition. This would be expected from averaging sulfur isotope compositions of large numbers of pyrite samples over wide swaths of geologic time: the average sediment value of stable isotope compositions must ultimately approach the average value of the source seawater sulfate, if the system is effectively closed to large net sulfur inputs or removals. However, the individual analyses display a large range of values, between −55 and +80‰, characteristic of microbiological sulfate reduction.

By contrast with sedimentary and hydrothermal pyrite, the sulfur isotopic compositions of framboids have been rarely reported. It would be expected that the $\delta^{34}S$ for sedimentary framboids is negative since the pyrite-S is derived from microbial sulfate reduction.

Canfield et al. (1992) mechanically separated the early framboidal and later pyrites from sediments off Long Island Sound. They found that framboids were concentrated in the 4–21 μm pyrite size fraction and used the sulfur isotopic compositions of this fraction as a proxy for the framboid sulfur isotopic composition. They reported that, at 4.5 cm depth below the sediment-water interface, the framboids had similar $\delta^{34}S$ values to the bulk pyrite (around −30‰). They found that the later-formed pyrites had $\delta^{34}S$ values 10‰ to 15‰ heavier than the earlier framboidal pyrite and ≤40‰ heavier than pyrites associated with magnetite. The change with depth resulted from the increased proportion of non-framboidal pyrite. The very fine pyrite in the clay fraction had $\delta^{34}S$ values similar to those of the framboidal pyrite. The result suggests sulfate-reduction in a partly limited system where the later reactant sulfide becomes isotopically heavier as the lighter fraction is removed and sequestered in pyrite framboids. Zhang et al. (2011) separated framboid concentrates from 250 Ma Permian-Triassic boundary clays from south China and found that $\delta^{34}S$ values varied between − 18.7‰ and −30.2‰.

With the introduction of microanalytical methods, such as nano-scale secondary ion mass spectrometry (NanoSIMS), it has become possible to analyze intra-grain and inter-grain variations in the sulfur isotope composition of sedimentary pyrite. Kohn et al. (1998) reported large variations in sulfur isotopic compositions of sedimentary pyrite, including framboids, from the top 20 cm of sediments from off the coast of California. They analyzed 18 framboids and their $\delta^{34}S$ isotopic compositions averaged −24‰ and ranged from −5‰ to −41‰. However, they found no variation with depth in the sediment. Zhao et al. (2018) found that $\delta^{34}S$ values for framboids from Carlin style ore deposits in China displayed very light S (−53.2‰ to −44.0 ‰).

By contrast with pyrite grains and nodules, where $\delta^{34}S$ compositions commonly increase toward the margins (e.g., McKibben and Riciputi, 1998), no sulfur isotopic variations have been reported within individual framboids. This is consistent with the process of burst nucleation in framboids described in section 11.1 of Chapter 11. It does suggest that framboids pick up a more accurate measure of the sulfur isotopic composition of the prevailing dissolved sulfide and are likely to retain this over geologic time.

9.4. Controls on the Trace Element Concentrations in Framboids

It is obvious that the trace element concentrations in framboids represent something about the environment in which the framboids were formed. Framboidal pyrite forms by burst nucleation (section 11.1 of Chapter 11) and rapid crystal growth (section 12.3 of Chapter 12) and therefore will sample the fluid in which it was formed. This was neatly demonstrated by Gregory et al. (2014) who compared the trace element contents of pyrite, including framboids, in a contaminated estuary sediment with those from an uncontaminated estuary. A Zn smelter had been built on the shores of the contaminated estuary in 1917.

Their data are summarized for selected elements in Table 9.12. It can be seen that the Co content reflects the sediment composition, whereas the high Ni in

Table 9.12 Selected Mean Trace Element Contents of Pyrite Framboids and Bulk Sediments and Their Ratios, from Uncontaminated and Contaminated Estuaries in Tasmania

		Co	Ni	Cu	Pb	Zn	As	Mo
Uncontaminated	Framboid	5	7	29	1	43	239	16
	Sediment	14	441	28	19	59	20	5
	Framboid/ Sediment	0.36	0.02	1.04	0.05	0.73	11.95	3.20
Contaminated	Framboid	12	5	73	244	304	849	3
	Sediment	26	33	401	1493	7317	263	3
	Framboid/ Sediment	0.46	0.15	0.18	0.16	0.04	3.23	1.00

Recalculated from Gregory (2013).

the uncontaminated sediments is not picked up by the framboid. This is consistent with crystal field theory which predicts that Co is compatible with the pyrite structure whereas Ni is not (Burns, 1970). Likewise, the framboid Cu content does not reflect the contaminated sediment value. The framboid Pb, Zn, and As contents seem to track the sediment value, at least within a magnitude, although arsenopyrite was detected as a discrete phase in the more contaminated sediments. The Mo content is lower in the contaminated sediment and probably results from Mo being mainly sequestered in organic matter rather than pyrite.

These results, together with the data discussed throughout this chapter, suggest that pyrite framboids may pick up the local environmental trace element variations. However, interpretation of the results is currently complex. The original sequestration of trace elements is likely to be in part determined by the pyrite crystal chemistry, and there may be a limit to how much of any given trace element can be sequestered by pyrite. The pyrite may be in competition for trace elements with other sedimentary components, which is well illustrated with Mo distributions. There seems to be strong evidence that trace element contents in framboidal pyrite decrease during early diagenesis due to equilibration. This is likely to be enhanced during late diagenesis and early metamorphism, and it is not altogether clear how individual trace elements behave over geologic time. These effects are likely to be increased in hydrothermal framboids.

10

Pyrite Framboid Formation Chemistry

Framboids are overwhelming constituted by pyrite. Although other minerals may display the framboid texture, they are extremely limited in distribution. Pyrite formation is therefore key to understanding framboid formation, although framboids formed by other minerals may test these ideas, as discussed in Chapter 8.

In this context, *formation* is an ill-defined term since the formation of pyrite in framboids is divided into two processes: nucleation and crystal growth. The nucleation and crystal growth of framboids are considered in more detail in Chapters 11 and 12. The processes involved in pyrite formation must therefore explain the chemistry of both pyrite nucleation and crystal growth. It is worth noting at this stage that crystal growth is the process producing the overwhelming mass of the material and that, consequentially, when a natural sample or experimental reaction product is analyzed, the pyrite process which is being interrogated is mainly crystal growth. It is through this prism that previous reports on the formation of pyrite must be viewed.

10.1. Pyrite Chemistry

The framboid texture itself places considerable constraints on the chemistry of mineral formation in framboids. In particular, framboids contain up to tens of thousands of equidimensional, equant, and equimorphic microcrystals which must have formed at the same time (see section 11.1 in Chapter 11) throughout the individual framboid space. This in turn requires that the chemistry of the mineral formation process in framboids facilitates the nucleation and crystal growth of the microcrystals at the same time. In this chapter, the chemistry of pyrite and the molecular and thermodynamic bases for its nucleation and crystal growth in solution are discussed in order to provide a foundation for understanding framboid formation.

Framboids. David Rickard, Oxford University Press. © Oxford University Press 2021.
DOI: 10.1093/oso/9780190080112.003.0010

10.1.1. Pyrite Solubility

The solubility of pyrite was revised by Rickard and Luther (2007) in the light of contemporary thermodynamic data. They found the solubility of pyrite, written in terms of Equation 10.1, is $10^{-14.2}$.

$$FeS_2 + H^+ = Fe^{2+} + HS^- + S^0 \qquad\qquad 10.1$$

This suggests that the ion activity product $\{Fe(II)\}\{\Sigma S(-II)\}$ at pH = 7 is 10^{-21} where $\{\Sigma S(-II)\}$ is the total dissolved sulfide activity (Equations 10.2 and 10.3):

$$\{\Sigma S(-II)\} = \{H_2S\} + \{HS^-\} \qquad\qquad 10.2$$

$$\{H_2S\} = \{HS^-\} \text{ at pH} = 7 \qquad\qquad 10.3$$

The equilibrium solubility equations illustrate the extreme insolubility of pyrite in aqueous solutions at standard temperature and pressure or SATP (25°C, 0.1 MPa). For example, where the aqueous Fe concentration is at nanomolar levels, just picomolar concentrations of dissolved sulfide are necessary for pyrite microcrystals to grow. These concentrations are difficult to analyze and show the extreme stability range of pyrite in the natural environment.

The solubility product is important in framboid formation since it defines the saturation level in the solution in equilibrium with pyrite; that is, the minimum dissolved iron and sulfide concentrations necessary for pyrite crystals to grow. The degree of supersaturation of a solution with respect to pyrite, Ω, is usually presented as the ratio of the ionic activity product in solution to the equilibrium solubility product (Equation 10.4).

$$\Omega = \frac{\left(\{Fe(II)\}\{\Sigma S(-II)\}\right)_{solution}}{\left(\{Fe(II)\}\{\Sigma S(-II)\}\right)_{equilibrium}} \qquad\qquad 10.4$$

The consequence of the low solubility product for pyrite is that most natural solutions in low oxygen environments are supersaturated with respect to pyrite. Indeed, as discussed in section 11.3.1 of Chapter 11, extreme relative supersaturations commonly occur and these extreme supersaturations power framboid nucleation. The widespread supersaturation with respect to pyrite in low oxygen environments explains its abundance, and the extreme supersaturations explain the ubiquity of framboids.

10.1.2. Pyrite Formation Chemistry

There are innumerable recipes for synthesizing pyrite (e.g., Ohfuji and Rickard, 2005; Rickard, 2012; Rickard and Luther, 2007; Xian et al., 2016), but the kinetics and mechanisms of only two basic processes have been established.

(1) The polysulfide pathway: This is the reaction between FeS and polysulfide to form pyrite, which was first defined by Rickard (1975) and confirmed by Luther (1991).

(2) The H_2S pathway: The kinetics and mechanism of the oxidation of FeS by H_2S to form pyrite were first established in aqueous solutions by Rickard (1997) and Rickard and Luther (1997a). The reaction was well known in the chemical literature since it was originally described by Berzelius in (1845). It was revisited and popularized by Wikjord et al. (1976) and Huber and Wächtershauser (1997).

Both of these reactions have been shown to produce framboidal pyrite (e.g., Butler and Rickard, 2000b; Sweeney and Kaplan, 1973).

10.2. The Polysulfide Reaction

10.2.1. Polysulfide Reactants

Much of the early work on pyrite formation with polysulfide was predicated on the basis that the most abundant polysulfide species in aqueous solutions at STP were $S_4(-II)$ and $S_5(-II)$ species. This led to mechanistic interpretations which involve cleaving of these longer chain polysulfides such that $S_2(-II)$ moieties were created (e.g., Luther, 1991). Although this mechanism is valid, it is likely to play a relatively minor role on pyrite formation. The reason is that the stability constants for $S_n(-II)$ species were put on a far firmer footing by the work of Kamyshny et al. (2004).

The implications of this revised data set were considered by Rickard and Luther (2007), who showed that the dominant polysulfide species in aqueous solution between pH 4 and 11 was HS_2^-, a protonated disulfide (Figure 10.1). Rickard and Luther (2007) calculated that HS_2^- can contribute as much as 1% of the total dissolved sulfide over much of this pH space. A caveat here is that this is true in the presence of solid sulfur. The stability of the stable phase, orthorhombic cyclooctasulfur, S_8, is well known to be limited to acidic solutions. For example, Rickard and Luther (2007) showed that it was limited to pH <4.5

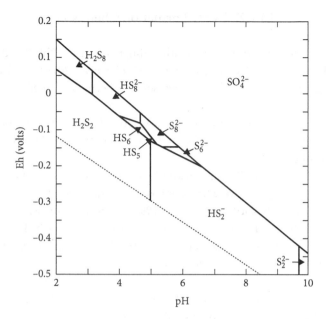

Figure 10.1. Stability regions of polysulfide species in aqueous solutions at SATP in terms of pH and Eh space with stable sulfide species removed.

Reprinted with permission from Rickard, D., & Luther, G. W. 2007. *Chemistry of Iron Sulfides. Chemical Reviews.* Fig. 10. Copyright (2007) American Chemical Society.

at 10^{-3} M total dissolved sulfur activities. However, the solubility of solid sulfur is highly dependent on its state which varies from orthorhombic S_8, to various hydrophilic and hydrophobic sulfur colloids, often of biologic origin and dissolved sulfur (Avetisyan et al., 2019; Kafantaris, 2019; Kleinjan et al., 2005). Most experimenters are familiar with the problem of colloidal sulfur precipitating and then persisting in pH regions well outside its ideal stability zone. The stability diagram for HS_2^- is then likely to describe its approximate relative abundance in marine sulfidic sediments as well as the more acidic freshwater systems in which S_8 is stable.

10.2.2. The Polysulfide Reaction Mechanism

The overall reaction can then be written as Equation 10.5, where FeS refers to any surface bound or aqueous iron (II) monosulfide.

$$FeS + S_2 \left(\text{-II} \right) = FeS_2 + S \left(\text{-II} \right)$$
10.5

The reaction is essentially an exchange reaction in which the S(-II) in FeS is exchanged with aqueous S_2(-II). The reaction was proved by Butler et al. (2004), who demonstrated isotopically that the resultant pyrite S was entirely derived from the aqueous polysulfide (Equation 10.6).

$$Fe^{32}S + {}^{34}S_2(-II) = Fe^{34}S_2 + {}^{32}S(-II)$$
10.6

The reaction has been widely used in framboid syntheses (see Ohfuji and Rickard, 2005), including the original experimental synthesis by Sweeney and Kaplan (1973).

Luther (1991) originally suggested that the mechanism of the Rickard's (1975) polysulfide pathway for pyrite formation involved the formation of an $[Fe(SH).S_x]^-$ intermediate where S_x refers to a polysulfide group of indeterminate stoichiometry. In the light of subsequent data on polysulfide stability described earlier, this indeterminate reaction intermediate can be replaced by the specific species $[FeS.HS_2]^-$ and the Luther mechanism can be generalized to cover both pyrite nucleation and crystal growth in aqueous solutions (Equations 10.7–10.9).

$$Fe(II) + S(-II) \rightarrow FeS$$
10.7

$$FeS + HS_2^- \rightarrow [FeS.HS_2]^-$$
10.8

$$[FeS.HS_2]^- \rightarrow FeS_2 + S(-II)$$
10.9

In these reactions, S(-II) refers to aqueous sulfide species like HS and H_2S, FeS includes both aqueous FeS clusters and =FeS surface moieties, and the mechanism applies to both pyrite nucleation and crystal growth chemistry. In this model, reactions 10.8 and 10.9 represent a classical Eigen–Wilkins substitution reaction mechanism (Eigen and Wilkins, 1965). This substitution mechanism has been independently verified by the isotopic results of Butler et al. (2004), as shown in Equation 10.6.

10.2.3. Aqueous FeS as a Reactant

Evidence for non-protonated, aqueous FeS was first described kinetically by Pankow and Morgan (1979) and thermodynamically by Davison (1991). Theberge and Luther (1997) reported that aqueous FeS occurs as clusters of hydrated FeS molecules with indeterminate stoichiometries, analogous to the well-known FeS clusters which form the active centers of FeS proteins in organic systems. There is some evidence to support the idea that the dimer, $Fe_2S_2(H_2O)_4$, may be the most abundant species. Aqueous FeS clusters have been shown to be widespread in the natural environment and to play a key role in the transport of Fe in sulfidic systems.

The reaction between aqueous FeS, represented by the monomer FeS^0, and polysulfide, represented by HS_2^-, the most abundant polysulfide species in most aqueous solutions at SATP, is shown in Equation 10.10. The equilibrium constant for this reaction is 10^{16}.

$$FeS^0 + HS_2^- = FeS_2 + HS^- \qquad\qquad 10.10$$

Rickard et al. (2001) proved that aqueous FeS clusters play a key role in pyrite formation. They showed that suppressing the clusters also suppressed pyrite formation. They used aldehydic carbonyl as a suppressant, and this leads to the idea that some conflicting data on pyrite formation chemistry might be due to the presence or absence of traces of organic inhibitors like aldehydic carbonyl in the system (Thiel et al., 2019).

The relationship between aqueous FeS and particulate FeS phases such as mackinawite, FeS_m, was established by Rickard et al. (2006). Rickard et al. (2006) confirmed the earlier observations of Davison (1991) that at pH >5.3 in aqueous solutions at SATP the dissolution of FeS_m is independent of pH. This is also consistent with the original measurements of the kinetics of FeS_m dissolution by Pankow and Morgan (1979) who demonstrated a pH-independent dissolution regime at pH >5.7. The consequence of these observations is that the products of FeS_m dissolution in the pH-independent regime cannot include significant concentrations of protonated species such as $FeSH^+$ or $Fe(HS)_2$. The main product must be a non-protonated FeS complex. Rickard and Luther (2006b) showed that FeS cluster chemistry could be modeled thermodynamically by addressing the monomer FeS^0 as a minimal representation of the FeS cluster composition (Equation 10.11).

$$FeS_s = FeS^0 \qquad\qquad 10.11$$

I use the formula FeS^0 to represent the monomeric form of the aqueous FeS clusters, and FeS_{aq} to refer to the aqueous FeS in general. Likewise, FeS_s refers to any particulate FeS phase, including phases which have previously been described as amorphous FeS, disordered mackinawite, mackinawite, and nanoparticulate FeS.

The relative abundance of FeS_{aq} in aqueous solutions closely approximates the equilibrium values. The equilibrium reaction (Rickard et al., 2006) is described by Equation 10.12.

$$FeS^0 + H^+ = Fe^{2+} + HS^- \qquad \log K = 2.2 \qquad\qquad 10.12$$

A significant FeS_{aq} abundance might be defined as where $\{FeS^0\} > \{Fe^{2+}\}$, that is, where $\{Fe^{2+}\}/\{FeS^0\} \leq 1$.

Figure 10.2 shows the plot of where the FeS^0 concentration equals the dissolved Fe(II) concentration in terms of pH and the total dissolved sulfide concentration. The total dissolved sulfide concentration is the sum of the H_2S and HS^- concentrations. It is calculated from Equation 10.13.

$$\{\Sigma S\text{-II}\} = \{HS^-\}\{H^+\}/10^{-7} + \{HS^-\} \qquad\qquad 10.13$$

where 10^{-7} is the equilibrium coefficient of reaction 10.14

$$H_2S = HS^- + H^+ \qquad\qquad 10.14$$

The activities are converted to approximate concentrations assuming that the activity coefficients $\gamma H_2S = \gamma HS^- \simeq 1$. The actual values in seawater are probably 1.03 for H_2S and 0.74 for HS^- (Millero, 1986; Rickard and Luther, 2006b). The approximation of unit activity then introduces an error of around 25% in the concentration estimate for seawater, but this is a reasonable uncertainty for sulfide chemical analyses in a natural solution. The error is further ameliorated by the

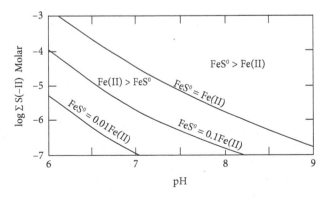

Figure 10.2. Locus of $FeS^0_{aq} = Fe(II)_{aq}$ in terms of $\log \Sigma$ S(-II) (Molar) and pH showing the broad regimes of dominance of FeS_{aq} even at moderate total dissolved sulfide concentrations.

logarithmic scale, of course. Similar errors are intrinsic in the $FeS^0/Fe(II)$ computation. The activity coefficient for FeS^0 is expected to approximate unity as a neutral species, but the activity coefficient of aqueous $Fe(II)$ in seawater is relatively uncertain and may be as low as 0.3 (Rickard and Luther, 2006b). The error of 70% in the assumption of unit activity coefficient for $Fe(II)$ is still within the expected analytic uncertainty for dissolved $Fe(II)$ in sulfidic natural environments.

The results show that FeS^0 becomes the dominant dissolved $Fe(II)$ species at total dissolved sulfide concentrations at the micromolar level in seawater (pH $\simeq 8.2$), at the tens of micromolar level in sulfidic sediments (pH $\simeq 7.5$) and requires millimolar levels of sulfide in freshwater systems (pH $\simeq 6.5$). At pH $\lesssim 6$, FeS^0 has limited stability. Note that FeS^0 constitutes a significant proportion of total dissolved Fe^{2+} (i.e., >1%) at micromolar total dissolved sulfide concentrations at pH >6.5.

FeS_{aq} has been reported from lacustrine, fluvial, wetland, estuarine, and marine sediments, according to the review in Rickard (2012) and from vadose waters, hydrothermal vents, flooded mines, and sewage treatment works, according to Luther et al. (2008) and Hansen et al. (2014). The occurrence of FeS_{aq} may give counterintuitive results for $Fe(II)$ solubility in sulfidic waters. For example, Schlosser et al. (2018) showed that dissolved $Fe(II)$ concentrations actually increased with increasing dissolved H_2S concentrations in upwelling waters of the Peruvian Trench. Schlosser et al's (2018) results are interesting since they showed that the presence of FeS_{aq} in systems where the total dissolved $Fe(II)$ and S(-II) concentrations never reached the mackinawite solubility product. As expected, FeS_{aq} was present in the absence of mackinawite.

10.2.4. Protonated Iron Sulfide Complexes as Reactants

Aqueous FeS clusters dissociate in acid solutions to form Fe^{2+} and H_2S (Equation 10.15).

$$FeS^0 + 2H^+ = Fe^{2+} + H_2S \qquad \log K = 9.2 \qquad \qquad 10.15$$

Where FeS_{aq} is written as the monomer, FeS^0, the acid dissociation equilibrium constant is $10^{9.2}$ (Rickard, 2006). Rickard and Luther (2007) surveyed the literature on the reported stability of aqueous $Fe(II)$ sulfide complexes and concluded that only $FeHS^+$ was likely to be significant in natural solutions. They noted that evidence for higher order iron sulfide complexes, although probable chemically, had only been provided by curve fitting during titration analyses and, unlike $FeSH^+$, these had not been detected voltammetrically.

$$Fe^{2+} + HS^- = FeSH^+ \qquad\qquad 10.16$$

The formation constant for $FeSH^+$ (Equation 10.16) of $10^{5.2}$ is derived from the analysis of published equilibrium constants by Rickard and Luther (2006a, 2007) and is consistent with experimental observations by Luther et al. (1996), which showed the presence of $FeSH^+$ in aqueous solutions down to pH around 5.

The relative distributions of FeS_{aq} and $FeSH^+$ are shown in Figure 10.3. The data show that $FeSH^+$ is likely to dominate sulfidic solutions in acidic solutions. The reaction between $FeSH^+$ and HS_2^- (Equation 10.17) to form pyrite has an equilibrium constant of $10^{15.2}$.

$$FeSH^+ + HS_2^- = FeS_2 + H_2S \qquad\qquad 10.17$$

This compares with the equilibrium constant of 10^{16} for the reaction with FeS_{aq}. Within the errors of the measurements of the equilibrium constants, this is a small difference. The major divergence experimentally in the two systems is simply that the FeS_2 formed in acidic conditions includes progressively more marcasite—the orthorhombic dimorph of pyrite—until, at pH $\lesssim 5$, the product is commonly all marcasite. Although marcasite is metastable with respect to pyrite, it can exist for millions of years at STP without equilibrating.

Microscopic aggregates of marcasite, which may at first glance look similar to framboids, are readily synthesized in acid solutions. Figure 10.4 shows that the orthorhombic marcasite microcrystals fail to form framboids primarily because

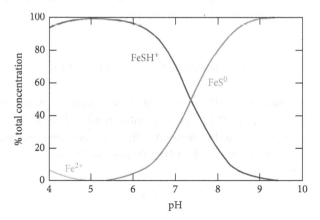

Figure 10.3. Relative distribution of FeS_{aq} and $FeSH^+$ in terms of pH.

Reprinted with permission from Rickard, D. & Luther, G. W. 2007. *Chemistry of Iron Sulfides*. *Chemical Reviews*. Fig. 21. Copyright (2007) American Chemical Society.

Figure 10.4. Marcasite pseudoframboids: (a) general view showing occasional apparent spheroidicity; (b) detailed view showing microcrystalline aggregates. Scanning electron micrograph images by Ian Butler.

of their elongated habit rather than the equant microcrystals required for the framboid texture. Otherwise the chemistry of marcasite formation appears to be similar to that of pyrite, with similarly low solubility (e.g., Schoonen and Barnes, 1991b) leading to large supersaturations prior to nucleation. Syntheses of marcasite require pH <5 and the rate of formation becomes rapid at pH <4 (e.g., Allen et al., 1912; Schoonen and Barnes, 1991a). Various theories have been proposed to explain why marcasite forms in acidic solutions rather than pyrite (e.g., Kitchaev and Ceder, 2016; Luther et al., 2003; Murowchick and Barnes, 1986). The important aspect of this from the point of view of framboids is that framboids are unlikely to form in more acidic solutions, where the dominant iron disulfide phase is marcasite rather than pyrite.

10.2.5. Aqueous Fe(II) as a Reactant

The dominant aqueous Fe(II) species in most natural aqueous solutions is hexaaquairon(II), $[Fe(H_2O)_6]^{2+}$, conveniently rendered as Fe_{aq}^{2+}. A balanced pyrite-forming reaction can be written for the reaction between aqueous Fe(II) and a dissolved S(-II) species such as HS_2^- (Equation 10.18).

$$Fe_{aq}^{2+} + HS_2^- = FeS_2 + H^+ \qquad 10.18$$

The equilibrium constant of $10^{13.84}$ for Equation 10.18 is derived from thermodynamic data listed in Rickard and Luther (2007). The large equilibrium constant suggests that this reaction is thermodynamically very favorable. At pH = 7, $\{Fe^{2+}_{aq}\}\{HS_2^-\}$ is $10^{-20.9}$. $\{HS_2^-\}$ can be directly related to $\{HS^-\}$ through reaction 10.19.

$$S^0 + HS^- = HS_2^- \qquad\qquad 10.19$$

S^0 represents stable orthorhombic sulfur ($1/8S_8$) and the equilibrium constant of $10^{-1.76}$ for this reaction is from Kamyshny et al. (2004). Substituting other sulfur forms makes no significant difference overall to this discussion. Combining reactions 10.18 and 10.19 produces Equation 10.20 for the overall pyrite formation reaction.

$$Fe^{2+}_{aq} + HS^- + S^0 = FeS_2 + H^+ \qquad\qquad 10.20$$

The equilibrium constant of $10^{12.08}$ for Equation 10.20 suggests that pyrite will be precipitated through reaction with dissolved Fe(II) and S(-II) concentrations less than 1mM.

The problem is that experimentally the reaction between aqueous Fe(II) and S_2(-II) does not readily produce pyrite partly because this involves a change in the spin state of the Fe from high spin in aqueous Fe(II) to low spin in pyrite. Instead, high spin aqueous Fe(II) is immediately precipitated as FeS, in which the Fe remains in the high spin configuration, and which rapidly begins to develop the structural characteristics of mackinawite, FeS_m (Equation 10.21).

$$Fe^{2+} + HS_2^- = FeS_m + S^0 + H^+ \quad \log K = 5.3 \qquad\qquad 10.21$$

The reaction to produce FeS_m is thermodynamically favored but obviously far less so than the equivalent reaction for the formation of the stable phase pyrite (Reaction 10.20).

An additional, and more prosaic, reason for the formation of FeS instead of FeS_2 is simply that, in aqueous polysulfide solutions, S(-II) is necessarily present and is the dominant dissolved sulfur species. For example, we can write three reactions describing S_2(-II) equilibria in aqueous solutions at SATP:

$$\frac{1}{8}S_8 + HS^- = S_2^{2-} + H^+ \quad \log K = -11.46 \qquad\qquad 10.22$$

$$\frac{1}{8}S_8 + HS^- = HS_2^- \quad \log K = -1.76 \qquad\qquad 10.23$$

$$\frac{1}{8} S_8 + HS^- + H^+ = H_2S_2 \qquad \log K = 3.24 \qquad\qquad 10.24$$

Simple inspection of reactions 10.22–10.24 shows that in natural solutions where HS^- is the dominant S(-II) species, the dominant S_2(-II) species, HS_2^-, makes up less than 2% of the total dissolved sulfide. In acidic solutions where H_2S is the dominant S(-II) species, H_2S_2 becomes the dominant species at pH less than 4. In other words, in all natural aqueous solutions ranging in pH from 4 to 11, the disulfide ion only exists in the presence of excess sulfide. The kinetics of the reaction between aqueous Fe(II) and S(-II) have been determined (Rickard 1995) and show that FeS forms in less than 1 ms at SATP. In other words, FeS will be formed almost instantaneously in an aqueous Fe (II) and S_2(-II) solution, whereas it may take some time to build up the extreme supersaturations required for pyrite, and especially framboid, nucleation (see section 11.3.1 of Chapter 11).

However, it does not answer the question of whether the direct reaction between Fe(II) and S_2(-II) in aqueous solutions to form pyrite is possible, or whether it always leads to the dissociation of S_2(-II) and the formation of FeS. The direct reaction seems to be favored by Peiffer et al. (2015), although the detailed mechanism has not been established. The thermodynamics obviously strongly favor pyrite formation for this reaction. The equilibrium constant for the pyrite formation reaction 10.18 is $10^{13.8}$, suggesting that pyrite formation is favored in all aqueous solutions at SATP. By contrast, the thermodynamics of the same reaction, written for FeS^0 formation (Equation 10.7), show that this reaction is less favored and is highly dependent on solution conditions, especially pH. On the other hand, the kinetics do not favor direct pyrite formation because of the change in Fe(II) spin state.

Further insights into the direct reaction between Fe(II) and S_2(-II) have been reported by researchers synthesizing pyrite nanoparticles for the electronics industry. A variety of organic reactants have been investigated where the S_2(-II) is attached to organic radicals. For example, Rhodes et al. (2017) investigated a large number of organosulfur reagents and found that pyrite was only formed directly, without an FeS intermediate, with diallyl disulfide, 4,5-dithia-1,7-octadiene, which displays the weakest C-S bond strength. In all other cases, the S-S bond was cleaved by the incoming Fe(II) ion and FeS or Fe_3S_4 were formed instead of FeS_2. The S-S bond dissociation energy is calculated to be around 265 kJ mol^{-1} (Rhodes et al., 2017; Steudel, 1975), which makes it one of the strongest common homonuclear bonds, only exceeded by H-H and C-C single bonds. Even so, the C-S bond strength in diallyl disulfide is lower than the S-S bond strength, permitting the release of persulfide, S_2(-II), rather than the dissociation of the S-S moiety into organic sulfides, R-S. The implication here is then that the

S-S bond in $S_2(-II)$ is normally split by reaction with Fe(-II) to produce FeS as an aqueous, surface, or nanoparticulate phase, and sulfur. This is reminiscent of the result of Rickard et al. (2001), who demonstrated that aldehydic carbonyl inhibited aqueous FeS formation and pyrite did not form.

The equilibrium constant for the reaction with FeS^0 is about two magnitudes greater than that for Fe^{2+}. This shows that under similar reaction conditions, the pyrite formation via the reaction of polysulfide with the Fe(II) bonded to S is thermodynamically preferred over the direct reaction with aqueous Fe^{2+}. Chadwell et al. (1999, 2001) reported tetrasulfide and pentasulfide complexes of Fe(II), FeS_4, FeS_5, $Fe_2S_4^{2+}$ and $Fe_2S_5^{2+}$. They showed that the mononuclear forms had negligible stabilities in aqueous solutions at SATP, whereas the polynuclear forms could have a significant stability. The data might suggest that the reaction between aqueous Fe^{2+} and S_2^{2-} is likely to produce an Fe_2S_2 complex rather than pyrite.

10.2.6. Iron Sulfide Minerals as Reactants

R. A. Berner originally proposed that FeS was involved in pyrite formation in sediments (Berner, 1970). In fact, many of the reaction schemes advanced for pyrite formation by Berner (1970) presaged the various mechanisms for pyrite formation that were later discovered. By contrast, some later researchers suggested that two metastable iron sulfide minerals, mackinawite (tetragonal FeS_m) and greigite (cubic Fe_3S_{4g}), were essential reactants for pyrite formation. These workers believed that pyrite could not form without the intervention of these phases, and they called them *necessary precursors*. These minerals are sometimes mistakenly referred to as "monosulfides" by soil scientists (e.g., Bush et al., 2004). The sequence was applied to pyrite framboid formation by Wilkin and Barnes (1997). This process was suggested since (1) FeS_m precipitated rapidly from aqueous solutions; (2) the only route for Fe_3S_{4g} formation was oxidation of FeS_m; and (3) Fe_3S_{4g} is metastable with respect to pyrite and transforms rapidly to pyrite and pyrrhotite above 200°C. The process was implicated in framboid formation since (1) greigite framboids had apparently been identified (see section 8.2.2 of Chapter 8); (2) magnetic pyrite framboids had been reported, and these were assumed to contain relic Fe_3S_{4g}; and (3) the organization of framboidal microcrystals was assumed to require a magnetic force, such as might be provided by magnesioferrite ($MgFe_2O_4$) or greigite, the attractive van der Waals forces being thought to be insufficient to account for the aggregation process (Taylor, 1982; Wilkin and Barnes, 1997).

As pointed out by Rickard and Luther (2007), the presence of FeS_m in experimental pyrite-forming systems is neither happenstance nor because of some requirement for FeS_m as a precursor in pyrite formation. The reason is simply that, in order to produce measurable quantities of pyrite, especially in conventional batch systems, the dissolved iron and sulfide concentrations need to be so high that nanoparticulate FeS_s which has some properties similar to the mineral mackinawite, precipitates. However, as soon as pyrite starts to nucleate, the solubility product of mackinawite is breached and nanoparticulate FeS_s rapidly dissolves (cf. Pankow and Morgan, 1979; Rickard et al. 2007).

The key properties of mackinawite, greigite, and pyrite are summarized in Table 10.1. The transformation from mackinawite through greigite to pyrite requires that two-thirds of the Fe(II) is oxidized to Fe(III) while the S(-II) remains reduced, and then the Fe(III) is reduced to Fe(II) while the S(-II) is oxidized to S_2(-II). At the same time, there is a spin transition between the high-spin Fe in mackinawite and greigite and the low-spin Fe in pyrite. In addition, although there is structural homology between mackinawite and greigite, no such structural conformability exists between greigite and pyrite. This is consistent with widespread experimental studies which demonstrated that magnetic precursors were not necessary for framboid formation (e.g., Butler and Rickard, 2000b; Graham and Ohmoto, 1994; Sunagawa, 1993).

Apart from being highly improbable and chemically inelegant, many experimental observations demonstrate that this mackinawite-greigite-pyrite sequence of transformations does not occur in pyrite formation and is not compatible with the characteristics of framboids (see Rickard and Luther, 2007, for review). Additionally, observations from the natural environment are also incompatible with this scheme. Even Large et al. (2001), who assumed that the iron sulfide nanoclusters they described were constituted by greigite and that they grew into pyrite framboids, found considerable problems with this theory.

Table 10.1 Comparison of Chemical and Structural Characteristics of Pyrite, Mackinawite, and Greigite

	Mackinawite	Greigite	Pyrite
Composition	FeS	Fe_3S_4	FeS_2
Structure	Tetragonal $P4/nmm$	Cubic $Fd3m$	Cubic $Pa3$
	Fe sheets with S intercalations	Inverse spinel	Low symmetry cubic
Fe oxidation state	2+	$1 \times 2+, 2 \times 3+$	2+
Fe spin state	high spin	high spin	low spin
S oxidation state	2–	2–	1–

The general problems with the idea that framboids grew from mackinawite and greigite precursors include the distribution of FeS_m and Fe_3S_{4g} in natural systems. Mackinawite is rarely observed in marine sediments where pyrite and framboids are abundant (e.g., Rickard and Morse, 2005). It is possible that it was present but has subsequently dissolved, but the lack of any evidence for its presence makes the idea reminiscent of the arguments for the phlogiston theory. Greigite is almost exclusively found in freshwater sediments and not in marine sediments. One reason is that its metastability range is limited to more acid solutions (Rickard and Luther, 2007). Where mackinawite and greigite are observed in association with pyrite, it is often the case that the formation of these minerals does not precede pyrite formation but occurs after pyrite formation (e.g., Burton et al., 2011). As shown by Rickard and Morse (2005), evidence for the presence of mackinawite and greigite in marine sediments is rare, and assumptions that the presence of pyrite is evidence in itself for their presence is false. Indeed, Howarth (1979) wrote that pyrite forms rapidly in wetland sediments only where iron monosulfides are undersaturated. This conclusion has been widely validated in soils and wetland sediments (e.g., Burton et al., 2011; Ferreira et al., 2007; Huerta-Diaz et al., 1998; Noel et al., 2014).

The other environmental problem is, of course, that pyrite and pyrite framboids form at high temperatures where mackinawite, and consequently greigite, are not formed. Mackinawite is not precipitated from aqueous solutions much above 100°C and greigite is limited to less than about 200°C. The situation is complicated by the fact that both mackinawite and greigite can be used as reactants in the synthesis of pyrite. Both minerals include surface =FeS moieties which facilitate pyrite formation (section 11.5.1). However, as pointed out by Rickard and Luther (2007), otherwise there is no particular difference between these minerals as reactants in pyrite formation and a host of other Fe minerals such as iron (oxyhydr)oxides which, in sulfidic environments, produce surface =FeS. Indeed, the lists of reactants used experimentally for pyrite synthesis emphasizes this point (e.g., Ohfuji and Rickard, 2005; Rickard, 2012; Rickard and Luther, 2007; Xian et al., 2016).

Crystal growth of pyrite commonly occurs via the addition of monomers to the surface (section 12.2 of Chapter 12), and any theory of pyrite formation chemistry must also account for the rapid growth of pyrite crystals. As discussed in section 12.2 of Chapter 12, crystal growth mainly takes place at molecular defect sites on the pyrite surface as, for example, at steps, dislocations, vertices, and edges. One of the characteristics of these sites on the pyrite surface is the relative concentration of sulfide monomers compared to conventional stoichiometric disulfide sites (e.g., Murphy and Strongin, 2009). The polysulfide reaction and H_2S mechanisms (sections 10.2.2 and 10.3.1) which show that pyrite can be formed through the reactions of surface =FeS sites therefore also explain how pyrite crystals grow (section 12.2 of Chapter 12).

10.2.7. Nanoparticulate FeS

The first formed precipitate through the reaction between aqueous Fe(II) and S(-II) is a nanoparticulate phase which has been variously described as mackinawite (FeS$_m$), amorphous FeS, (FeS$_{am}$), disordered mackinawite, and nanoparticulate FeS (FeS$_{nano}$). I use the abbreviation FeS$_s$ to distinguish this particulate phase from aqueous FeS clusters with no a priori assumptions as to its exact structure or composition. It differs from the more crystalline material in being electroactive. This makes it difficult to distinguish from aqueous clusters with current technical methods. For example, the size of the most abundant aqueous FeS cluster, Fe$_2$S$_2$.4H$_2$O, is about 0.5 nm or toward the lower end of the size range of FeS$_s$ nanoparticles (Rickard and Luther, 2007).

Matamoros-Veloza et al. (2018) and Ohfuji and Rickard (2006) confirmed Wolthers et al.'s (2003) original report that the precipitated FeS phase is around 2 nm in size when first probed. The first observed 2 nm FeS$_s$ phase contains about 150 FeS molecules (Ohfuji and Rickard, 2006). However, this FeS$_s$ nanoparticle size is observed after at least 20 minutes of FeS$_s$ crystal growth (Ohfuji and Rickard, 2006) and nanoparticulate FeS$_s$ precipitates in less than 1 ms (Rickard, 1995). The initial size of the FeS$_s$ nanoparticles is therefore likely to be less than 2 nm. The limiting size of an FeS$_s$ nanoparticle is probably around 1 nm, which would contain around 20 FeS molecules. The first observed structure shows FeS layers separated by 0.5–0.7 nm. This would suggest that the 2 nm particles are close to the minimum size for a particle to display this structure. Smaller particles would not contain enough FeS molecules to form separate layers, which characterize even the embryonic FeS$_m$ structure. The conclusion is that the first formed FeS$_s$ precipitate cannot display an approach to an infinite lattice which defines mineral crystal structures, such as mackinawite.

Some limitations on the chemical properties of FeS$_s$ can be obtained by examining the data for the crystalline phase, mackinawite, FeS$_m$. Since FeS$_m$ is both larger and more crystalline than FeS$_s$, it is more stable and it provides a lower limit to the solubility of FeS$_s$. Pankow and Morgan (1979) found that, in the pH-independent regime for FeS$_m$, dissolution, the kinetics could be described by Equation 10.25.

$$\frac{-dFeS_m}{dt} = k_2 \qquad\qquad 10.25$$

where the rate constant, k_2, is 3×10^{-7} mol m^{-2} s^{-1}. Rickard and Luther (2007) calculated that this would be equivalent to around 10^{-4} mol s^{-1} using Ohfuji and Rickard's (2006) estimate of the specific surface area for FeS$_m$. The rate of FeS$_m$ dissolution formation is very fast since this rate suggests that 2 nm diameter FeS$_m$ particles dissolve to form FeS$_{aq}$ in less than a picosecond. The rate of dissolution of smaller nanoparticulate FeS$_s$ is likely to be even greater than this.

As noted earlier, the dissolution of FeS$_s$ is congruent at pH >5.3 (Equation 10.26).

$$FeS_s = FeS_{aq} \qquad\qquad 10.26$$

The net result of solutions where the {FeII}{S-II} ion activity product exceeds the solubility product of FeS$_s$ is a soup of FeS$_s$ nanoparticles and aqueous clusters as the FeS$_s$ nanoparticles flicker in and out of existence in response to small variations in the saturation level.

10.3. The H$_2$S Reaction

Pyrite framboids have been synthesized by the reaction between H$_2$S and FeS (Butler and Rickard, 2000a):

$$FeS + H_2S = FeS_2 + H_2 \qquad\qquad 10.27$$

The overall reaction (Equation 10.27) is thermodynamically favorable at low temperatures ($\Delta G_r^\circ = -29$ kJ mol^{-1}, at 25°C and 100 kPa total pressure, where FeS = FeS$_m$). In this reaction, H$_2$S acts as an oxidizing agent with respect to FeS. Butler et al. (2004) confirmed this mechanism isotopically. They demonstrated that the δ^{34}S of the product FeS$_2$ from the reaction equaled a 1:1 mixture of the δ^{34}S of the reactant FeS and H$_2$S (Equation 10.28).

$$Fe^{34}S + H_2{}^{32}S = Fe^{34}S^{32}S + H_2 \qquad\qquad 10.28$$

The idea that H$_2$S is a good oxidizing agent, potentially more potent than O$_2$ itself, appeared counterintuitive to geochemists raised on the notion that sulfidic systems are synonymous with ill-defined reducing environments. The lowest unoccupied molecular orbital (LUMO) for H$_2$S is around -1.1 eV and therefore

H_2S is a better electron acceptor (i.e., oxidizing agent) than O_2 where the LUMO is only -0.47 eV (Drzaic et al., 1984; Rickard and Luther, 1997b). The reaction of FeS with H_2S to form pyrite stems from an original observation by Berzelius (1845) which was revisited by Huber and Wächtershauser (1997), Rickard and Luther (1997a, 1997b), Wikjord et al. (1976), and, inadvertently, by Wilkin and Barnes (1996) in their "iron-loss pathway" (see Butler et al., 2004).

10.3.1. The H_2S Reaction Mechanism

The mechanism of the H_2S pathway was determined from the kinetics by Rickard and Luther (1997b). The mechanism involves the formation of an inner-sphere complex between FeS and H_2S, followed by electron transfer between S(-II) and H(I) to produce S2(-II) (Equations 10.29 and 10.30).

$$FeS + H_2S \rightarrow \left[FeS \rightarrow SH_2 \right] \qquad\qquad 10.29$$

$$\left[FeS \rightarrow SH_2 \right] \rightarrow FeS_2 + H_2 \qquad\qquad 10.30$$

Again, FeS refers to both aqueous FeS species and the surface =FeS moiety, and the mechanism explains both pyrite nucleation and crystal growth chemistry. The reaction has been proven independently by the isotopic mass balance study by Butler et al. (2004) and the microbial study by Thiel et al. (2019). H_2S becomes the dominant dissolved sulfide species in acidic solutions where pH <7. Under these conditions, both particulate and aqueous FeS become less stable and $FeSH^+$ becomes the dominant species (section 10.2.4). There is a narrow window down to pH 6, where FeS_{aq} has a significant presence (Figure 10.3).

$$FeSH^+ + H_2S = FeS_2 + H_2 + H^+ \quad \log K = 13.9 \qquad\qquad 10.31$$

The equilibrium constant for the reaction of H_2S with $FeSH^+$ (Equation 10.31) is pH dependent. At pH = 7, log K is 6.9, and at pH = 6 it is 7.9. Thus, although log K for the overall reaction appears not to be significantly different from the equilibrium constant for the polysulfide reaction with $FeSH^+$ (Equation 10.17), in reality it is far lower. The conclusion is that the reaction between H_2S and $FeSH^+$ to form pyrite is less significant than the reaction between H_2S and FeS_{aq} in natural systems. At pH values where it might become significant, marcasite forms in place of pyrite and the framboid texture is not developed (section 10.2.4).

This H_2S mechanism is important since S(-II) species are involved in all reported aqueous syntheses of pyrite. However, it is likely to dominate framboidal pyrite formation at pH around 7 and below. H_2S is less nucleophilic than HS^- since the highest molecular orbital (HOMO) for H_2S is relatively stable (ca -10 eV) (Drzaic et al., 1984; Rickard and Luther, 1997b). H_2S can act as an oxidizing reagent with respect to FeS; HS^- cannot. The rate is thus directly dependent on the H_2S concentration (Rickard, 1997; Rickard and Luther, 1997a). Since pK_1 for H_2S is close to 7, the relative concentration of H_2S declines logarithmically at pH >7. Thus, at pH 6, H_2S makes up ca. 90% of the total dissolved S(-II) whereas, in normal seawater at pH 8.1, H_2S makes up only ca. 9% of the total dissolved sulfide.

10.3.2. Microbial Mediation

One of the problems discussed by Rickard and Luther in their original study of the kinetics and mechanism of the H_2S reaction was that stoichiometric amounts of H_2 were not recovered. Thiel et al. (2019) showed that lithotrophic microorganisms mediated the reaction if metabolically coupled to methane-producing bacteria. The coupled reactions are shown in Equations 10.32 and 10.33 and the net reaction balance in Equation 10.34.

$$4FeS + 4H_2S = 4FeS_2 + 4H_2 \quad \Delta G^\circ_r = -41\,kJ\,mol^{-1} \qquad 10.32$$

$$CO_2 + 4H_2 = CH_4 + 2H_2O \quad \Delta G^\circ_r = -131\,kJ\,mol^{-1} \qquad 10.33$$

$$4FeS + 4H_2S + CO_2 = 4FeS_2 + CH_4 + 2H_2O \quad \Delta G^\circ_r = -295\,kJ\,mol^{-1} \quad 10.34$$

In this reaction sequence (Equations 10.32–10.34), the H_2 produced by the pyrite-forming reaction 10.32 is utilized by methanogens to produce methane in reaction 10.33. The effect is to enhance pyrite formation in reaction 10.32 by the Le Chatelier effect. The net process (Equation 10.34) is energetically extremely favorable. Coupling of pyrite formation to methanogenesis was proposed by Holmkvist et al. (2011) to be a critical part of the deep biosphere. The observations of Lin et al. (2016) that framboidal pyrite formation was enhanced within the sulfate-methane transition zone of marine sediments are in accordance with this process. The mechanism was also the basis for Huber and Wächtershauser's (1997) original suggestion as the basis for autocatalytic metabolism and the origin of life.

10.3.3. Iron (III) (oxyhydr)oxide

The presence of solid reactants in framboid formation is precluded by the requirement for simultaneous nucleation of pyrite microcrystals throughout the framboid volume. Thus, although the reaction between iron (III) (oxyhydr)-oxides and sulfide may be interesting for pyrite formation in general, it is not directly related to framboid formation. The kinetics and mechanism of the reaction between goethite (α- FeOOH) and sulfide in aqueous solutions were first determined by Rickard (1974) and subsequently refined by Pyzik and Sommer (1981), Dos Santos Afonso and Stumm (1992), Yao and Millero (1996), and Kumar et al. (2018). The products of this reaction are FeS and colloidal sulfur. The rate is dependent on pH, the total dissolved sulfide, temperature, and the surface area of the goethite, and the kinetics are consistent with the formation of a sulfide complex at the (oxyhydr)oxide surface, followed by electron transfer between the sulfide and Fe(III).

Peiffer and coworkers have expanded the experimental study to include lep-idocrocite (γ-FeOOH) and ferrihydrites (perhaps $Fe_{10}O_{14}(OH)_2$) (e.g., Hellige et al., 2012; Peiffer et al., 2015; Wan et al., 2017). They showed that one of the reaction products was Fe(II) and that this could enter into solution, providing an aqueous Fe(II) source for pyrite formation. The results further illuminate potential processes behind the reactive iron shuttle that transports iron from continental margins to deeper parts of the basin (e.g., Canfield et al., 1996; Lyons and Severmann, 2006). As such it is indirectly significant for pyrite framboid formation sediments since it describes a process for the transport of dissolved Fe(II) to the site of framboid formation, which is a necessary attribute of the framboid nucleation and crystal growth processes (Chapters 11 and 12). However, it is probably more directly responsible for the formation of pyrite single crystals, as discussed in section 11.5.1. The reaction of HS^- with Fe(III) (oxyhydr)oxides is thought to be a major source of polysulfide formation in sediments (Avetisyan et al., 2019) and disulfide is formed during this reaction, bonded to the Fe(III) (oxyhydr)oxide surface (Wan et al., 2014).

10.4. Pyrite Framboid Syntheses

The first experimental synthesis of pyrite framboids was reported by Rickard (1966), who described large synthetic pyrite framboids (<100 μm) formed on an iron plate which had been exposed to the actions of sulfate-reducing bacteria at ambient temperatures in aqueous solution (Figure 10.5).

These first synthetic framboids were defined with an optical stereomicro-scope and thus the resolution is limited (Figure 10.5). The framboids were small

Figure 10.5. Optical microscopic image of first recorded synthesis of framboids (Rickard, 1966).

spheroids of pyrite and they can be seen to be constituted by individual pyrite microcrystals and thus they were true framboids. The pyrite was defined by XRD and electron microprobe analysis. The chemistry of the synthesis was complex, involving the obligate anerobic, sulfate-reducing bacteria, a chemically defined medium, and metallic iron. However, Rickard's (1966) analysis suggests that they were formed by the polysulfide pathway. The synthesis is interesting since it is the only reported synthesis of framboids using defined sulfate-reducing bacteria— the major sulfide providers in sedimentary pyrite framboid formation.

A possible further experiment involving sulfate-reducing bacteria to produce pyrite framboids was reported by Vietti et al. (2015). This experiment added de-fleshed domestic pig bones to samples of natural muds and seawater that had been incubated anaerobically for two weeks to simulate marine carcass falls. After 1 week, pyrite framboids were discovered on the bones. The sizes of the framboids after three weeks were around 4 μm and their constituent microcrystals were 0.7 μm in size on average (Figure 10.6). They noted that the framboids formed on the bone were not distinguishable from framboids already in the sediment. The experimental conditions, including the oxic overlying water and the presence of sulfur bacteria in the sediment biome, suggests that these framboids were formed by the polysulfide pathway. An unremarked image shown by Vietti et al. shows what appears to be framboids with hollow microcrystals (2015, Figure 1d). The authors merely comment that this framboid had anhedral microcrystals. However, it is similar to the framboids with hollow microcrystals described in section 4.5.6 of Chapter 4.

Figure 10.6. Framboids formed on pig's bone in a simulated marine carcass fall experiment.

Scanning electron micrograph images by Laura Vietti.

Experimental syntheses of pyrite framboids were reviewed by Ohfuji and Rickard (2005). They found 11 reports of apparent framboid syntheses but that many of these actually described pyrite forms which are not strictly framboidal, according to the definition of framboids (see section 1.1 of Chapter 1). Since that time there has been an explosion of reports of nano- and micro- pyrite syntheses mainly directed at producing controlled pyrite morphologies for energy storage, photovoltaic, and environmental protection applications. A review of syntheses of various pyrite morphologies in this context was published by Xian et al. (2016). Surprisingly, perhaps, Xian et al. only reported one synthesis of framboids (Wu et al., 2004), although they listed six reports of the synthesis of pyrite microspheres.

In the present discussion, I have divided these synthetic forms into framboids, micro- and nano- framboids, microspheres, and microcrystal clusters.

Although apparent framboids had been synthesized earlier (e.g., Berner, 1969; Rickard, 1966; Sunagawa et al., 1971), the first clearly defined synthetic framboids were reported by Sweeney and Kaplan (1973). Not only were the synthetic framboids well defined (Figure 10.7), they were also large, normally up to

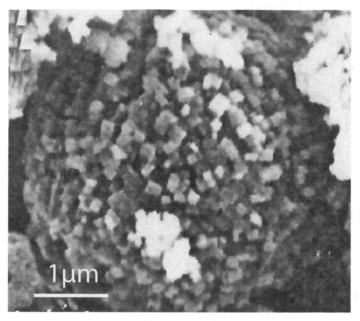

Figure 10.7. Framboid synthesized by Sweeney and Kaplan (1973).
Image reprinted by courtesy of the Society of Economic Geologists.

10 μm in diameter, but with occasional framboids up to 50 μm in diameter. The framboids were characterized by cubic microcrystals and were not ordered. The report was important in the history of the study of framboids since it confirmed the proposition by Sunagawa et al. (1971) that framboids could be formed inorganically without the intervention of bacteria or organic matter. The problem has been that since this initial report, no one has been able to reproduce the Sweeney and Kaplan (1973) synthesis, although many groups made extensive attempts.

Sunagawa et al. (1971) had earlier synthesized framboid-like pyrite forms at 250°C after 20 hours (Figure 10.8). They reported that the smooth spheres of pyrite were constituted by uniform microcrystals 0.5–1.0 μ m in size, although these cannot be discerned in the published images. They were equivocal about these being true framboids (e.g., 1971, p. 13), notwithstanding the title of their paper. However, their synthesis was more or less repeated by Graham and Ohmoto (1994), who reported large, ordered pyrite framboids (Figure 10.9).

Graham and Ohmoto (1994) synthesized framboids at high temperatures (200°–350°C) in hydrothermal bombs. Their system included an excess of rhombic sulfur, which took a globular form at these temperatures. The synthesis is interesting since these are the only synthesized framboids to display organized microcrystals (Figure 10.9). The microcrystals are arranged in a hexagonal close

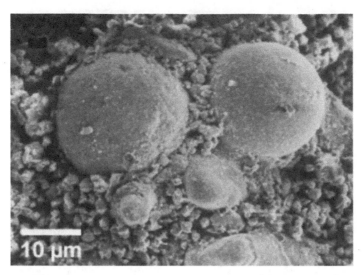

Figure 10.8. Framboids (?) synthesized hydrothermally by Sunagawa et al. (1971). Reproduced courtesy of The Society for Resource Geology.

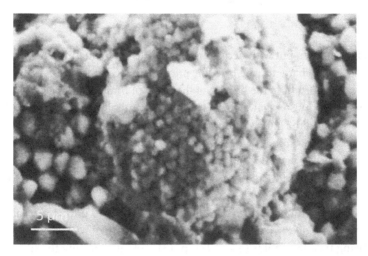

Figure 10.9. Framboid synthesized at >250ºC by Graham and Ohmoto (1994), showing organized microcrystals.

From Graham, U. M., and Ohmoto, H., 1994. Experimental-study of formation mechanisms of hydrothermal pyrite. *Geochimica et Cosmochimica Acta*, 58(10): 2187–2202, figure 10d. Image reprinted by courtesy of the Society of Economic Geologists.

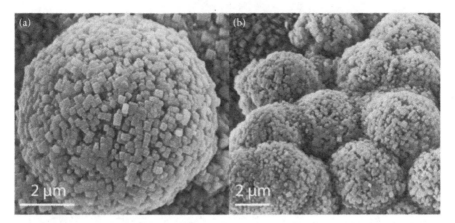

Figure 10.10. Framboids synthesized with the H₂S reaction: (a) individual framboids showing cubic crystals (b) cluster of framboids.
Scanning electron micrograph images by Ian Butler.

packed array reminiscent of the icosahedral framboids described in section 3.3 of Chapter 3. This result may be significant since, as discussed in section 13.3.2 of Chapter 13, framboid-style microcrystal organization is entropy-driven and can be facilitated by confinement of microcrystals in globular droplets.

The synthesis of large framboids was reported by Butler et al. (2000). The synthesized framboids were also constituted by cubic microcrystals and were disordered (Figure 10.10). Even clumps of framboids were synthesized (Figure 10.10). However, by contrast with the Sweeney and Kaplan synthesis which followed the polysulfide pathway for pyrite formation, the Butler et al. framboids were synthesized by the H₂S pathway in an anoxic, continuous flow chemostat under rigorous chemical conditions (Butler and Rickard, 2003; Wolthers et al., 2006). The Butler et al. (2000) syntheses demonstrated that framboid formation was dependent on hydrodynamic factors. In particular, framboids formed in stagnant zones in the system where the major nutrient transport was diffusion-controlled. This is consistent with framboids being formed abundantly in fine-grained sediments where advective flow is limited.

The Sweeney and Kaplan (1973) and Butler et al. (2000) syntheses also demonstrate that framboid formation is apparently independent of the pyrite formation pathway, since both the polysulfide pathway and H₂S pathway produced framboids. However, there is a caveat here. The reactants in pyrite syntheses involving polysulfide include dissolved sulfide and therefore the H₂S pathway cannot be excluded as a potential reaction in these conditions. By contrast, the experiments involving the H₂S pathway excluded S(0) and other oxidizing reagents and thus can be sure to strictly result from the reaction between H_2S and FeS species.

10.4.1. Microframboids and Nanoframboids

Microframboids and nanoframboids describe small, typically <2 μm, spherular pyrite forms which appear to consist of smaller particles. They are discussed in section 2.8 of Chapter 2. Nanoframboids of various metal sulfides were synthesized by Farrand (1970) and are illustrated in Figure 10.11(b). They are similar to the nanoframboids described by Yücel et al. (2011). Butler and Rickard's (2000a) synthesis of microframboids using the H_2S pathway was repeated by Gartman and Luther (2014) under hydrothermal conditions. However, nanoframboids do not appear to have been reported from the geological record. By contrast, microframboids overlap with the smallest sizes of ancient framboids (section 2.4). The point is important since it has been demonstrated that ancient framboids do not show an exponential size distribution which would include framboids with sizes into the nanoparticle range. They display a distinct cutoff in the micrometer range. The conclusion, as reported in section 2.8, is that nanoframboids probably dissolve, oxidize, or crystallize soon (in geological terms) after formation.

These nano- and microframboids have also been called protoframboids (Butler and Rickard, 2000b; Large et al., 2001). It is apparent now (see section 2.8 of Chapter 2) that these small framboids do not grow into larger ones. The absence of a continuous size distribution, discussed in section 2.8,

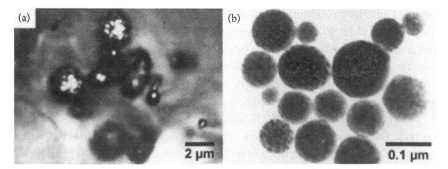

Figure 10.11. Examples of synthesized (a) microframboids and (b) nanoframboids.

(a) Berner, R.A., 1969. Synthesis of framboidal pyrite. *Economic Geology*, 64(4): 383–384, figure 1, by courtesy of the Society of Economic Geologists.

(b) Transmitted electron micrograph from *Mineralium Deposita*, 5(3): 237, figure 6. Framboidal sulphides precipitated synthetically. Farrand, M. J. Copyright (1970) with permission from Elsevier.

proves this. Furthermore, framboid formation is shown in section 12.3 of Chapter 12 to originate within a space similar in size to that occupied by the final framboid. That is, although the microcrystals grow, the framboid diameter does not change significantly with time. This is the basis on which the idea that framboid size can be used as paleoenvironmental indicators is predicated. The conclusion is that these nano- and microframboids are discrete pyrite textures.

10.4.2. Microspheres

Pyrite microspheres are a common experimental reaction product. They are characterized by a spheroidal form but do not appear to be constituted by individual microcrystals or particles. They display massive or radiating internal structures. They are common constituents of ancient deposits particularly in the so-called colloform textures of hydrothermal ore deposits. They are also common in Proterozoic sediments (especially >1.6 Ga), where they appear to replace framboids as the dominant spheroidal pyrite form.

Pyrite microspheres are often difficult to distinguish from framboids, as discussed earlier in the case of the Sunagawa et al. (1971) framboids. Wang and Morse (1996) described an even more misleading texture: pyrite microspheres coated in pyrite microcrystals. The point being that these microspheres appeared to display a framboidal texture when viewed externally. Gartman and Luther (2013) also described these textures and referred to them as pseudoframboids.

Gartman and Luther (2013) described the synthesis of pyrite microspheres with an external "ropy" texture. Close examination of these microspherules shows that this texture was due to the constituent microcrystals being elongated and flattened. They are reminiscent of the marcasite microspheres described in 10.2.3. Gartman and Luther (2013) described 50 nm holes in these microspheres (Figure 10.12(a)). Wu et al. (2004) also produced microspherules of pyrite with holes by a solvothermal method (Figure 10.12(b)). They described the product as *honeycomb pyrite*. Both of these textures were produced by the polysulfide pathway. Although they are not identical to the ancient framboids with hollow microcrystals described in section 4.5.6 of Chapter 4, they may cast some light on the formation processes of this texture. In particular, both syntheses followed the polysulfide pathway.

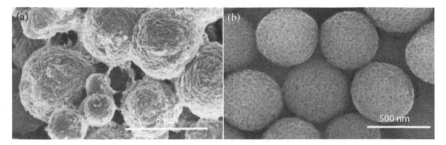

Figure 10.12. Scanning electron microscope images of pyrite microspheres with hollow microcrystals.

(a) From *Geochimica et Cosmochimica Acta*, 120: 447–458, figure 3a. Comparison of pyrite (FeS$_2$) synthesis mechanisms to reproduce natural FeS$_2$ nanoparticles found at hydrothermal vents. Gartman, A., and Luther, G. W., Copyright (2013) with permission from Elsevier.

(b) From *Small*, 10(22): 4754–4759 figure 2a. Synthesis of honeycomb-like mesoporous pyrite FeS$_2$ microspheres as efficient counter electrode in quantum dots sensitized solar cells. Xu, J., Xue, H. T., Yang, X., Wei, H. X., Li, W. Y., Li, Z. P., Zhang, W. J., and Lee, C. S. Courtesy of John Wiley and Sons.

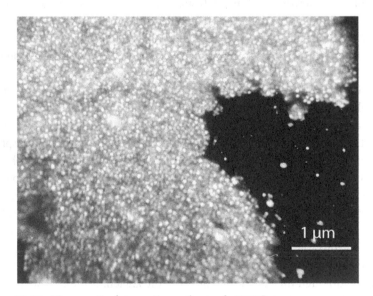

Figure 10.13. Nanoparticulate pyrite synthesized at 25°C.

10.4.3. Nanoparticulate Pyrite and Pyrite Clusters

The most common product of experimental pyrite formation at ambient temperatures is nanoparticulate pyrite, which we colloquially called pyrite dust (Figure 10.13). This is also found abundantly in deep ocean hydrothermal vents

Figure 10.14. Microcrystal clusters are loose aggregates of pyrite microcrystals which may show some degree of sphericity.

(a) From *Geochimica et Cosmochimica Acta*, 55(10): 2839–2849. Pyrite synthesis via polysulfide compounds. Luther, G. W., Copyright (1991) with permission from Elsevier.

(b) from Sunagawa, I., Endo, Y. and Nakai, N., 1971. Hydrothermal synthesis of framboidal pyrite. *Mining Geology* 2: 10–14, courtesy of The Society for Resource Geology.

and sediments where it can show patches of ordered nanocrystals (see section 2.6 of Chapter 2). It does not appear to be reported from the geological record and thus it may dissolve with time, as with the nano- and microframboids.

At the other end of the scale are intergrown pyrite crystals which are abundant in the geological record of course. Between these two extremes, clusters or aggregates of pyrite microcrystals have been frequently reported as experimental products at all temperatures.

These aggregates often display some degree of sphericity (Figure 4.6 in Chapter 4) and are best described as pyrite microcrystal clusters. Similar irregular aggregates of small pyrite crystals are commonly observed associated with pyrite framboids, in particular, in recent sedimentary environments (e.g., Cornwell and Morse, 1987; Jiang et al., 2001; McKay and Longstaffe, 2003). There does appear to be a continuum between framboids and clusters of pyrite microcrystals. The pyrite microcrystal clusters appear to be loosely aggregated pyrite microcrystals which have not formed tight spheroids and do not show any internal organization. In contrast with framboids, there is no requirement for all the microcrystals in a cluster to form at the same time. Indeed, close inspection of many clusters show successive growth of microcrystals. There is therefore a fundamental difference between the formation processes of framboids and clusters, which in turn reflects the differences in the environments in which they were formed. In particular, the hydrodynamic regime which is required for framboid formation (Butler et al., 2000) is not necessary for clusters: that is, framboids form in stagnant, diffusion-dominated systems, whereas clusters evidence advective flow environments.

Table 10.2 Syntheses of Pyrite Framboids and Framboid-Like Textures by the
Polysulfide Reaction

	pH	T °C	Duration	Product	Size (μm)
1.	7.9–8.0	65	2 weeks	Microframboids	1
2.	6–6.5	200–300	5–70 hours	Framboids (?)	<50
3.	—	60–85	9–11 days	Framboids	1–60
4.	6.5–7.5	25	~3 months	Microspheres	10–50
5.	5.5–8	100	4–48 hours	Clusters	5
6.	>3.8	200–350	2 hours–8 weeks	Framboids	2–100
7.	ca. 7	70	125–330 hours	Microspheres, clusters	<7
8.	5–8	25	3–24 months	Microspheres, clusters	10
9.	<7	25	~1 year	Microspheres, clusters	10
10.	6.2–12.5	145	12 hours	Microspheres	3
11.	nd	Ambient	1–3 weeks	Framboids	4
12.	>7–3	90–280	6–48 hours	Framboids	3

nd: not determined

Sources: 1. Berner (1969); 2. Sunagawa et al. (1971); 3. Sweeney and Kaplan (1973); 4. Kribek (1975); 5. Luther (1991); 6. Graham and Ohmoto (1994); 7. Wilkin and Barnes (1996); 8. Wang and Morse (1996); 9. Morse and Wang (1997); 10. Gartman and Luther (2013); 11. Vietti et al. (2015); 12. Wu et al. (2004).

10.5. Comparison of the Polysulfide and H$_2$S Pathways for Framboid Formation

As discussed in section 10.1.2, there are two clearly defined reaction pathways for pyrite formation in aqueous solutions: the polysulfide pathway and the H$_2$S pathway. Gartman and Luther (2013) found that syntheses via the H$_2$S pathway mimicked natural hydrothermal vent nanoparticles more closely than syntheses via the polysulfide pathway. They found nano- and microframboids only in experiments with the H$_2$S pathway. In contrast, several experiments involving the polysulfide pathway produced large framboids (e.g., Graham and Ohmoto, 1994; Sweeney and Kaplan, 1973).

Reported framboid syntheses apparently via the polysulfide pathway are listed in Table 10.2. Where the pathway is not explicitly stated, I have determined the

Table 10.3 Syntheses of Framboids by the H_2S Reaction

	pH	T °C	Duration	Product	Size (µm)
1.	7–8	25	~1 year	Nano- and micro- framboids	0.1–10
2.	6	60–100	5–45 days	Microframboids and clusters	2–5
3.	Not reported	145	4.5–12 hours	Nanoframboids	<0.6
4.	6–7	40	<37 days	Framboids	<8

Sources: 1. Farrand (1970); 2. Butler and Rickard (2000b); 3. Gartman and Luther (2013); 4. Butler et al. (2000).

apparent pathway mainly on the reported presence or absence of elemental or orthorhombic sulfur in the reaction. The syntheses occurred at temperatures from ambient to 350°C, demonstrating that the formation of the framboid texture is independent of temperature. The problem with the apparent polysulfide pathway is that H_2S is always present in the reaction systems. This means that experiments with both S(0) and S(-II) have both $S_2(-II)$ and H_2S present as potential reactants, and it has not been possible to discriminate between the two pathways in these systems.

Reactions with the H_2S pathway (Table 10.3) have the advantage that the reaction only occurs through the reaction between H_2S and FeS and polysulfide (and therefore particulate sulfur) is not involved as a reactant. In the most rigorous experimental system with a continuous flow chemostat (Butler et al., 2000) large, clearly defined framboids were produced. This is the only reported synthesis of large clearly defined framboids with the H_2S reaction. The other syntheses produced micro- and nanoframboids. It is interesting to note, however, that the syntheses of micro- and nanoframboids are not limited to hydrothermal conditions but occur over the range of temperatures from 25°–145°C.

The results suggest, however, that framboid formation is not pathway dependent. This in turn implies that both pathways involve a similar key stage in pyrite formation. This appears to be the formation of aqueous FeS species. The rigorously controlled reaction in a continuous flow chemostat reported by Butler et al. (2000) occurred in conditions which were certainly undersaturated with respect to particulate FeS formation. This means that framboids form from solution. The corollary is that the polysulfide reaction involved in framboid formation follows a similar solution pathway (cf. Luther, 1991). This is consistent with the framboid formation process described in terms of nucleation and crystal growth in Chapters 11 and 12.

11

Nucleation of Framboids

The formation of framboids involved three processes: (a) nucleation, (b) crystal growth, and (c) self-assembly. Nucleation refers specifically to the process of the formation of a new phase. It is discussed in this chapter with particular regard to the formation of pyrite framboids. However, the conclusion regarding the nucleation of pyrite in general also elucidates the processes involved in framboid formation. Nucleation is also used to describe the molecular basis of crystal growth and this is discussed in Chapter 12.

Nucleation in solution can be conveniently divided into homogeneous nucleation, where the new phase develops from the dissolved species, and heterogeneous nucleation, where the new phase develops on a surface.

11.1. Burst Nucleation: The LaMer Theory

The nucleation of pyrite in framboids provides even more rigorous constraints for the processes involve in pyrite formation. Chemical engineers in the early 1950s produced a qualitative description of a process of nucleation and growth which included a stage called burst nucleation (LaMer, 1952; LaMer and Dinegar, 1950). The theory has become known as the LaMer model (Figure 11.1). LaMer kinetics are characterized by (1) a lag phase before nucleation becomes significant; (2) burst nucleation where the rate of nucleation increases exponentially and may be completed in seconds; (3) a short growth phase where nucleation becomes again insignificant. The growth phase is dominated by crystal growth, which is limited by the supply of nutrients, which is not replenished, and growth is rapidly extinguished. The result of LaMer kinetics is a large number of similarly shaped and sized colloidal or nano-sized particles.

The fact that all the microcrystals in a framboid must have been formed at the same time was first noted by Read (1968) in an unpublished thesis and was reported in the literature by Rickard (1970). In fact, this is another qualitative characteristic of burst nucleation (see later discussion; cf. Baronov et al., 2015). The involvement of LaMer-style burst nucleation in framboid formation was originally suggested by Wilkin and Barnes (1997).

The lag phase and burst nucleation of the LaMer process are characteristic of materials like pyrite which require large supersaturations, Ω, to nucleate. The

Framboids. David Rickard, Oxford University Press. © Oxford University Press 2021.
DOI: 10.1093/oso/9780190080112.003.0011

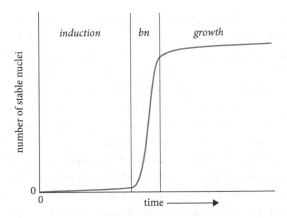

Figure 11.1. Elements of the LaMer theory for the formation of uniform colloidal particles. An initial induction period with insignificant nucleation rates is followed by a short period of burst nucleation (bn) where the rate increases exponentially followed by a growth dominated period where the formation of new nuclei is again insignificant.

induction phase of the process can be regarded as the metastable phase zone width (Kashchiev, 2011): that is, the period in which the system is in a metastable state. For pyrite nucleation from aqueous solutions, the lag phase can be quite extensive (Schoonen and Barnes, 1991). This is consistent with the requirement in pyrite nucleation for large supersaturations to be built up in solution for pyrite to nucleate.

11.2. Homogeneous and Heterogeneous Nucleation of Framboids

It has been widely assumed that pyrite only nucleates heterogeneously and that homogeneous nucleation of pyrite from aqueous solution does not occur. One reason for this assumption was that pyrite appeared reluctant to nucleate from experimental aqueous solutions (e.g., Schoonen and Barnes, 1991) except at very high dissolved Fe and S concentrations. At these high experimental concentrations of dissolved Fe and S, nanoparticulate FeS_s forms rapidly and begins to develop the mackinawite (FeS_m) structure (e.g., Matamoros-Veloza et al., 2018; Rickard, 1995; Wolthers et al., 2003). It thus appeared that FeS_m was a necessary precursor to pyrite formation in aqueous solutions and that its role was to facilitate pyrite nucleation. It should be explained that the use of the descriptor "precursor" is widely used in the materials science literature as a synonym for "reactant." However, in the geochemical literature it was mistakenly supposed that mackinawite was a *necessary* precursor to pyrite and that it transformed to pyrite through a complex series of solid-state reactions often involving the magnetic thiospinel, griegite, Fe_3S_{4g}. This is, of course, not valid. The apparent frequent intervention of FeS_m in the formation of pyrite from aqueous solutions is mainly an experimental artifact (e.g., Rickard and Luther, 2007). In order to synthesize sufficient pyrite for analysis, huge supersaturations are required in batch systems, and this saturation often approaches or exceeds the ion activity product of FeS_m at SATP. Mackinawite has been implicated simply because this was the first phase in the FeS_s precipitate which was identifiable by the conventional powder X-ray diffraction analyses available to the early researchers: in fact, this phase developed from the original nanoparticulate FeS_s over time as the particles grew to approximate the infinite lattices which are intrinsic to powder XRD theory.

We can follow Kashchiev (2011) and define supernuclei as nuclei which are stable and have the potential to grow. The supernuclei are thus the embryos on which the pyrite microcrystals grow. In heterogeneous nucleation it is obvious that the number of supernuclei that can be formed is limited by the number of heteronuclei, such as nanoparticles, in the system. When all the heteronuclei have been titrated against the supernuclei, no more supernuclei can be formed by heterogeneous nucleation. The concentration of heteronuclei required for the formation of, for example, a 10 μm diameter framboid containing 1 μm microcrystals can be estimated. The framboid volume is 5×10^{-16} m³ and it contains around 740 microcrystals (section 4.2 in Chapter 4). The concentration of microcrystals is then $> 10^{15}$ L^{-1}. The number of heteronuclei must be at least this concentration for heterogeneous nucleation to be involved in framboid formation. By contrast, in homogeneous nucleation, every molecule in solution provides a potential nucleation site (Kashchiev, 2011).

A further consequence of burst nucleation is that the solution volume in which nucleation occurs must be similar to volume of the final framboid. On a microscopic scale, the formation of individual supernuclei depletes virtually all the dissolved Fe and S in a volume of solution as it bursts into existence. The mass of pyrite in a supernucleus is given by its critical radius, and the concentrations of dissolved Fe and S in solution are constrained by the critical supersaturation for pyrite nucleation. The result is that the volume of solution required for each pyrite nucleus to form is comparable to the framboid microcrystal dimensions. The final framboid volume is then similar to the volume in which the framboid originally nucleated.

The involvement of burst nucleation in framboid formation places quite severe restraints on the processes involved in pyrite formation in general and the development of this texture in particular. The idea that pyrite formation proceeds by the successive solid-state transformation of the minerals mackinawite and greigite to pyrite is excluded since the process cannot lead to the simultaneous nucleation of pyrite. Likewise, pyrite microcrystal nucleation within a pre-formed spherical aggregate of FeS_s is precluded since the diffusion of pyrite nutrients into the solid would result in a concentration gradient being set up from the outside of the spherical aggregate to its center. This would in turn mean that pyrite would initially nucleate in the outer parts of the sphere first and only later in the center. In other words, the pyrite microcrystals in the framboid would not all nucleate at the same time. If this were the case, then framboids would display larger microcrystals at the outside of the framboids and smaller microcrystals in the center. In fact, the intrinsic characteristics of the framboid texture—similarly shaped and equal-sized microcrystals—evidences homogeneous nucleation of pyrite in solution. Heterogeneous nucleation of pyrite occurs (see section 11.5.1) but is the main route for the production of single crystals rather than framboids.

11.3. Classical Nucleation Theory

Although criticism of classical nucleation theory (CNT) has become a staple of the university teaching curriculum, innumerable reports of the consistency of the results of the theory with experimental results has proven that it provides a good approximation to reality. In CNT the rate of formation of spherical nuclei, R_N, can be described by Equation 11.1 which relates the rate of formation to a variety of basic physical and chemical parameters, including γ, the surface energy (J m^{-2}), v_m, the molecular volume (m^3 molecule^{-1}), k, Boltzman's constant (1.38×10^{-23} J K^{-1}), T, the temperature (K), and Ω, the supersaturation given by the ratio of the ion activity product (IAP) to the solubility product (K_{sp}).

$$R_N = A \, \exp\left[\frac{-\left(16\pi\gamma^3 v_m^{\,2}\right)}{\left(3k^3T^3\left(\ln\Omega\right)^2\right)}\right] \qquad\qquad 11.1$$

where A is the pre-exponential factor, which is essentially a kinetic quantity that considers the concentration of nucleation sites, the frequency of the attachment of monomers to the nucleus, and the Zeldovich factor, a measure of the probability that the critical nucleus will go on to form a particle and not redissolve. The pre-exponential factor varies between 10^{13} and 10^{41} m^{-3} s^{-1} (Kashchiev, 2011), with the smaller values being characteristic of systems with seeds or active centers and/or a low frequency of monomer attachment. Most calculations of the pre-exponential factor give values around $10^{33 \pm 3}$ cm^{-3} s^{-1} (Pina and Putnis,

2002). Simple inspection of Equation 11.1 shows that the rate of nucleus formation is not especially sensitive to uncertainties in the pre-exponential factor.

11.3.1. Critical Supersaturation

For pyrite we can define the supersaturation, Ω, by Equation 11.2.

$$\Omega = \frac{\left(Fe^{2+}\right)\left(S_2^{2-}\right)}{K^0} \qquad 11.2$$

K^0 is the solubility product for the congruent reaction (Equation 11.3) and, according to Rickard and Luther (2007), has a value of 10^{-24} (at SATP, standard
ambient temperature and pressure: 25°C and 0.1 MPa)

$$FeS_2 = Fe^{2+} + S_2^{2-} \qquad 11.3$$

The supersaturation required for pyrite to nucleate is conveniently defined in
terms of the critical supersaturation, Ω^*, which experimentally defines a point at
which nuclei begin to form at a measurable rate.

We have experimental data on the magnitude of Ω^* for heterogeneous nucleation at SATP for pyrite. The values range between 10^{11} and 10^{14} (Harmandas et al.,
1998; Rickard et al., 2007). These values were collected from experimental systems in which heterogeneous nucleation of pyrite occurred on pyrite seeds and on
organic surfaces. The actual Ω^* value for homogeneous nucleation of pyrite from
aqueous solutions is unknown and is likely to be at least at these levels. However,
some idea of the critical supersaturation for the homogeneous nucleation of
pyrite can be obtained from theoretical considerations since the critical supersaturation is related to the solubility via the surface energy. The greater the solubility, the smaller the surface energy and the lower the critical supersaturation.
The pre-exponential term, A, in Equation 11.1 is dependent on the concentration of growth units in the solution at a given supersaturation: in sparingly soluble
substances like pyrite this value is small and A is small (Putnis, 2010). Large critical supersaturations are characteristic of sparingly soluble materials like pyrite.

The mackinawite solubility product can be assumed to be a limiting value for
the maximum concentrations of Fe(II) and S(-II) in solution. If the product of
the Fe(II) and S(-II) activities exceeds the mackinawite solubility product, then
mackinawite, FeS_m, precipitates. In fact, it takes some time for the mackinawite
structure to develop in aqueous solution at SATP. The initial precipitate is
nanoparticulate and develops the tetragonal mackinawite structure over periods

of up to two years in solution (Rickard, 1969). This initial precipitate, which includes particles less than 2 nm in size (Ohfuji and Rickard, 2006), is obviously more soluble than mackinawite and therefore the solubility of mackinawite is a limiting value for the concentrations of aqueous Fe(II) and S(-II).

Harmandas et al. (1998) measured a critical supersaturation of $10^{14.7}$ for heterogeneous nucleation of pyrite on pyrite seeds at dissolved Fe(II) and S(-II) concentrations below the mackinawite solubility limit, which suggests that the critical supersaturation for the homogeneous nucleation of pyrite in the absence of FeS is likely to be $>10^{15}$.

$$FeS_2 + 2H^+ = Fe^{2+} + H_2S + S^0 \qquad 11.4$$

The maximum value is given by simple solubility limitations. Thus where log $\Omega = 20$, the product of the concentrations of aqueous Fe^{2+} and H_2S, $[Fe^{2+}][H_2S]$, for pyrite according to Equation 11.4 approaches 1 M at pH 7 and is unlikely in natural systems. As discussed in section 10.2.3 of Chapter 10, the approximation of unit activity coefficients creates an error of less than one magnitude for this approximation. A more reasonable maximum dissolved Fe and S concentration is 10^{-3} M, which would produce a supersaturation with respect to pyrite of 10^{18} at pH 7. As the solution becomes more acid, Equation 11.4 shows that the IAP for any given supersaturation is even greater. Thus, the critical supersaturation for pyrite nucleation is $10^{11} > \Omega^* < 10^{18}$.

As pointed out in section 10.2.3 of Chapter 10, in neutral to alkaline environments, aqueous FeS clusters dominate dissolved Fe(II) speciation at micromolar total dissolved sulfide concentrations. In these environments, the solubility of FeS_s is determined by aqueous FeS clusters rather than aqueous Fe(II) and pH. The limiting solubility of FeS_s can be estimated with reference to the more insoluble macroscopic crystalline phase mackinawite, Equation 11.5, for which log $K = -5.7$.

$$FeS_m = FeS^0 \qquad 11.5$$

From the viewpoint of framboid nucleation, the process involves the increase of aqueous sulfide until the concentration of aqueous FeS clusters become significant. The region where this occurs is shown in Figure 10.2 in Chapter 10 in terms of pH and total dissolved S(-II) concentration, calculated according to Equation 10.11. Figure 10.2 shows that the proportion of FeS^0 increases logarithmically with total dissolved sulfide concentration until it equals and surpasses the Fe^{2+} concentration. At pH >6.5, the concentration of FeS^0 becomes significant (i.e., $>1\%$ of dissolved Fe^{2+}) at micromolar total dissolved sulfide concentrations.

Aqueous FeS is an intermediate in pyrite framboid nucleation. Indeed, the suppression of aqueous FeS cluster formation has been demonstrated to inhibit pyrite formation (Rickard et al., 2001) and, as described in Chapter 10, Fe-S moieties are necessary precursors for pyrite formation.

11.3.2. Surface Energy of Pyrite Nuclei

Equation 11.1 shows that the rate of nucleus formation in CNT is highly sensitive to the surface energy term, γ. The definition of γ is one of the major weaknesses of CNT. In particular, the surface energy of a cluster of a few pyrite molecules is not easy to define. The general approach of approximating the surface energy of nuclei to the bulk surface energy (e.g., Privman et al., 1999) is difficult to justify, except on the grounds that it appears to give a good approximation to experimental results. The surface energy is almost certainly not size independent and at best refers to the mean surface energy. In Equation 11.1 the shape of the nucleus is assumed to be spherical. The idea of shape for a few pyrite molecules may be intrinsically irrelevant and, although different shape factors can be incorporated in Equation 11.1, they cannot normally be better justified. Furthermore, incorporation of different shape factors does not cause substantial variations in the results.

The rate of formation of pyrite nuclei is exceptionally sensitive to the value of γ in Equation 11.1 which approximates to a step function within the accuracy of the system. As can be seen from Figure 11.2, a change in the supersaturation of just one magnitude leads to a change in the rate of formation of nuclei of several magnitudes. This places quite severe constraints on the variation in magnitude of the parameters in Equation 11.1. This sensitivity can be utilized by making the approximation that the initiation of nucleation in a nucleation burst process can be approximated to a rate of 1 nucleus s^{-1}. Equation 11.1 can then be solved for γ:

$$\gamma^3 = 3 \ln A k^3 T^3 \left(\ln \Omega \right)^2 16 \pi \upsilon_m^{\;2} \qquad\qquad 11.6$$

A graph of the solution to Equation 11.6 for pyrite supersaturations $0 > \log \Omega < 20$ at SATP is shown in Figure 11.3.

As discussed earlier, the critical supersaturation for pyrite in aqueous solution at SATP ranges between 10^{11} and 10^{18} and solving Equation 11.6 suggests $0.49 < \gamma < 0.68$ J m^{-1} for this range of Ω^* values. This is lower than the computed surface energies measured for bulk pyrite (ca. 1 J m^{-1}) but larger than some measured values for bulk pyrite (ca. 0.05 J m^{-1}). A mean surface energy of ca. 0.6 J m^{-2} for pyrite nuclei in aqueous solution at SATP seems to be a reasonable approximation.

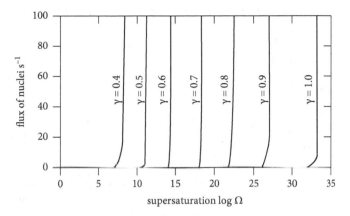

Figure 11.2. Graphical solutions for Equation 11.1 for various values of the surface energies of pyrite nuclei, γ, in J m⁻¹.

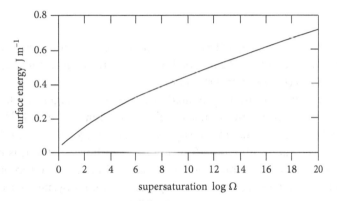

Figure 11.3. Logarithm of pyrite supersaturation versus pyrite surface energy in aqueous solutions at SATP. The logarithm of the critical supersaturation for pyrite ranges between 11 and 18, suggesting a mean surface energy for pyrite nuclei of around 0.5 J m⁻¹.

11.3.3. Pyrite Critical Nucleus Size

During nucleation, pyrite clusters form and dissociate until a minimum nucleus size is produced where the cluster can grow irreversibly to form stable nuclei. This minimum or critical size can be described in terms of the critical radius r^* of a spherical nucleus. In CNT, the critical radius is given by the relationship in Equation 11.7.

$$r^* = \frac{2\gamma \upsilon_M}{RT\ln\Omega}$$

11.7

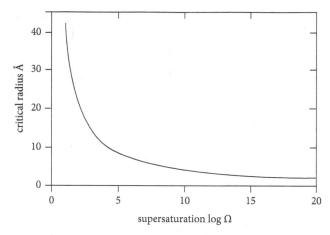

Figure 11.4. Critical radius, r^*, for pyrite nuclei (Å) plotted against the supersaturation, log Ω, according to Equation 11.7.

where γ is the surface energy (J m^{-2}), v_M is the molecular volume (m^3 mol^{-1}), R is the universal gas constant (8.3147 J K^{-1} mol^{-1}), T is the temperature (K), Ω is the supersaturation (IAP/K_{sp}), and r^* is in m.

The critical radius for pyrite is plotted against log Ω in Figure 11.4. The critical supersaturation for pyrite nuclei ranges between 10^{11} and 10^{18} at SATP. Equation 11.7 suggests that in this supersaturation range, the pyrite critical nucleus reaches a radius of about 10 Å. Figure 11.4 shows that at this point the curve is becoming asymptotic and large changes in log Ω result in only small changes in r^*. In the range of critical supersaturation for pyrite nuclei at SATP discussed earlier, the variation is 2.3 <r^* <3.8 Å. Although these variations have been ascribed to the presence of varying surfaces, these calculations suggest that the variations in these ranges of Ω are within the experimental uncertainties.

The volume of an FeS$_2$ molecule is around 40 Å^3, which is equivalent to a 2.2 Å diameter sphere. In Figure 11.4, the pyrite critical radius becomes asymptotic between 2 and 3 Å or a sphere with a 4–6 Å diameter. It actually reaches this critical size at log Ω >~16. Bulk pyrite has a cubic unit cell containing the equivalent of 4 FeS$_2$ molecules, a volume of 159 Å^3, and a cell parameter of 5.4 Å. Although it cannot be assumed that the form of the critical nucleus is identical with the habit of the bulk unit cell, the similarity of the dimensions of the pyrite critical nucleus and the bulk pyrite unit cell is noticeable.

Since the size of the pyrite critical nucleus is an inverse function of the supersaturation, the number of monomers required for the nucleus to reach critical stability rapidly increases at lower supersaturations, making pyrite nucleation more unlikely. For example, at log Ω = 1, the critical nucleus requires that over

7000 FeS_2 molecules assemble, assuming a pyrite unit cell volume of 159 $Å^3$ and the equivalent of 4 FeS_2 molecules per unit cell. The critical nucleus size required for pyrite nucleation is consistent with the rate of nucleation being a limiting condition for pyrite precipitation, which has been widely observed and reported from experimental syntheses.

11.4. Framboid Size and Supersaturation

One consequence of CNT which is significant for framboid formation is the relationship between the solution volume V and Ω^*. Kozisek et al. (2011) showed that in all cases, Ω^* is a monotonically decreasing function of the volume of the nucleation solution. Since, in framboids, the volume of the nucleation solution can be approximated to the size of the framboid itself, Kozisek et al.'s results imply that Ω^* varies with framboid size: the bigger the framboid, the smaller is Ω^*. The consequence of this is that the framboid size reflects the local supersaturation. Populations of smaller framboids reflect formation from systems with higher supersaturation, and these populations have been correlated with framboid formation in the water column itself (Wilkin et al., 1996).

The critical supersaturation Ω^* is related to the volume V through the relationship in Equation 11.8 (Kashchiev, 2011).

$$\Omega^* = \left[B/\ln\left(k_n C_0 V\right)\right]^{1/2} \qquad\qquad 11.8$$

where C_0 is the concentration of nucleation sites (m^{-3}), k_n is a kinetic factor, and B is a thermodynamic parameter. If B, k_n, and C_0 can be regarded as constants, then Equation 11.8 shows that the critical supersaturation decreases monotonically with increasing V. Applying this to pyrite framboids, where V_f is assumed to approximate the volume of the framboid, the supersaturation can be related to the framboid diameter, D (Figure 11.5).

The relative increase in the critical supersaturation for framboidal pyrite can be estimated from the assumption that the smallest framboids have a diameter of 2 µm. Equation 11.8 then shows that the increase in critical supersaturation with framboid diameter becomes asymptotic as the diameter increases and is maintained at less than one magnitude throughout. Thus, for example, if the critical supersaturation for a 2 µm framboid is 10^{11}, that for the largest framboid will not exceed 10^{12}. The exponential relationship between the critical supersaturation and framboid diameter shown in Figure 11.5 demonstrates that the greatest increase in critical supersaturation occurs between 2 and 5 µm. That is, all else being constant, framboids ≤5 µm in diameter need significantly higher

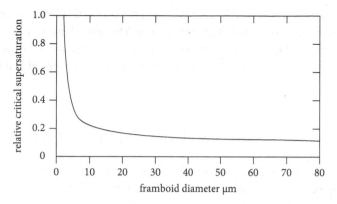

Figure 11.5. Relative critical supersaturation versus framboid diameter in μm. The critical supersaturation is standardized to Ω^* for the smallest, 2 μm diameter, framboid.

supersaturations to form. Conversely, framboid size is a function of relative supersaturation in the nucleating solution.

11.5. Nucleation of Framboids

11.5.1. Heterogeneous Nucleation of Pyrite

Many researchers have commented on the coexistence of euhedral pyrite with framboids in sediments and sedimentary rocks. Framboids usually make up a minor fraction of the total pyrite in these systems, with estimates ranging from 1% to 10% of the total pyrite (Rickard, 2012b). There is no doubt that framboids and euhedral pyrite are formed roughly contemporaneously in sediments (Figure 11.6). Then the question has been posed as to what causes the formation of the different forms of pyrite at the same time in different regions of the sediment? As discussed in section 10.3.3 of Chapter 10, the reaction of HS^- with Fe(III) (oxyhydr)oxides produces surface =FeS. It also is thought to be a major source of polysulfide formation in sediments (Avetisyan et al., 2019) and disulfide is formed during this reaction, bonded to the Fe(III) (oxyhydr)oxide surface (Wan et al., 2014). The resulting formation of surface FeS_2 moieties through the reaction between =FeS and $S_2(-II)$ leads to the heterogeneous nucleation of pyrite. The conclusion is that a major pathway for the heterogeneous nucleation of pyrite is the reaction between dissolved sulfide and Fe(III) (oxyhydr)oxides.

Obviously, surface =FeS groups that can react to form pyrite are not restricted to iron (oxyhydr)oxides. Any iron mineral in a sulfidic environment may

Figure 11.6. Pyrite euhedra coexisting with small organized framboids from the ca. 380 Ma Rammelsberg deposit.

Scanning electron micrograph by Ian Butler.

develop surface =FeS groups and several, including the metastable iron sulfides mackinawite and greigite, are known to have these surface groups. More significantly, as described in Chapter 12, pyrite itself has surface =FeS groups and these constitute the major sites for pyrite crystal growth.

It seems probable that nucleation controls the ultimate pyrite form. Certainly, framboids evidence burst nucleation in solution, producing multiple pyrite nuclei at the same instant. By contrast, euhedral pyrite crystals evidence the formation of isolated nuclei.

The conclusion is that pyrite framboids result from homogeneous nucleation and pyrite aggregates and single crystals from heterogeneous nucleation. Homogeneous nucleation is dominant in stagnant environments where aqueous Fe(II) and S(-II) concentrations can build up to critical supersaturation levels. Heterogeneous nucleation occurs on preexisting surfaces where the nucleation is enhanced by the presence of active sites. In framboid formation, further crystal growth is constrained by the rate of diffusion of aqueous Fe(II) and S(-II) to the framboid site. Single crystals and crystal aggregates, by contrast, develop in systems dominated by both diffusive and advective flow: the key requirement being the continued replenishment of Fe(II) and S(-II). In the case of isolated single pyrite crystals, it appears that the concentration of dissolved Fe(II) and S(-II) remains under the supersaturation level necessary to nucleate further pyrite critical nuclei but above the solubility product for pyrite.

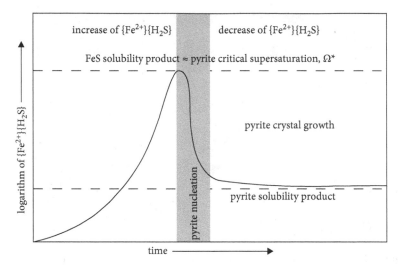

Figure 11.7. Nucleation processes in framboid formation in terms of the logarithm of the aqueous ion activity product $\{Fe^{2+}\}\{H_2S\}$ and time. The ion activity product increases until it reaches the point where FeS_{aq} forms and pyrite nucleates. Burst nucleation results in a decrease in $\{Fe^{2+}\}\{H_2S\}$ and pyrite crystal growth follows.

11.5.2. Nucleation of Pyrite Framboids

The nucleation of pyrite framboids is shown schematically in Figure 11.7. Burst nucleation of pyrite occurs when $\{Fe(II)\}\{S(-II)\}$ approaches the critical supersaturation level, Ω^*, which is at least 10^{11} times the pyrite solubility product, K_{sp}. At these Fe(II) and S(-II) concentrations, aqueous FeS forms and this instigates the burst nucleation of pyrite, which in turn results in a rapid decrease in the concentrations of dissolved Fe(II) and S(-II). The decrease of the $\{Fe(II)\}\{S(-II)\}$ product means that pyrite can nucleate no longer because the $\{Fe(II)\}\{S(-II)\}$ product is less than Ω^*, the critical supersaturation level for pyrite.

In conclusion, the processes involved in the nucleation of the pyrite microcrystals in framboids at ambient temperatures include (1) the increase in the concentration of dissolved Fe and S and the formation of FeS_{aq}; (2) burst nucleation of pyrite on reaction of polysulfide or H_2S with FeS_{aq}; (3) decrease in the concentration of dissolved Fe and S, so that the pyrite supersaturation limit is not reached and no more pyrite is nucleated; and (4) crystal growth of pyrite (see Chapter 13).

12

Framboid Microcrystal Growth

12.1. Origin of Pyrite Microcrystal Habits

The large variety of different crystal habits displayed by pyrite, as described in Chapter 4, is due fundamentally to its crystal structure, which produces a large number of possible crystal faces with variable surface energies. The surface energy is proportional to the solubility, with higher energy faces being less stable and more soluble than low energy faces. The surface energies of pyrite faces are not known precisely, despite having been the subject of various experimental and, more particularly, molecular modeling efforts. These theoretical computations are easier to carry out than the experimental measurements, but ground truthing of the theoretical models is difficult. The computed surface energies tend to be empirical; although relative computed values have been used with merit, the precision and accuracy of absolute values are unknown. In particular, there are challenges for molecular modeling to encompass important factors in the surrounding environment, such as pH and saturation state. Furthermore, the application of the values obtained for macroscopic crystals and nanocrystals to framboidal microcrystals is uncertain.

The original experimental investigation by Murowchick and Barnes (1987) of the causes of varying crystal habits in pyrite has not been surpassed. Although many other syntheses of, especially, nanocrystalline pyrite have been reported since that date, few have looked at the effects of saturation and temperature on pyrite crystal habit. Most of these syntheses have involved complex organic molecular precursors (e.g., Yuan et al., 2015) and deconvoluting the effects of supersaturation and temperature on the crystal habit is not straightforward.

Murowchick and Barnes (1987) originally reported that as the degree of supersaturation increased, the pyrite habit changed from cube → octahedron → pyritohedron (Figure 12.1).

12.1.1. Pyrite Surface Energies

A summary of computed surface energies for pyrite was compiled by Rosso and Vaughan (2006b), Arrouvel and Eon (2019), and Kitchaev and Ceder (2016) (Table 12.1).

Framboids. David Rickard, Oxford University Press. © Oxford University Press 2021.
DOI: 10.1093/oso/9780190080112.003.0012

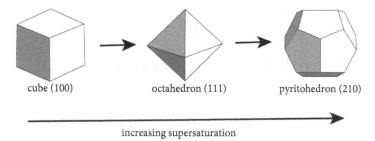

cube (100)   octahedron (111)   pyritohedron (210)

increasing supersaturation

Figure 12.1. The development of pyrite habit with increased supersaturation in hydrothermal conditions according to the experimental results of Murowchick and Barnes (1987).

Table 12.1 Surface Free Energies, γ, in J m^{-2}, for Pyrite Surfaces in Vacuum, $\gamma\frac{vac}{hkl}$, and in Contact with Pure Water, $\gamma\frac{H_2O}{hkl}$

Surface		$\gamma\frac{vac}{hkl}$ J m^{-2}	$\gamma\frac{H_2O}{hkl}$ J m^{-2}
(100)	Cubic	1.38	1.11
(111)	Octahedral	1.80	1.71
(210)	Pyritohedral	1.82	1.07
(110)	Dodecahedral	2.14	1.26

From Kitchaev and Ceder (2016).

Even if the absolute values are uncertain, at least for vacuum conditions at 0 K, it is widely accepted that surface energy increases from cubic (1 0 0) through octahedral (1 1 1) to pyritohedral (2 1 0) and dodecahedral (1 1 0) surfaces (Alfonso, 2010; Kitchaev and Ceder, 2016; Rosso and Vaughan, 2006b; Sunagawa, 1957), which is consistent with the experimental observations of Murowchick and Barnes (1987). Thus, the cube face is the most stable and most abundant form, and this seems to be intuitively correct. The higher surface energies displayed by the octahedral faces suggest that these are less stable. Kitchaev and Ceder (2016) found that pyritohedral surfaces become relatively more stable in water (Table 12.1). Alfonso's (2010) calculations suggested that (2 1 0) and (1 1 1) surfaces become progressively more stable as the chemical potential of sulfur increases.

The estimation of surface energies for pyrite is complicated by molecular imperfections on the surfaces such as dislocations and step edges. Indeed, the founders of modern crystal growth theory, Burton, Cabrera, and Frank, originally concluded that crystal growth could not occur at low supersaturations in the absence of dislocations (Burton et al., 1949).

I use the designation =M as shorthand for the surface bound moiety, M. Both experimental and computational results suggest that species such as $=S^{2-}$ and low-coordinated =Fe occur at edge locations (e.g., Hung et al., 2002; Rosso and Vaughan, 2006a). The low-coordinated =Fe sites are likely to be spin-polarized and thus more reactive toward paramagnetic species such as O_2 and adsorbates. However, the exact nature of the speciation at pyrite surfaces is the subject of considerable debate since the experimental conditions can produce empirical results, and the applicability of computational predictions so far to real-world crystal growth conditions may be uncertain (e.g., Murphy and Strongin, 2009). It should also be noted that no information has been published on the surface characteristics of framboidal microcrystals.

The results can be used to compute the equilibrium shape of pyrite crystals. This is often called the Wulff shape, after the Russian scientist George Wulff, who proposed the basic equation relating surface energy to the growth of crystal faces based on J. Willard Gibbs's original proposal that the equilibrium shape of a crystal is that which minimizes the total surface free energy; that is, the sum of the individual products of surface area and interfacial energy. Thus, the ratio of the surface energies of the various pyrite crystal surfaces will determine the ideal shape. For example, a ratio of (1 1 1)/(1 0 0) surface energies of 1.2 will result in the formation of truncated cubes or truncated octahedra (Arrouvel and Eon, 2019). The computed surface energies were originally derived from calculations in vacuum at 0 K. Under these conditions, the ordering of common pyrite Wulff crystal shapes based on stoichiometric surface energies is truncated cube < cube < truncated octahedron < octahedron < truncated pyritohedron < pyritohedron < dodecahedron (Figure 12.2).

Since then, algorithms have been developed which permit *ab initio* estimates of the surface energies in water (Figure 12.3) (Barnard and Russo, 2009b; Kitchaev and Ceder, 2016) and the effects of sulfur fugacity. (Barnard and Russo, 2009a; Zhang et al., 2015).

The effect of sulfur fugacity is incorporated into the computations through the differential molecular configurations of the various pyrite faces. Thus, there is only one terminal layer for {1 1 0} and {2 1 0} and this is constituted by both =Fe and =S atoms and is therefore stoichiometric. This means that the surface energies of these faces are constant with respect to sulfur fugacity. In contrast, there are three terminal layers configurations for {1 0 0} and

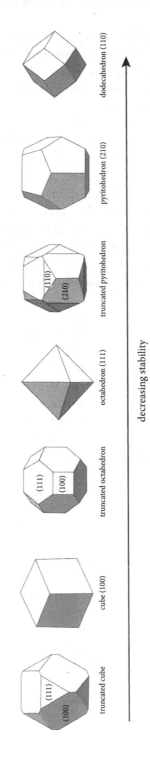

truncated cube   cube (100)   truncated octahedron   octahedron (111)   truncated pyritohedron   pyritohedron (210)   dodecahedron (110)

decreasing stability

Figure 12.2. Relative stabilities of equilibrium pyrite forms in vacuum at 0 K according to computations by Barnard and Russo (2009).

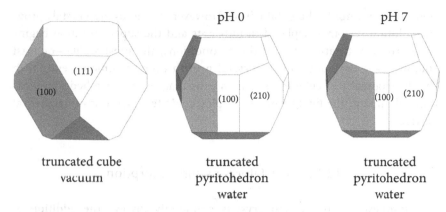

Figure 12.3. Effect of water on pyrite Wulff shapes from computations by Kitchaev and Ceder (2016).

{1 1 1} surfaces: (1) stoichiometric {1 0 0} and {1 1 1} surfaces with both =Fe and =S atoms; (2) non-stoichiometric {1 1 1}:Fe surfaces with only =Fe atoms; and (3) non-stoichiometric {1 0 0}:S and {1 1 1}:S surfaces with only = S atoms. This allows computations to be made on the effect of S fugacity on the Wulff shapes of pyrite crystals. Barnard and Russo (2009) suggest that Fe-fugacities can be included in the computations by assuming that S-poor conditions are necessarily Fe-rich, and vice versa. Including sulfur fugacity in the computations suggest that cubic nanocrystals are predicted under low sulfur fugacities. The relative infrequency of cubic pyrite microcrystals in framboids (section 4.5.1 of Chapter 4) might therefore suggest that in most situations, pyrite framboid formation occurs in sulfur-rich conditions. This is consistent with the idea that in sediments, sulfide production through microbiological sulfate reduction is the key parameter determining framboid formation in sediments.

In conditions where the Fe and S fugacities are stoichiometric, pyritohedral nanocrystals are predicted. Again, the relative scarcity of pyritohedral microcrystals in framboids (section 4.5.2) might suggest that stoichiometric conditions are less common in framboid-forming environments.

One of the major features of pyrite nucleation is the huge supersaturations required to initiate pyrite nucleation, which is discussed in more detail in section 11.4. The result is that the initial stage of pyrite crystal growth occurs in solutions with large supersaturations and the less stable face develops first. This is simply an expression of Ostwald's step rule. The first formed habit is therefore commonly the octahedron. However, as the pyrite crystals grow, the solution becomes depleted in nutrients and the supersaturation begins to approach saturation. Under these conditions, the most stable—or least soluble—face develops and the crystal growth event commonly ends with the development of cube faces. This means that, even if octahedra are initially abundant, the final form will display cubic terminations of octahedral vertices.

12.1.2. Spectator Ions and Adsorption

The syntheses of pyrite nanocrystals commonly involve the addition of surfactants which interact preferentially with particular pyrite surfaces, thus producing differently shaped crystals. These molecules have been called "spectator ions" by Kitchaev and Ceder (2016) since they influence crystal growth but are not represented in the chemical formula of the bulk pyrite. For example, polyvinylpyrrolidone (PVC) strongly interacts with {1 1 1} producing octahedra. PVC molecules have a long polymer chain which provides large steric hindrance which prevents particle agglomeration (Wang et al., 2010). The oxygen atoms of PVP have two lone electron pairs which strongly bind to some pyrite crystal faces. Non-ionic solvents such as ethanol (Liu et al., 2013) and ethanolamine (Yuan et al., 2015) stabilize cubic surfaces. The implication is that it is possible that naturally occurring organic molecules, with which framboids are commonly associated, might have similar effects on microcrystal form. However, it would appear that such naturally occurring organic molecules would not possess long polymer chains, since their steric hindrance might prevent the pyrite microcrystals from self-assembling into framboids.

In contrast with these organic spectator ions, inorganic species not only influence pyrite growth forms, but may also be incorporated into the bulk pyrite as trace elements. The role of these inorganic species on pyrite crystal growth forms is less clear since far less experimental work has been directed to syntheses involving inorganic species. This is in contrast to pyrite surface adsorption studies which have been mainly targeted at organic species (Murphy and Strongin, 2009; Rickard, 2012c). An exception is arsenic. It has long been known that arsenic is preferentially adsorbed on (1 1 1) surfaces through reactions with =Fe sites (Hayashida and Muta, 1952; Sunagawa and Takahishi, 1955). The net result is the inhibition of growth in ⟨1 1 1⟩ directions and the association with octahedral pyrite crystals in As-rich environments (Arrouvel and Eon, 2019). There is little information about the effects of other trace elements on pyrite crystal growth. Murowchick and Barnes (1987), for example, remarked that Co, Ni, and Se do not appear to affect pyrite crystal habits.

12.2. Growth Mechanisms and Microcrystal Habit

There are three basic processes in crystal growth: monomer addition, crystallization by particle attachment, and Ostwald ripening (De Yoreo et al., 2015). These three growth regimes are neither necessarily successive nor exclusive. They can overlap and any one may dominate at any time and, as discussed later, all three processes may be variously involved in the development of pyrite microcrystals in framboids.

The microcrystals in any individual framboids must have grown at the same time, over the same time period, and in the same environment. If the microcrystals form at different times, then there is little likelihood that they will produce an equal size distribution. Not only must the microcrystals be formed at the same time, but they also must grow for the same length of time. The result is that the growth of microcrystals in any framboid took place over a specific time period characteristic of that framboid. Likewise, similarity in microcrystal habit in individual framboids suggests that the microcrystals all developed in very similar environments.

Molecular mechanisms of crystal growth from solution are not well-defined (e.g., Shtukenberg et al., 2013), although rapid progress is being made through advances in, for example, atomic force microscopy. Even so, understanding of pyrite crystal growth is still based on classical theory.

12.2.1. Screw-Dislocation Growth

Burton, Cabrera, and Frank (1949) launched the modern era of crystal growth theory with the idea that screw dislocations on a crystal surface continually extrude steps to which molecules can attach (Figure 12.4). The ends of the spirals resolved the paradox of fast growth from solutions at low

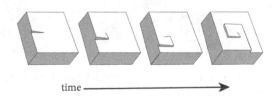

time ——————————————→

Figure 12.4. Screw dislocation growth.

supersaturation since the molecules situated in these locations are characterized by incomplete or dangling bonds. Indeed, Burton et al. (1949) concluded that crystal growth cannot proceed at low supersaturations in the absence of dislocations.

Seager (1952) first reported screw dislocations on (1 0 0) faces of pyrite. However, no investigation of the surface growth forms of pyrite microcrystals has been reported. The best we can do at present is to re-examine some of the SEM images obtained in routine observations of framboids. This is unsatisfactory since these images are taken at relatively low resolution and growth forms are often not clearly observed. An example is given in Figure 12.5, where a blow-up of a cubic microcrystal from Figure 4.6 (Chapter 4) shows possible spiral growth patterns. The images are blurred, being at the limits of the resolution of the original image. However, the intimated pattern is very similar to the spiral growth pattern reported from macroscopic pyrite crystals by Endo (1978, Plate 33-3). Reviewing the image in Figure 4.6 in detail shows that spiral dislocation growth appears to have been prevalent throughout this framboid. Likewise, a review of the image of the large microcrystals in the framboid from Figure 4.11 show that these appear to show triangular hillocks. Kevin Rosso (personal communication, 2019) points out that the screw dislocations giving rise to the features in figure 12.5a run along [1 0 0]. The trisoctahedral form in figure 12.5b is then emerging by fast spiral growth along a [1 0 0] screw dislocation. That is, defect-enabled growth may be a mechanism for the expression of higher energy faces such as (1 1 1).

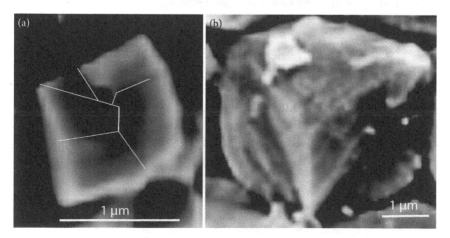

Figure 12.5. Blurred details of microcrystal growth forms: (a) spiral growth on cubic microcrystals from figure 4.6; (b) triangular growth of pyrite microcrystals from Figure 4.11, giving an apparent trisoctahedral form to the microcrystal.

12.2.2. Surface Nucleation Growth

Surface nucleation growth is also referred to as 2D nucleation growth (De Yoreo and Vekilov, 2003), layer growth or single nucleation growth (e.g. Cubillas and Anderson, 2010). It was originally proposed by Sunagawa (1957) for the growth of hydrothermal pyrite on (1 0 0) surfaces and observed by Endo (1978). Endo (1978) concluded that surface nucleation growth was the dominant growth process for hydrothermal pyrite.

Surface nucleation growth involves the attachment of the monomer to a site on the crystal surface and the continued nucleation to form an island of monolayer height: a 2D nucleus (Figure 12.6). The problem with surface nucleation for pyrite is that it suffers from the same problem as 3D-nucleation discussed in Chapter 10: there is a critical nucleus size and a free energy barrier to nucleation. Thus, high supersaturations are required to initiate surface nucleation growth of pyrite microcrystals by this method (De Yoreo and Vekilov, 2003).

Since the outermost face of any pyrite microcrystal reflects the point where the concentration of dissolved constituents equals or becomes less than the pyrite solubility product, we can assume that, generally, the outermost face is formed under low supersaturation conditions. Evidence for surface nucleation growth of pyrite framboid microcrystals is consequently rare, since growth at low supersaturations classically requires dislocations. A possible example of surface nucleation growth may be that reported by Sawlowicz (2000) from a framboid in the ca. 14 Ma, Miocene shales in the Fore-Carpathian Basin, Poland (Figure 12.7). Here framboidal microcrystals occur with flat sub-circular decorations possibly representing surface nucleation islands. A further possible example is given by the detailed image (Figure 12.5(b)) of a microcrystal from the framboid shown in Figure 4.11. This again shows a texture very similar to those imaged in macroscopic crystals by Endo (1978) and, consistent with his observation, the triangular pyramid has the opposite orientation to the {1 1 1} face. Endo (1978) reported that most pyrite {1 1 1} faces display these regular triangular growth patterns. This is consistent with the comment by Rosso mentioned previously.

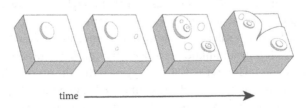

time ⟶

Figure 12.6. Surface nucleation growth.

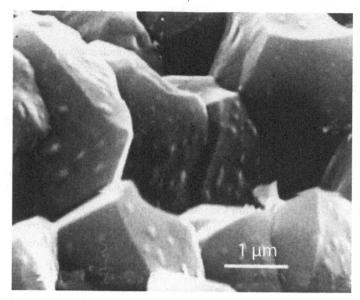

Figure 12.7. Framboidal microcrystals showing sub-circular decorations.
Scanning electron micrograph images by Zbigniew Sawlowicz.

12.2.3. Crystallization by Particle Attachment (CPA)

Monomer attachment rates are proportional to the solubility, so that as the solubility drops to the extremely low values exhibited by pyrite, the rates of monomer addition—at the same levels of supersaturation—may drop by factors of the order of 10^{10} or even more (De Yoreo et al., 2015). This means that crystallization by particle attachment (CPA)—crystallization through the attachment of nanoparticles rather than monomers—becomes potentially more attractive in phases with low solubilities such as pyrite.

CPA is also known as *aggregative growth*, but this may be confusing in discussing framboid crystallization since the framboids themselves develop by aggregation or assembly of pyrite microcrystals (Chapter 14). CPA is a general term which includes processes such as oriented attachment (De Yoreo et al., 2015) where particles orient themselves for attachment to specific crystal faces. In the CPA model, pyrite microcrystals grow by aggregation of nuclei and subsequent crystallization by an unknown process that probably involves minimization of surface energy. Gong et al. (2013) suggested that the process involved four stages (Figure 12.8). First, the nucleation of a seed crystal: the shape illustrated in Figure 12.8 is the truncated cube, potentially one of the more stable Wulff shapes, according to Barnard and Russo (2009b). The seed crystals assemble by

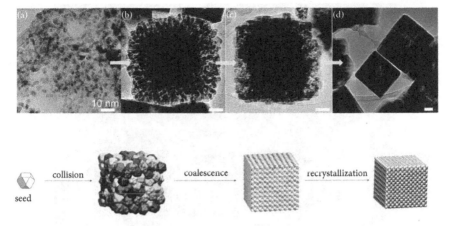

Figure 12.8. Sequence of TEM images showing the CPA growth process and a schematic summary: (a) formation of pyrite seeds; (b) seed collision; (c) mesocrystal formation; and (d) cubic crystal formation.

Images from Gong et al., 2013. *Scientific Reports* 3: 2092. Reproduced by permission.

oriented attachment along (100) facets. These coalesce preferentially to form a cubic mesocrystal which then recrystallize to produce the cubic crystal. A similar process was observed by Lucas et al. (2013).

CPA is well documented in metal sulfides, including Zn sulfide (Zhang et al., 2003) and Fe sulfide (Guilbaud et al., 2010) as well as pyrite (Gong et al., 2013; Li et al., 2011; Liu et al., 2013; Lucas et al., 2013; Zhu et al., 2015). Xian et al. (2016) reported the synthesis of quasi-octahedral pyrite mesocrystals through CPA (oriented attachment), classical nucleation growth, and Ostwald ripening. The microtexture displayed by these synthesized octahedral mesocrystals (Xian et al., 2016, Figure S3) is very similar to that observed by Butler (1994) (Figure 12.9) in cubic crystals in synthetic framboids. It is interesting to note that Xian et al. (2016) refer to this as a rough surface consisting of nanoparticles up to 5 nm in diameter, which provides a link between the observed rough surfaces of some pyrite framboid microcrystals, described earlier, and CPA. Mesocrystals are metastable phases which transform to single crystals (Xian et al., 2016), and thus the observation of mesocrystals in framboids, especially those of geologic age, is unlikely.

12.2.4. Ostwald Ripening

Ostwald ripening or coarsening is a conventionally accepted mechanism for the conversion of smaller nanoparticles into larger ones. By definition, therefore,

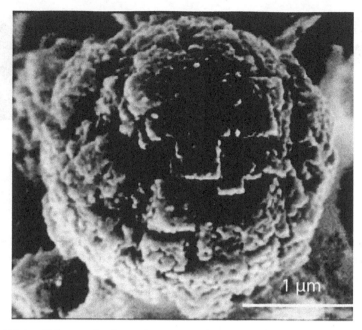

Figure 12.9. CPA growth of a synthetic quasi-framboid. The individual pyrite microcrystals are constituted by smaller pyrite particles which results in a lack of uniformity in the microcrystal form.
Scanning electron micrograph image by Ian Butler.

it is important in systems where the crystals are of different sizes: that is, if the system is quenched at any time, Ostwald ripening normally produces a spectrum of crystal sizes. In framboids, the microcrystals are similarly sized and therefore the scope for Ostwald ripening is limited. This is not to say that it does not occur in framboid formation, and possible examples of Ostwald ripening occurring in very large framboids are described in section 4.2.2 of Chapter 4.

12.2.5. Interrelationship between Monomer Attachment, CPA, and Ostwald Ripening

The process of framboid formation involves the increase in the pyrite solubility product, $\{Fe^{2+}\}\{S_2^{2-}\}$, to very large supersaturations, followed by burst nucleation of pyrite nanoparticles or seeds. The burst nucleation occurs at supersaturations of at least 10^{11}. One of the interesting features of the logarithmic scale for describing concentrations is that a reduction of the concentration of one magnitude, for example, results in a real-world decrease of 10% in the concentration.

In CPA the number of critical nuclei in the systems is not directly related to the number of microcrystals in the resultant framboid. In fact, it is much greater, since a 1 μm pyrite cube may consist of 10^3 supernuclei, rather than being the result of crystal growth of just one supernucleus. The result is that the number of critical nuclei being formed in a CPA system is much greater than that in a diffusion-limited growth process. Ostwald ripening then results in a reduction in the number of particles.

Wang et al. (2014) introduced the concept of the nucleation function, which is a plot of the nucleation rate versus time (Figure 12.10). In this plot, the areas under the curves are the number of nuclei; the width of the curve is the time window of nucleation. It can be seen that the narrower the time window for the same number of nuclei formed, the smaller are the nuclei. This means that with an equal number of monomers generated, the more rapid nucleation process will produce a smaller mean particle size.

The growth trajectories of a large number of experimental systems producing individual nanocrystals have been observed with real-time visualization of nucleation and growth events in TEM since the original report of Zheng et al. in 2009. All these show that CPA is widespread but rarely exclusive, and monomer addition also occurs. The results of these studies showed that monomer addition

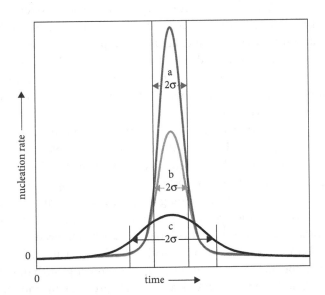

Figure 12.10 The nucleation function according to Wang et al. (2014). The nucleation rates are described by Gaussian profiles characterized by their 2σ widths. The areas under the curves are the numbers of nuclei generated. Curves b and c have the same 2σ widths but curve c reaches a greater nucleation rate.

produces nanocrystals that grow continuously until reaching a saturation size and maintain similar morphologies. In contrast, CPA nanocrystal growth is discontinuous and stepwise as nanocrystals with similar dimensions coalesce in a short time period. During growth pauses, the nanoparticles rearrange by dissolution/reprecipitation to subspherical forms. Also, during these periods structural rearrangement and recrystallization occurred, ultimately producing a single crystal nanoparticle. Similar processes have been observed for sulfides. For example, Evans et al. (2011) showed monomer-attachment, coalescence, and Ostwald-ripening for PbS particles.

Although there is a spectrum of processes between end-member growth by monomer addition and CPA, these processes may influence the final form of the framboid. The particular problem has been explaining the extreme internal ordering displayed by some framboids. The discovery that framboids are aggregates of crystals rather than single crystals led to the idea that small impurities in the aggregating system would result in a breakdown of potential ordering. The monomer addition and CPA processes provide a different interpretation. End-member monomer growth produces more extreme monodispersed populations of microcrystals. By contrast, CPA appears to result in quite irregular microcrystals, which naturally lead to more disorganized geometries. Synthetic pyrite aggregates displaying evidence of the CPA growth of microcrystals also show extreme variations in microcrystal size, leading to a border-line definition of the texture as framboidal—rather than a simple aggregate of pyrite crystals (Figure 12.9). Gradations between these end-member processes lead to various degrees of similarity between microcrystal habit and size: indeed, framboids and groups of pyrite microcrystals often show varying degrees of ordering and disorder within the same mass.

Wang et al. (2014) reviewed the kinetics and mechanism of CPA of nanocrystals. They pointed out that the CPA is only possible after classical nucleation and some growth has already occurred: there must be some particles to aggregate. The third regime is Ostwald ripening. Wang et al. (2014) noted that neither CPA nor Ostwald ripening necessarily occurs, and, where they do occur, there may be some overlap between the three regimes. For example, the initial classical nucleation and growth regime may be followed by a second induction and growth period associated with CPA. Zheng et al. (2009b) described differences in the nanoparticles produced by monomer growth only and monomer growth followed by CPA. Nanocrystals growing by monomer addition grow continuously as single crystals. In contrast, nanocrystal growth by CPA is discontinuous. During these growth pauses the particle rearranged its morphology, a process involving both structural rearrangement and recrystallization.

12.2.6. Supersaturation and Pyrite Framboid Microcrystal Growth Forms

The relationship of pyrite crystal form to degree of supersaturation has been described empirically by Wang and Morse (1996) and Sunagawa (1987) in terms of differential growth mechanisms for octahedral and cubic faces (Figure 12.11).

The classical sequence for pyrite framboid microcrystal formation is illustrated in Figure 12.12. The monomer concentration builds up in a restricted volume by diffusion until the concentration reaches the critical supersaturation for pyrite, which is at least 10^{11} times the pyrite solubility constant K_{sp} at SATP. At that point, burst nucleation occurs and the monomer concentration declines. Initially, at relatively high supersaturations, octahedral microcrystals are formed by surface nucleation growth. If the monomer concentration is depleted to the level of the pyrite solubility constant during this stage, framboids with octahedral microcrystals are produced. Commonly, the diffusion of monomers continues and ion-by-ion growth produces the most stable surface, (1 0 0), which begins to form by screw dislocation growth. This results in the formation of framboids with truncated octahedral microcrystals. This may be the most common form of framboid microcrystal and certainly it lends itself ideally to close packing, as discussed in section 5.2 of Chapter 5. If crystal growth continues at low supersaturations, framboids with cubic microcrystals are produced. As discussed in section 4.5.1, these seem to be relatively uncommon, especially in sediments, and it seems that the continuous supply of monomers by diffusion sufficient to maintain the development of cubic facets on the framboidal microcrystals is not the normal situation. As noted earlier, the rate of diffusion of monomers to the pyrite surface must be greater than the rate of pyrite crystal growth for this situation to occur. The depletion of monomer concentrations

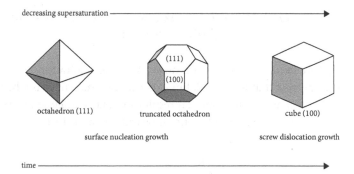

Figure 12.11. Development of framboid microcrystals over time and relationship with surface growth forms in a system with decreasing supersaturation.

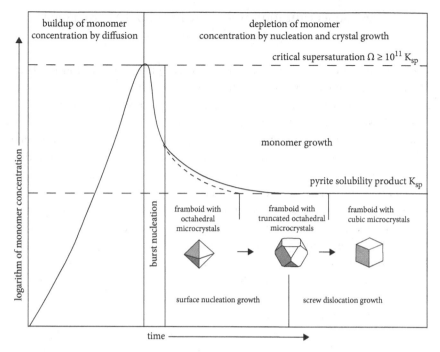

Figure 12.12. Interrelationships between saturation and framboid microcrystal form for a classical LaMer sequence in terms of the logarithm of the monomer concentration and time.

during the growth of the octahedral faces suggests that, at least during this phase, the rate of monomer supply via diffusion is less that the rate of surface nucleation growth.

12.3. Origin of Framboid Sizes and the Rate of Pyrite Framboid Formation

The rate of pyrite crystal growth is unknown. Various qualitative observations over the years suggest that it is rapid, but experimental data have not been collected. The best we have is the empirical result from Harmandas et al. (1998) in which experimental data was fitted to the empirical Equation 12.1.

$$R_p = ks\Omega_p^n \qquad\qquad 12.1$$

where R_p is the rate of pyrite crystal growth (mol m^{-2} s^{-1}), k is a crystal growth rate constant, s is a function of the active growth sites on the crystal surface, Ω_p^n

is the supersaturation, and n is the apparent order of the crystal growth process. Harmandas et al. (1998) found that the rate varied between approximately 10^{-8} and 10^{-6} mol m^{-2} s^{-1} for supersaturations between $10^{7.0}$ and $10^{7.4}$. These are of course empirical figures. What is more significant was their finding that the apparent reaction order of the growth process was 3.5 ± 0.5. This strongly suggests that the rate is diffusion controlled.

This is consistent with an estimation of the diffusion rate of dissolved Fe and S in aqueous solution. The diffusion constant for Fe^{2+}_{aq} in pure water is 3.31 + 0.15T°C and, because the solubility product for pyrite is so low, the relative supersaturation closely approximates the ion activity product $\{Fe^{2+}\}\{S_2^{2-}\}$ so that the concentration of Fe^{2+}_{aq} is approximately half the relative supersaturation. This means that, in the Harmandas et al. (1998) experiments, the limiting diffusive flux of aqueous Fe^{2+} to the pyrite surface was 10^{-6} mol m^{-2} m^{-1}, or the same order as the measured pyrite growth rate. In other words, pyrite crystal growth was rapid and limited by the rate of diffusion of dissolved components to the pyrite surface.

12.3.1. Crystal Size Distribution (CSD) Theory

The idea that the sizes of crystals in a system might be a probe into nucleation and crystallization processes was introduced into chemical engineering mainly through the work of Randolph and Larson (1988). It is an empirical theory that is based on crystal population density. The original idea was based on a simple mass balance in a batch crystallization chamber that assumed that the number of crystals is conserved as they grow: an equation can be written where the number of crystals within a particular size range is determined by the rate at which crystals reach this size range and leave the system in a particular time interval. It was rapidly taken up by petrologists, led by the pioneering work of Marsh (1988) and Eberl et al. (1998), and has been mainly used by igneous petrologists during this millennium. The attraction of the theory to geologists is that it does not require any knowledge of nucleation and growth kinetics, and the necessary input can simply be obtained by measuring crystal sizes in a rock. However, the application of CSD theory to natural systems involves a number of assumptions: (1) that there is a simple removal process by which crystals can leave the system in order that the number of crystals can be conserved as they grow; (2) that the growth rate is independent of crystal size; (3) that the number of crystals with zero length is nonzero and constant; and (4) that the system has reached steady state and the number of crystals with any size range is constant. The solution to the mass balance is then in the form of an exponential relationship between crystal population density and size. The shape of the curve of plots of size versus number of crystals may be used to provide information about the growth mechanism of the crystals and has even be used to estimate the rate of crystal growth. However, Lasaga (1997) pointed out that solutions to the CSD equations can be obtained if the growth rate and the boundary conditions are known (cf. Špillar and Dolejš, 2013).

CSD theory was applied to pyrite framboids by Wilkin et al. (1996) as a possible explanation for their thesis that the ratio of the microcrystal diameter to the framboid diameter could be used as a paleoenvironmental indicator. However, the CSD model does not appear to be strictly relevant to framboids. In particular, the evidence for the widespread removal of pyrite crystals from the system is lacking and, considering the extreme stability of pyrite in natural systems, it seems unlikely. It is also the case that framboid formation is dependent on the hydrodynamics of the system, as well as the physico-chemistry of nucleation. Each framboid represents an essentially isolated system where the rate of introduction and removal of new material to the site of framboid formation are restricted. This means that the number of crystals is constant in each individual framboid system. The sizes of large numbers of framboids may not be interrelated in terms of mass balance in the system as a whole.

Wilkin et al. (1996) described a number of crystal size versus number curves that clearly showed exponential relationships both for the framboids and for the microcrystals within individual framboids. The consequence is that these curves must derive from variations in kinetic parameters—although Pan (2001) argued that the relationship was intrinsic to the measurements themselves and therefore represented a circular argument. Lasaga (1998) showed that a similar exponential relationship would be obtained if both nucleation and growth were included in a system that closely approximates burst nucleation followed by quenched growth.

12.3.2. Diffusion-Controlled Growth

In the classical LaMer model, the burst nucleation event depletes the local system of nutrients to such an extent that the monomer concentration falls below the nucleation threshold. Subsequent microcrystal growth is then determined by diffusion of nutrients to the surface. Pyrite framboids tend to form in environments with little advective mixing, such as the interstices of fine-grained clastic sediments or in stagnant water columns. This has been observed experimentally by Rickard (2012b) and Wang and Morse (1996a). In these systems the rate of pyrite crystal growth is greater than the rate of diffusion of nutrients to the site of framboid formation, and microcrystal growth is limited. In framboids the outermost microcrystals are similar in size to the interior microcrystals: if a continual supply of dissolved Fe and S were available, then the outermost microcrystals would continue to grow. This is seen in framboidal textures, often given exotic names such as *sunflower pyrite* (Figure 1.13, Chapter 1), where there are overgrowths of pyrite on the framboids either due to continued nutrient supply or a new supply after a break in pyrite formation. In so-called *colloform pyrite* (Figure 3.5, Chapter 3), which is often formed in hydrothermal systems, the mass of overgrowth pyrite itself limits access of further dissolved Fe and S to the surface and the framboids appear to be buried in massive pyrite.

In a classical system where the growth of microcrystals is determined by monomer addition, the number of critical nuclei generated in the period of burst nucleation determines the number of growing particles (section 11.2). In framboids, therefore, where the classical model fully operates, this computation may be reversed, and the number of microcrystals in a framboid approximates the number of critical nuclei generated in the burst nucleation period.

The relative significance of diffusion-controlled growth in the development of the monodisperse microcrystals observed in framboids can be evaluated by considering a theoretical system in which the concentration of monomers in the bulk solution, c_b, is constant. The concentration at the particle surface is c_0, the equilibrium concentration. The rate of increase in particle radius, r, with time, t, is then

$$\frac{dr}{dt} = \frac{\left[D(1+r/\delta) v_0 (c_b - c_0)/r \right]}{\left[1 + D(1+r/\delta)/k_i r \right]} \qquad 12.2$$

where D is the diffusion coefficient, δ is the width of the diffusion boundary layer around the particle, v_0 is the molar volume of the monomer and k_i is the rate constant for the interface reaction.

In a surface reaction-controlled process, the interface reaction becomes rate controlling ($D >> k_i$) and the rate of growth is independent of particle size. In the diffusion-limited reaction ($D << k_i$), the growth rate is a function of the particle size: as the particle size increases the growth rate decreases. Therefore diffusion-controlled growth has a much stronger tendency to produce the type of monodispersed microcrystals observed in framboids.

12.3.3. The Rate of Pyrite Framboid Formation

The time taken for a pyrite framboid to form in sedimentary environments was reported by Rickard (2019), who estimated that the time for an average sedimentary 6 μm diameter framboid to form is 5 days.

In the framboid-forming system, the limiting bulk solution in the sediment or water column can be regarded as approaching a constant value. This means that the rate of formation of the framboids can be closely approximated by a steady state solution to the three-dimensional equation for spherical diffusion (Berner, 1971; Carslaw and Jaeger, 1959; Lasaga, 1998):

$$J_s = J/A = 2\varphi\pi D D R_b \left(c_b - c_0\right)/\left(R_b - 1\right)$$ 12.3

(see Table 12.2 for symbols, definitions and units in Equation 12.3).

One of the interesting aspects of this system is that the flux of dissolved nutrients to the framboid is independent of the radius of the bulk solution volume. This is because pyrite is relatively insoluble (i.e., $c_b - c_0 \sim c_b$) and the radius of the bulk $R_b \gg 1$ μm (i.e., $R_b - 1 \sim R_b$), and Equation 12.3 can be simplified to

$$J_s = 2\varphi\pi D D c_b$$ 12.4

In the model, the formation of framboids is diffusion-limited and the smallest diffusion coefficient of the dissolved Fe and S species limits the maximum flux. Boudreau (1996) gave $D = 3.31 + 0.15T°C$ in pure water. In sediments,

Table 12.2 Symbols, Definitions, and Units Used in the Diffusion Equations (e.g., Equation 12.3)

Symbol	Definition	Units
J	Flux	$g\,\mu m^{-2}\,s^{-1}$
J_s	Flux per unit area of the framboid	$g\,s^{-1}$
A	Surface area of the framboid	μm^2
D	Diffusion coefficient	$\mu m^2\,s^{-1}$
D_s	Whole sediment diffusion coefficient	$\mu m^2\,s^{-1}$
D	Framboid diameter	μm
R_b	Radius of the bulk solution	μm,
c_b	Concentration of monomers in the bulk solution	moles μm^{-3}
c_0	Concentration of monomers at the framboid surface	moles μm^{-3}
φ	Sediment porosity	dimensionless
Θ	Tortuosity	dimensionless

diffusion is represented by the whole sediment diffusion coefficient D_s which takes into account the tortuosity, Θ, a measure of the relative path length of the diffusion process (Berner, 1971). Since $\Theta > 1$, $D_s < D$. Various suggestions for the value of D_s have been published, but the variation is relatively insignificant compared with uncertainties in other areas of the diffusion equation. Thus, for example, Raiswell and Anderson (2005) used 0.75 as a factor describing the properties of the first 30 cm of fine-grained sediments, which gave D_s $(Fe^{2+}) = 3.53 \times 10^{-6}$ cm^2 s^{-1}.

The time taken for average-sized sedimentary framboids to form is shown in Figure 12.13, calculated from Equation 12.4. The time here is actually the period between burst nucleation and the final assembly of the microcrystals into the complete framboid; the point is that it may take some time for the concentrations of dissolved Fe and S to build up to a sufficient supersaturation that pyrite nucleates. This period of time—equivalent to the lag phase of nucleation in classical kinetics—is indeterminate.

The computations are for the limiting condition where the concentration of dissolved Fe and S in the bulk solution is controlled by the solubility of FeS_m (mackinawite). Two situations are shown: diagenetic, for framboids formed in fine-grained clastic sediments with porosity of around 20%, and the other, syngenetic, for framboids formed in the water column.

The time taken for framboids with diameters up to 80 µm to form is shown in Figure 12.14. As shown in the figure, framboids take between a few hours and a few years to form in sediments. Most framboids are within this size range, but rare framboids have been reported with diameters >80 µm. However, the

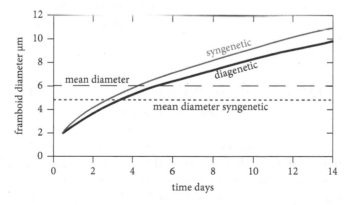

Figure 12.13. Logarithm of time in seconds versus framboid size in µm for water column (syngenetic) and sediment (diagenetic) at STP.

Reprinted from *Earth and Planetary Science Letters*, 513: 64–68. How long does it take a pyrite framboid to form?. Rickard, D., 2019. Copyright (2019) with permission from Elsevier.

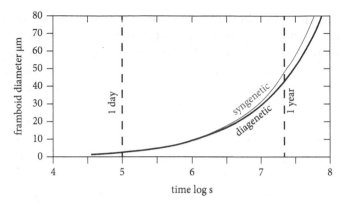

Figure 12.14. Time (log s) taken for framboids up to 80 μm in diameter to form in sediments (diagenetic) and in open water (syngenetic). The curve is exponential so very large framboids (> 80 μm in diameter) do not take much longer than a few years to form.

Reprinted from *Earth and Planetary Science Letters*, 513: 64–68. How long does it take a pyrite framboid to form?. Rickard, D., 2019. Copyright (2019) with permission from Elsevier.

exponential nature of the curves suggests that these larger framboids will not take significantly longer to form than the 2.2 years needed for 80 μm diameter framboids: the time taken for the largest framboid, about 250 μm in diameter (section 2.1 of Chapter 2) is about 3 years.

The different times taken for the formation of syngenetic and diagenetic framboids may contribute to the smaller average diameters of syngenetic framboids (section 2.5.1 in Chapter 2). In the relatively dynamic environment conditions of open water, the stable conditions necessary for the formation of any individual framboid are likely to be relatively short-lived, resulting in smaller framboids.

12.3.4. The Rate of Polyframboid and Framboid Cluster Formation

The results of the model explain the common occurrence of pyrite framboids in clusters and within dead organisms and fossils, such as shells and plant remains. If each framboid required a fixed volume of solution in which to form, then we might expect to see framboids distributed at a distance from each other dependent on the size of the system necessary to contain the mass of dissolved Fe and S contained within the framboid. Once the critical supersaturation for pyrite is reached within the shell, the size of the shell itself—which effectively constitutes the volume of the bulk solution in the model—is irrelevant. A feature of the model is that the rate of formation is independent of the size of the framboid-forming system, which is consistent with the observation of framboids commonly occurring in groups. In fact, burst nucleation only depletes a solution volume similar to that of the individual framboid. Subsequent pyrite crystal growth is relatively rapid and determined by the diffusion gradient caused by the difference between the extremely low solubility of pyrite and the concentrations of dissolved iron and sulfur in the bulk solution.

Many polyframboids, for example, have been formed within the shells of organisms, such as foraminifera (see section 7.1.7), which provides the long-term stable environment required for their formation. The model reveals how these polyframboids are able to form since the formation of the individual framboids is independent of shell volume.

The framboids in the polyframboid pictured in Figure 3.14 (Chapter 3), for example, vary between 18 and 25 µm in diameter. This means they took between 63 and 162 days to form, according to Equation 12.4. This is longer than the average sedimentary framboid of ca 6 µm diameter takes to form (around 5 days) and suggests a long-term stable environment, such as might be provided by the interior of a foraminifera (section 7.1.7 of Chapter 7), for example. The longer times taken to form also explain several of the framboid boundaries which have been obviously impinged by contact with an adjacent framboid. However, the time period is not unreasonable for the burial of an individual foraminifera with the internal pyritization through polyframboid formation.

12.4. Molecular Mechanism of Pyrite Crystal Growth

As discussed in sections 10.2.2 and 10.3.1 in Chapter 10, pyrite crystal growth occurs through the same process as pyrite nucleation from solution: the reaction of $S_2(-II)$ or H_2S with FeS moieties, where the FeS moieties in pyrite crystal growth are the =FeS sites at the pyrite surface. The overall reaction (Equation 12.5) with polysulfide is an exchange reaction between the dissolved persulfide, written generically as $S_2(-II)$ to include such species as HS_2^- as well as S_2^{2-}.

$$=FeS + S_2(-II) = =FeS_2 + S(-II) \qquad 12.5$$

The S(-II) produced reacts with neighboring =Fe sites to propagate more =FeS sites. The net effect is that all non-stoichiometric sites on the pyrite surface react to produce more =FeS$_2$. The location of these non-stoichiometric sites at surface defects, such as vertices, edges, and dislocations, is consistent with the observed growth characteristics of pyrite. For example, it explains how screw dislocations develop. This is consistent with the kinetics of pyrite crystal growth which is fast and diffusion-controlled rather than surface reaction–controlled. It is also consistent with the basic attributes of framboid microcrystals which require (1) that all the microcrystals within a framboid develop at the same time in the same environment, (2) that framboids develop in diffusion-dominated, stagnant systems, and (3) that burst nucleation is followed by diffusion-limited, microcrystal growth.

12.4.1. Effect of Temperature

Much of the early data on the origin of pyrite crystal habits in natural systems derives from studies of pyrite crystals in hydrothermal ore deposits (reviewed in Murowchick and Barnes, 1987).

Murowchick and Barnes (1987) found experimentally that pyrite habit changed as the temperature increased from 250° to 450°C from a cube through an octahedron to a pyritohedron (Figure 12.15). This was in concert with increases in supersaturations, and it would appear that a major effect of temperature on pyrite formation in aqueous solutions is an increase in supersaturation for the same dissolved Fe and S concentrations as the temperature, and thus solubility, rises.

The effect of temperature on the rate of framboid formation is not well constrained. This is because the concentration of monomers in solution is not limited by the solubility of FeS_m since mackinawite does not form in hydrothermal solutions. If the rate is diffusion-controlled, then the effect on the monomer flux is through the variation of the diffusion coefficient, D, with temperature, T. This tends to follow the Arrhenius equation, and the Arrhenius activation energy for diffusion is generally low relative to the activation energies

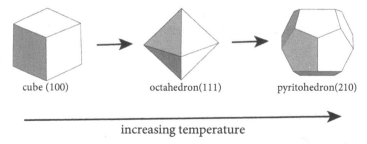

cube (100)   octahedron(111)   pyritohedron(210)

increasing temperature

Figure 12.15. Effect of increasing temperature on pyrite crystal form.

of chemical reactions, varying between about 12 and 27 kJ mole^{-1} (Rickard, 1975). Even so, the diffusion coefficient commonly increases by more than a magnitude over the temperature range from ambient to hydrothermal conditions. Boudreau (1996), for example, estimated that $D(Fe^{2+})$ increases from 3 x 10^{-6} cm^2 at 0°C to 33 × 10^{-6} cm^2 s^{-1} at 200°C and 63 × 10^{-6} cm^2 s^{-1} at 400°C. This means that rate of diffusion is over a magnitude faster in hydrothermal solutions than at ambient sedimentary environments. As a first approximation, this suggests that the rate of framboid formation is over a magnitude faster in hydrothermal systems than in sedimentary environments under similar monomer concentrations. In practical terms this means that an average 6 μm framboid forms in around 12 hours in hydrothermal systems, rather than 5 days in sediments.

This is significant since the hydrothermal environment is more unstable than the sedimentary environment, with large temperature and chemical gradients and a dominance of advective flow over stagnant, diffusion-dominated conditions which enhance framboid formation. The relatively rapid framboid formation rates in hydrothermal systems means that stable conditions need only be maintained for shorter periods for framboids to have the chance to form. Even then, it is more probable for pyrite macrocrystals to form in these conditions, and this conclusion is consistent with observations.

The combination of faster diffusion with a higher potential monomer concentration in the bulk hydrothermal solution leads to the possibility that the rate of pyrite microcrystal growth becomes surface reaction–controlled rather than diffusion dependent in hydrothermal solutions. This in turn suggests that surface nucleation growth is the dominant pyrite growth mechanism in hydrothermal systems. Surface nucleation growth was originally proposed by Sunagawa (1957) for the growth of hydrothermal pyrite on {1 0 0} surfaces and was observed by Endo (1978). Endo (1978) concluded that surface nucleation growth was the dominant growth process for hydrothermal pyrite. At ambient temperatures {1 0 0} faces develop at low supersaturations through screw dislocation growth, as described previously. The change to surface nucleation growth for these facets is consistent with the increased pyrite solubility so that the most stable faces are growing at greater monomer concentrations, thus facilitating surface nucleation growth. This suggests that the mechanism of crystal growth is largely independent of the supersaturation but directly dependent on the monomer concentration in solution. Consistent with this conclusion, computations suggest that the presence of water facilitates the development of truncated cubes at ambient temperatures evolving to cubes in hydrothermal solutions over 237°C (Barnard and Russo, 2009a).

12.4.2. Effect of pH

Although pyrite displays extreme stability in terms of pH space, variations in pH affect the rate of framboid formation through the effect of pH on the solubility of FeS_m and thus the value of c_b in Equation 12.4. In acid solutions the solubility of FeS_m becomes pH dependent (Rickard, 2006). In this region, the equilibrium concentrations of dissolved Fe^{2+} and H_2S are described by Equation 12.6 for which the equilibrium constant at SATP is $10^{3.5}$.

$$FeS_m + 2H^+ = Fe^{2+} + H_2S \qquad\qquad 12.6$$

In acidic solutions, the ionic activity product of Fe and S in solution is a function of the square of the proton concentration, meaning that the solubility increases rapidly with decreasing pH. As can be seen from Figure 12.16, the amount of dissolved Fe^{2+} in equilibrium with FeS_m in acidic solutions for any given total dissolved sulfide concentration is virtually unlimited: that is, FeS_m solubility may no longer limit the dissolved Fe^{2+} and H_2S concentrations.

The counter process here is that, as the pH becomes more acidic, the metastable dimorph marcasite tends to form in preference to pyrite. Interestingly, of course, microscopic spherical aggregates of marcasite microcrystals may be produced but, as mentioned in section 8.2.1 of Chapter 8, these do not meet one of the primary characteristics of framboids: orthorhombic marcasite tends not to

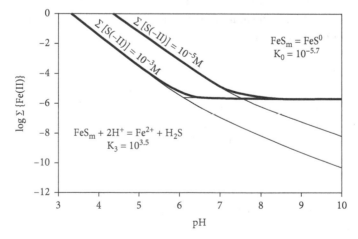

Figure 12.16. pH dependence of the solubility of FeS_m (bold lines) at SATP in terms of total dissolved Fe(II) concentrations, $\Sigma[Fe(II)]$, for total dissolved sulfide concentrations $\Sigma[S(-II)] = 10^{-3}$ and 10^{-5} M.

form equant microcrystals (Figure 8.2, Chapter 8). Marcasite is metastable with respect to pyrite and thus its formation at pH <5 has been a subject of considerable discussion (section 8.2.1). Kitchaev and Ceder (2016) suggested that phase selection between pyrite and marcasite is due to relative surface stabilities as a function of pH.

The acid region of pH space between marcasite formation (pH <5) and neutral conditions (pH = 7) is commonly found in freshwater systems. The conclusion that the rates of framboid formation in freshwater systems can be considerable is consistent with the observations of huge number of pyrite framboids, for example, in lake sediments (e.g., Vallentyne, 1963).

13

Framboid Self-Assembly and Self-Organization

The physical processes involved in the formation of framboids can be divided into (1) self-assembly and (2) self-organization. These two processes are basically unrelated. All framboids are aggregates and therefore self-assemble, but only a proportion are self-organized. The terms *self-assemble* and *self-organize* are often used synonymously in the literature, but I prefer *self-assembly* to describe the process of aggregation of framboidal microcrystals, and *self-organization* to describe the process whereby framboids achieve ordered or partially ordered arrays of their constituent microcrystals. As described in Chapter 5, for example, there is a gradation between organized, partially organized, and randomly organized framboids. In the most extreme cases, irregular masses of pyrite microcrystals may show patches where the microcrystals are organized (see Figure 13.5 later in the chapter). Rickard (1970) originally observed that framboids show varying degrees of order and disorder and, in an individual framboid, various domains of relative order and disorder may occur.

13.1. Self-Assembly: The DVLO Theory

The fundamental driving for the self-assembly of framboids is reduction in free energy, which reduces the surface free energies of the individual microcrystals, and the development of sub-spheroidal forms, which minimizes the total surface free energy.

I include a crib table here of symbols and abbreviations used in this chapter for ease of reference (Table 13.1).

Most theories describing aggregation of micro- and nano-particles are clustered around the Derjaguin-Landau-Verwey-Overbeek (DVLO) approach (Derjaguin and Landau, 1941; Verwey and Overbeek, 1948). The DVLO approach has had the widest application to date since many of the parameters involved can be estimated from experimental measurements.

The van der Waals energy, V_{vdW} (J m^{-2}), between parallel faces of two crystals can be described by Equation 13.1 where h (m) is the distance between the crystal faces and H is the Hamaker constant, which enters into the equation to

Framboids. David Rickard, Oxford University Press. © Oxford University Press 2021.
DOI: 10.1093/oso/9780190080112.003.0013

Table 13.1 Symbols and Abbreviations used in Chapter 13

Symbol	Definition	Units
CCC	Critical coagulation concentration	mol L^{-1}
DVLO	Derjaguin-Landau-Verwey-Overbeek	
e	Charge on the electron	91.6×10^{-19} C
F	Faraday constant	96 485 C mol^{-1}
F_{EDL}	Electrostatic double layer force	N m^{-2}
F_{vdW}	van der Waals force per unit area	N
H	Hamaker constant	zJ
h	Distance between the crystal faces	m
I	Ionic strength of the solution	mol L^{-1}
k_B	Boltzmann constant	$1.38064852 \times 10^{-23}$ m^2 kg s^{-2} K^{-1}
pH$_{IEP}$	Isoelectric point	pH
pH$_{zpc}$	Zero point of charge	pH
r	Radius	m
R	Gas constant	8.314 J mol^{-1} K^{-1}
V_{EDL}	Electrostatic double layer energy	J m^{-2}
V_{vdW}	van der Waals energy	J m^{-2}
Z	Interaction constant in equation 14.4	
z	Electrolyte valence	
ε_0	Permittivity in a vacuum	F^{-1} m^{-1}
ε_W	Permittivity of water	F^{-1} m^{-1}
ε	Relative permittivity	
ζ	Zeta potential	mV
κ	Inverse Debye length	m^{-1}
Ψ_0	Stern layer potential	mV

account for differences in the composition and structures of the materials (see section 13.1.1).

$$V_{vdW} \approx -\frac{H}{12\pi h^2}$$ 13.1

Equation 13.1 is derived from the equations describing the van der Waals force between two parallel plates (Israelachvili, 2011). It is strictly true for a plate of unit area interacting with the infinite area of a parallel plate. In practice, Equation 13.1 is a good approximation for situations where the crystal sizes are large compared with the separation distance. If the volume of the framboid was fixed *ab origo*, then the initial distances between the embryonic microcrystals was large relative to the size of the microcrystals, and Equation 13.1 does not strictly apply. As the microcrystals grow in a fixed volume, the intercrystal distance decreases and Equation 13.1 becomes a more accurate description of the van der Waals forces.

The force per unit area between two parallel plates, F_{vdW} (N), can be obtained by differentiating V_{wdW} with respect to h (Equation 13.2).

$$F_{vdW} = \frac{dV_{vdW}}{dh} = -\frac{H}{6\pi h^3}$$ 13.2

Equation 13.2 can be used to compare the van der Waals forces between cube and octahedra since all the basic parameters are the same. The van der Waals force between two cube faces is about three times that force between the faces of a similarly sized octahedra because of the relative difference in surface area. This suggests that octahedral microcrystals are less constrained than cubic microcrystals and can therefore more readily reorient into organized arrays. There is some evidence that octahedra or modified octahedra are often the microcrystal habit observed in well-organized framboids (see Chapter 4). In fact, the common truncated octahedra and cube forms of microcrystals can be approximated as spheres, and the equation for the van der Waals force between spheres of radius r could be used (Equation 13.3).

$$F_{vaW} = -\frac{H}{6h^2}$$ 13.3

Simple inspection of Equation 13.3 shows that the van der Waals force between two spheres increases proportionally to the size of the spheres and inversely as the square of the distance between the spheres. The net effect is that as the microcrystals grow in a fixed framboid volume, the van der Waals attractive force between adjacent microcrystals increases as a function of the cube of the intercrystalline distance and consequently as a similar function of time.

13.1.1. Hamaker Constant

Inspection of Equations 13.1 and 13.2 shows that the only material specific parameter involved in the determination of the van der Waals force is the Hamaker constant, H. It is unknown for pyrite and has been calculated for only a limited number of materials (e.g., Bergstrom, 1997; Faure et al., 2011). Israelachvili (2011) notes that H for semi-conductors like pyrite should be much greater than for dielectric and non-conducting materials. Measured and computed H values for metals in water vary between 10 and 40 zJ. Bergstrom (1997) calculated that H for a variety of materials in water varied between 0.3 and 11 zJ. The nearest analogies to pyrite, PbS and cubic ZnS, have H values close to 5 zJ. Similar values (5 to 10 zJ) were determined for iron oxides by Faure et al. (2011). Wilkin and Barnes (1997) assumed a value of 10 zJ for greigite.

13.1.2. Electrostatic Interactions

The electrostatic double layer is a simplified representation of the surface chemistry of particles in a medium such as water in which two layers of charge are defined. The first layer is the surface charge resulting from ions adsorbed onto the surface. The second layer is a diffuse layer, consisting of ions attracted to the charged surface layer. The energy V_{EDL} (J m^{-2}) between two microcrystal surfaces separated by a distance h (m) in water can be described by Equation 13.4 for two parallel plates (Israelachvili, 2011).

$$V_{EDL} = \frac{\kappa}{2\pi} Z e^{-\kappa h}$$

13.4

Where κ is the inverse Debye length (m^{-1}) and Z is an interaction constant given by Equation 13.5.

$$Z = 64\pi\varepsilon\varepsilon_0 \left(\frac{k_B T}{e}\right)^2 \tanh^2\left(\frac{z e \psi_0}{4 k_B T}\right)$$

13.5

In Equation 13.5, z is the electrolyte valence, e is the charge on the electron (91.6×10^{-19} coulombs), k_B is the Boltzmann constant ($1.38064852 \times 10^{-23}$ m^2 kg s^{-2} K^{-1}), and Ψ_0 is the potential at the Stern layer. The Stern layer potential is not strictly equivalent to the zeta potential, but since the zeta potential is measurable, it is commonly used as an approximation. Equation 13.5 can be simplified using the approximation for the series expansion in Equation 13.6.

$$\tanh(x) = x - \frac{1}{3}x^3 - \frac{2}{15}x^5 \ldots \approx x \left(where\, x < 1\right) \qquad 13.6$$

For a 1:1 electrolyte such as NaCl, $z = 1$ the electrostatic potential is then approximated by Equation 13.7.

$$V_{EDL} \approx 2\varepsilon\varepsilon_0 \kappa \Psi_0^2 e^{-\kappa h} \qquad 13.7$$

In Equation 13.7, ε is the relative permittivity and ε_0 the permittivity in a vacuum. For water the relative permittivity, ε is defined as the ratio of the permittivity of water, ε_w, to the vacuum permittivity. The permittivity has units of $F^{-1}\, m^{-1}$ whereas the relative permittivity is dimensionless. This representation is preferred to references to the dielectric constant since it has become progressively less clear exactly what that term describes (Delgado et al., 2005).

The force F_{EDL} ($N\, m^{-2}$) between two parallel crystal surfaces is then given by Equation 13.8.

$$F_{EDL} = -2\varepsilon\varepsilon_0 \kappa^2 \Psi_0^2 e^{-\kappa h} \qquad 13.8$$

13.1.3. The Debye Length and Microcrystal Self-Assembly

The thickness of the ionic atmosphere near the charged particle surface is described by the Debye length, $1/\kappa$. The magnitude of the Debye length is solely dependent on the properties of the solution and entirely independent of the nature of the surface. Conventional renderings of the reciprocal thickness of the double layer, κ, are described by equations like Equation 13.9.

$$\kappa = \left(\frac{2F^2\, I 10^3}{\varepsilon\varepsilon_0 RT} \right)^{1/2} \qquad 13.9$$

The relative permittivity of water in Equation 13.9 varies between 80 (20°C), 55 (100°C) to 35 (200°C) and ε_0 is $8.854 \times 10^{-12}\, F\, m^{-1}$; F is the Faraday constant

(96485 C mol^{-1}), R is the gas constant (8.313 J mol^{-1} K^{-1}), and I is the ionic strength of the solution (mol L^{-1}).

Equation 13.9 shows that the double layer thickness varies as the square root of the ionic strength of the solution. A plot of this equation for temperatures between 20°C and 200°C is shown in Figure 13.1. The plot shows that temperature over this range does not affect the Debye length significantly. However, as the ionic strength of the solution approaches that of seawater, the Debye length decreases to less than 10Å, or the size of a single molecule. In effect, the double layer collapses and the repulsive force becomes small, resulting in microcrystal self-assembly. The result is that the electrostatic force preventing framboid microcrystal assembly disappears in seawater. The only part of the DVLO force system occurring in pyrite microcrystals in seawater systems are the attractive forces.

The consequence of Equation 13.9 is, as illustrated in Figure 13.1, the collapse of the repulsive electrostatic double layer in higher ionic strength solutions. This in turn affects the kinetics of microcrystal self-assembly. Particle aggregation can be divided into two mechanisms: (1) fast aggregation, which is diffusion controlled and occurs in more concentrated solutions; and (2) slow aggregation, which is controlled by thermal activation and occurs in low salt concentrations. The demarcation between these two regimes is defined by the critical coagulation concentration (CCC).

H. Schulze and W. B. A. Hardy independently developed a rule for determining the CCC (Schulze, 1882; Hardy, 1899). Where z is the ionic valence, the Schulze-Hardy rule is given by Equation 13.10.

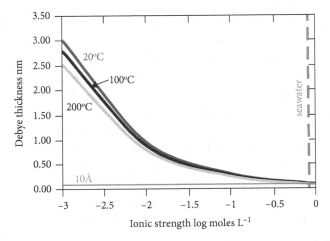

Figure 13.1 Debye thickness (nm) versus logarithm of the ionic strength (log mol L^{-1}) for particles in water at temperatures between 20 and 200°C.

$$CCC \propto \frac{1}{z^6}$$
                                                                    13.10

The Schulze-Hardy rule has quite dramatic consequences for the kinetics of framboid microcrystal self-assembly since it is a function of the 6th power of the ionic valence. Trefalt et al. (2017) revisited the Schulze-Hardy rule and showed that the behavior of the colloidal particles depended on whether the multivalent ions were counterions or co-ions. In particular, the dependence of the CCC on ionic valence varied inversely with the ionic valence to an inverse situation where the CCC varied directly with the ionic valence.

13.1.4. Pyrite Surface Charge

The CCC is independent of the nature of the solid surface apart from that of the overall surface charge. That is, counterions on one surface may become co-ions on another. Trefalt et al.'s contribution defines the relative contributions of electrolytes with varying stoichiometries to the critical coagulation concentration of variously charged surfaces. The consequence is that materials which have the same surface charge as pyrite will coagulate under similar conditions.

The isoelectric point, pH_{IEP} the pH at which the pyrite surface carries no net electric charge in water, was determined by Fornasiero et al. (1992) to be 1.2. Bebie et al. (1998) obtained a similar value of 1.4, and Weerasooriya and Tobschall (2005) reported 1.7 with potentiometric titrations.

Equation 13.8 shows that the repulsive force due to electrostatic layer effects is an exponential function of the zeta potential, ζ, that is the potential difference, in mV, existing between the surface of a solid particle immersed in a conducting liquid and the bulk of the liquid. In general, colloidal suspensions of particles with zeta potentials less than ca. 30 mV display incipient instability and a tendency to agglomerate. This general rule of thumb stems from an older idea that relates instability to the ζ^2/κ ratio (cf. Hunter, 1988) which is a consequence of the classic DVLO theory. Bebie et al. (1998) found that the ζ-potential for pyrite in aqueous solution was around -30 mV between pH 5 and 9. This suggests that pyrite colloids in aqueous solutions may display incipient instability and tend to agglomerate.

Although the pyrite surface in most natural solutions (pH $>>1.4$) is negatively charged, most perturbations to the chemical system will tend to reduce the zeta potential and therefore increase the instability of pyrite sols. Fornasiero et al. (1992) demonstrated that the inconsistent pH_{IEP} measurements reported for pyrite in the literature were primarily a function of the oxidation state of the surface. For example, pH_{IEP} values of around 6–7 are commonly reported for pyrite in air (e.g., Fuerstenau et al., 1968; Vilinska and Rao, 2011). The effect of dissolved metals on the ζ- potential of pyrite is also quite dramatic. For example, Bebie et al. (1998) showed that dissolved Fe^{2+} concentrations of 0.5 mM resulted in a pH_{IEP} of ~5 and a zero point of charge, pH_{zpc}, of ~8.

The interaction of the pyrite surface with organic molecules is of particular interest with respect to framboid formation in view of the frequent association of framboids with biofilm. It appears that the interaction of organics with the negatively charged pyrite surface occurs regardless of the charge of the organic species (Bebié and Schoonen, 1999, 2000). Organic interactions are dominated by chemical reactions at specific surface sites on the pyrite surface, and electrostatic forces are relatively unimportant. The net result is the reduction of repulsive electrostatic charge on the pyrite surface and a consequent increased tendency for the pyrite microcrystals to aggregate. In other words, the involvement of organic materials, such as biofilms, in pyrite framboid formation may not be simply mechanical, providing a structural support or constraining volume. The presence of organic macromolecules will reduce the repulsive electrostatic forces and promote microcrystal self-assembly.

The requirement for a limited window of surface charge to allow close packing to occur means that the process is sensitive to the chemistry of the solution in which framboid formation occurs. For example, the surface charge is pH dependent, and changes in pH have been used to force weakly charged silica spheres into alignment (Rugge and Tolbert, 2002). Likewise, the surface charge—and consequently self-assembly—can be facilitated by varying the surface charge through the presence of organic substances and even changes in the ionic strength.

The sensitivity of pyrite surface charge to environmental parameters has been discussed in some detail by Rickard (2012c). The problem that this sensitivity causes is that it seems impossible with the present state of knowledge to predict what the surface charge of pyrite will be in any particular natural system. On the other hand, we know that pyrite framboids form in a wide variety of natural environments, so we know that the actual surface charge on the pyrite microcrystals is fairly robust and normally varies between the limits required for microcrystal self-assembly.

13.1.5. Net DVLO Forces in Pyrite Framboid Self-Assembly

The consequences of the LaMer burst nucleation process described in section 11.1 of Chapter 11 are that (1) the volume of the final framboid is similar to the initial volume of the solution containing the pyrite supernuclei, and (2) the number of microcrystals in the framboid is equivalent to the original number of supernuclei. The result of these characteristics is that the size of the pyrite microcrystals, represented by r in Equation 13.3, and the intercrystalline distance, h, are not independent variables. As the pyrite microcrystals grow, they approach closer to their neighbors; that is, h decreases as r increases (Figure 13.2).

The van der Waals attractive force, F_{vdW}, is a function of the cube of the reciprocal distance between the microcrystals, $1/h^3$ (Equation 13.2). The electrostatic repulsive force, F_{EDL}, is an exponential function of the distance between the microcrystals (Equation 13.8). The DVLO force, or the net force, is then the sum of the repulsive electrostatic double layer repulsion and the van der Waals attraction, and this varies as a complex function of the distance separating the microcrystals. As the microcrystals grow in a confined volume, the van der Waals attractive force dominates, as shown in Figure 13.3.

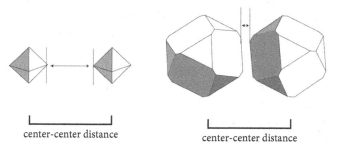

center-center distance   center-center distance

Figure 13.2. Decrease in separation distance for growing microcrystals with the same center-center distance according to Equation 13.3. The octahedral microcrystals develop cubic faces as the supersaturation decreases with time.

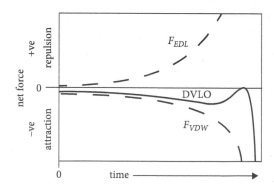

Figure 13.3. Net DVLO forces between microcrystals as a function of time and decreasing inter-crystal distance, with end-member electrostatic double layer repulsive force F_{EDL} and van der Waals attractive force F_{VDW} indicated. The net DVLO force is indicated for a pyrite-like material with low to moderate surface energy.

The development of net DVLO forces can therefore be illustrated with respect to time as the framboid microcrystals grow. The net DVLO force can be estimated from the Derjaguin approximation (Derjaguin, 1940), and the graph in Figure 13.3 is based on calculations in Ruths and Israelachvili (2011). The net effect is an increase in the attractive force between the microcrystals as the framboid develops. The final framboid is effectively dry, and the van der Waals interactions are much stronger in the absence of water. This results in a relatively strong inter-crystalline adhesive force, which helps explain why framboids are so robust and able to survive for millions of years of geologic time.

In fact, the development of DVLO forces illustrated in Figure 13.3 is an underestimate of the significance of adhesive forces in framboid formation. The results of the computations of net surfaces forces for pyrite microcrystals suggest that all the environmental parameters promote reduction in the net repulsive forces due mainly to electric double layer. For example, changes in ionic strength lead to a collapse in the thickness of the Debye layer and a decrease in the critical coagulation concentration, leading to rapid coagulation of the pyrite microcrystals. The presence of organic macromolecules in the system due to biofilms, for example, also promotes aggregation. Moreau et al. (2007) showed that extracellular proteins enhanced the aggregation of ZnS nanoparticles and reported that amino acids, particularly cysteine, were major actors in this process. Qi et al. (2019) further elucidated the mechanism, showing that the negatively charged amino acids facilitate the absorption between the extracellular proteins and positively charged metal moieties such as Zn(II). In both cases, the extracellular proteins were produced by sulfate-reducing bacteria which are, of course, the main source of sulfide in sedimentary pyrite framboids. The results contribute further to understanding the intimate relationship between biofilms and framboids discussed in section 7.2.1 of Chapter 7. The propensity for pyrite microcrystals to self-assemble is consistent with the widespread distribution of pyrite framboids and their independently computed formation rates.

13.2. Self-Organization

The internal organization of pyrite microcrystals has stunned microscopists ever since systematic studies of framboids in polished sections were initiated (e.g., Love and Amstutz, 1966; Kalliokoski and Cathles, 1969; Rickard, 1970). The eye-catching nature of these organized and partly organized framboids means that their abundances have probably been overestimated. Rickard (1970) estimated that up to 10% of the framboids he had studied were ordered or partly ordered, and Love and Amstutz estimated that 28% of the framboids they studied from the 390 Ma Devonian Chattanooga shale were ordered or partly ordered; the Chattanooga Shale has become the classical locality for the collection of ordered framboids. The formation of organized 3D arrays of pyrite microcrystals in framboids is one of the most extraordinary examples of abiologic self-organization observed in Nature. Even if only 10% of framboids show some degree of ordering, this suggests that over 10^{10} self-organized inorganic structures are being formed every second (cf. Rickard, 2015). This means that the process involved must be—at least from Mother Nature's standpoint— relatively mundane and an inevitable result of the physical chemistry of the framboidal system.

As discussed in section 6.3 of Chapter 6, the self-organization of microcrystals in framboids takes place by physical rotation of individual microcrystals during framboid formation. In this process, self-organization is a secondary feature of framboid formation after the self-assembly of pyrite microcrystals. Various hypotheses have been proposed to explain the processes that lead to self-organization in framboids, and these are discussed in the following. All have their limitations, and it may well be that different processes operate in different environments. One thing that is common to all these processes is that aggregated microcrystals must all be of the same size and shape for self-organization to occur. A secondary feature is that many framboids with dominantly randomly arranged microcrystals display organized regions, leading to the definition of partially organized framboids, as described in section 5.1 of Chapter 5. Third, irregular masses of pyrite microcrystals, so-called pyrite dust, can display regions where the microcrystals are organized, proving that, at least in this case, self-organization is not a necessary corollary of the development of a spheroidal form (see Figure 13.5, for example, later in this chapter).

13.2.1. Organic Templates

The proposal that framboids represented pyritized microorganisms is discussed in detail in section 7.1 of Chapter 7. However, framboids have been synthesized inorganically, and they form in natural environments where organisms cannot survive. Even so, the idea has taken a considerable hold on the imagination of researchers during the last 100 years since it was first proposed by Schneiderhöhn (1923), and it continues to crop up in the literature.

13.2.2. Spatial Constraints: Coacervates

The idea that framboids were formed in spatially confined sites was first suggested by Papunen (1966), who suggested that framboids were formed within spherical organic droplets which, at that time, were described as coacervates. The idea is discussed in detail in section 7.2 of Chapter 7. Obviously, not all framboids were formed in this way since framboids can be synthesized in inorganic systems and most framboids show no evidence of the contemporaneous presence of organic matter.

Strictly speaking, the coacervate hypothesis was not originally directed at explaining self-organization of framboids, but at explaining their apparent spherical shape and reconciling framboid formation with the organic residues many framboids contain. As noted in section 1.1.2 of Chapter 1, the apparent spherical shape of framboids was mainly due to the limitations of observations of framboid sections by optical microscopy; the introduction of high-resolution techniques for the observation of framboids in 3D, such as scanning electron microscopy, has revealed that many, if not most, framboids are faceted.

However, the coacervate theory has received a boost by the introduction of the solvent evaporation method for assembling colloidal particles (e.g., de Nijs et al., 2015; Wang et al., 2018). In this process, colloidal particles are dispersed in an apolar solvent. The dispersion is emulsified with water, and the solvent in the emulsion droplets slowly evaporate, causing the packing fraction of the colloidal particles to increase. In this process, the evaporation of the solvent and capillary action replace or augment the surface forces of the DVLO theory and permit the self-assembly of inert colloidal particles, such as polystyrene spheres, to be studied. The process appears to be somewhat analogous to Papunen's coacervate hypothesis, as well as being consistent with framboids being formed within biofilms (Large et al., 2001). Although the materials and methods of the syntheses are far removed from the natural environment, the results help elucidate the mechanism of self-organization in framboids, as discussed later in the chapter.

13.2.3. Spatial Constraints: Sediment Pores

The idea that spherical pores might occur in fine-grained sediments (e.g., Rickard, 1970), possibly caused by gas bubbles, was dismissed since framboids form in free water. The idea has received some new impetus by the discovery of methane bubbles at cold seeps in the ocean and inshore waters and the strong relationship between sulfate-reduction and methanogenesis. However, most pore spaces in sediments are highly irregular in shape. The sub-spheroidal shape of the sediment pores enclosing framboids is caused by micro-deformation of the sediment minerals around the pyrite framboid upon compaction and dewatering (Figure 13.4). This is consistent with the framboids being mainly formed during early diagenesis.

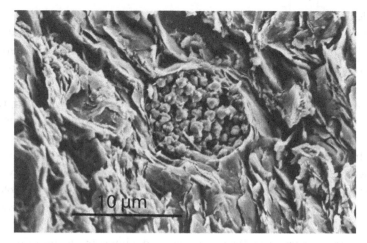

Figure 13.4. Clay minerals deformed around a pyrite framboid from the ca. 390 Ma Devonian Chattanooga Shale.
Scanning electron micrograph image by Ian Butler.

As described in section 4.5.7 of Chapter 4, there is apparently clear ev- idence for impeded growth in many natural sedimentary framboids (e.g., Figure 4.21, Chapter 4). In these framboids, the outer layers of microcrystals are only partially developed, forming a smooth outer surface to the framboid. At first glance, it appears that some of these partly formed microcrystals are flattened against the surrounding sediment minerals: it looks as though the outer layer of microcrystals either grew from the pore wall or grew against it. In fact, close observation of smooth framboid margins with partly formed outer layer microcrystals often clearly demonstrates a gap between the surrounding, com- monly clay, minerals and the pyrite (e.g., Figure 4.20b, Chapter 4).

Remnant organic sheaths around framboids are clearly related to the develop- ment of smooth framboid surfaces and consequent impeded growth of the outer layers of framboid microcrystals (e.g., Figure 4.21, Chapter 4). It seems probable that where framboids develop within organic matter, such as biofilms, there is a strong probability that smooth outer surfaces will result. The implication is that the organic envelope operates in the same way as the oil emulsion droplets in the syntheses of organized aggregates of polystyrene spheres. That is, it not only provides a confined space for microcrystal self-assembly but may, during intrinsic devolatilization, provide an additional force to bring the microcrystals together.

Assuming that the initial water-filled sediment pore in which the framboid nucleates and grows was not spherical, the suggestion is that, as the framboid microcrystals grew, compaction (with associated dewatering) of the sediment mechanically forced the sediment minerals around the framboid. That is, what we are observing is a coincidence between smooth-surfaced framboids with partially developed outer layer microcrystals and sediment microdeformation around the framboid on burial. The gap which is sometimes observed between smooth framboid surfaces and enclosing minerals may then be, partly, a trace of preexisting organic material.

These observations seem to suggest that smooth framboids tend to result from growth in organic matter envelopes: the coacervates of Papunen (1966) and Rickard (1970) and the biofilms of Large et al. (2001) and MaClean et al. (2008), for example. By contrast, fully formed outer layers of pyrite microcrystals evidence open space growth. Unfortunately, this idea has not yet been tested, and it would be interesting to further examine the relationship between organic matter and smooth framboids.

13.2.4. Topotactic Solid Phase Transformations

As discussed in section 10.2.6 of Chapter 10, some earlier workers wrongly supposed that sedimentary pyrite formed through a series of solid-state phase transformations involving, successively, mackinawite (tetragonal FeS_m) and greigite (Fe_3S_{4g}). The formation of sedimentary framboids by this mechanism was proposed in some detail by Wilkin and Barnes (1997). Close reading of their report shows that their analysis primarily concerned greigite rather than pyrite. They applied DVLO theory to a discussion of the formation of greigite framboids, and they concluded that greigite particles could not aggregate under van der Waals forces alone in freshwater systems. Greigite, being strongly ferrimagnetic, could aggregate by means of magnetic forces between the particles. Since they believed that greigite was a necessary precursor to pyrite, they supposed that their results were applicable to pyrite framboids.

In fact, the results of Wang et al. (2018) suggest that the inter-crystalline forces between the framboid microcrystals cannot be too strong. They must be strong enough to hold the particles together, but weak enough to permit rotation of the microcrystals into organized superstructures. This has been described as crucial to the development of internal ordering in these systems. This key requirement for weak inter-crystalline forces for framboid self-organization also precludes oriented attachment as a mechanism for self-ordering. Oriented attachment results in the fusion of the microcrystals and the formation of a crystal rather than a framboid. Indeed, oriented attachment might be involved in the recrystallization of framboids in metamorphic terrains, as described by Ostwald and England (1979), for example.

The results of the electron backscatter detection studies by Ohfuji et al. (2005) (section 6.3 of Chapter 6) also suggest that the mechanism for the development of self-organization of pyrite microcrystals in framboids through magnetic ordering of precursor greigite is unlikely. Since the magnetic moment of the greigite is crystallographically oriented, this process is incompatible with the crystallographic rotation of the pyrite microcrystals.

In fact, of course, greigite is not a necessary precursor to pyrite, and greigite framboids are rare in the natural environment. Greigite itself is a rare mineral, especially in marine systems, and tends to occur in freshwater sediments in association with plant material. In addition, pyrite framboids have been experimentally synthesized in the absence of greigite (e.g., Butler and Rickard, 2000), and greigite has not been shown to be involved in framboid syntheses (see section 10.2.6 of Chapter 10).

13.2.5. Compaction and Dewatering of Sediments

The idea that compaction and dewatering of sediments might play a key role in the development of framboid organization was first proposed by Amstutz et al. (1967). However, the orientation measurements conducted by Amstutz et al. (1967) were on framboids in ancient ca. 350 Ma old sediments which were completely dewatered, compacted, and lithified. They did not show a substantial degree of framboid orientation, and their conclusion that ordered framboids have a common preferred orientation in the shale was erroneous. Ohfuji (2004) and Ohfuji et al. (2005) examined framboids from the same shale as Amstutz et al. (1967) with electron backscatter detection (section 6.3 of Chapter 6). They found that crystallographic orientation data collected from the microcrystals that constitute ordered framboids confirm that there is no obvious preferred orientation of ordered framboids in the shale. In addition, regular arrangements of microcrystals are also observed in pyrite framboids from recent unconsolidated sediments (e.g., Bottcher and Lepland, 2000; Large et al., 2001; Stakes et al., 1999). Therefore, it seems that the self-organization of framboidal microcrystals occurs before late diagenesis and lithification of the surrounding sediments.

13.2.6. Microcrystal Alignment by Epitaxial Growth and Secondary Nucleation

One idea we had in the 1990s was that framboids were extreme examples of skeletal crystals where the individual microcrystals grew epitaxially on each other and nucleated successively through the structure (e.g., Butler, 1994). In fact, subsequent single crystal X-ray diffraction work by Butler himself, as well as the EBSD study by Ohfuji et al. (2005), proved that this was not the case (see

Chapter 6). A detailed crystallographic study on pyrite framboids reveals that the formation of the ordered microarchitectures is not crystallographically controlled, like some form of dendritic or skeletal growth, but probably occurs simply by reorientation of randomly aggregated microcrystals (section 6.4.1). This is analogous to the random crystal orientations of spherical nanoparticles (Naik and Caruntu, 2017).

13.3. The Self-Organization Process in Framboids

Much of our recent understanding of the processes involved in self-organization comes from studies of the syntheses of atomic clusters, nanocrystals, and inert colloidal spheres (e.g., Naik and Caruntu, 2017). There are necessarily some differences in the behavior of these particles during self-assembly compared with the pyrite microcrystals in framboids. Even so, the processes observed in experimental and simulated self-assembly of these particles are close analogies in the processes of self-organization of framboids.

Li et al. (2011) concluded that, in the case of monodisperse suspensions of micrometer-sized hydrated silica spheres, close packed arrays can be synthesized (a) if the free volume is restricted, (b) if the surface charge of the particles is sufficiently repulsive to discourage random aggregation but not large enough to prevent close packing, and (c) the low repulsive interactions between the particles allow them to rearrange to lower energy conformations as the free volume is decreased. Naik and Caruntu (2017) added that the particles needed to be of the same size and shape, as was identified as specific attributes of framboids by Ohfuji and Rickard (2005).

In their syntheses, Wang et al. (2019) showed that self-organization of the polystyrene particles occurs after self-assembly. This is consistent with the results obtained from electron backscatter diffraction studies of organized framboids (section 6.3 of Chapter 6). Organization appears to start in the center of the aggregate and near the rim at the same time, suggesting that icosahedral ordering is initiated even when the spheroid is in a mainly fluid state. The implication for framboids is that the constituent microcrystals cannot be tightly bound by surface or other forces. They must be able to rotate into positions that maximize the entropy of the framboid. This model suggests that this occurs as the microcrystals grow and their adjacent surfaces approach. The conclusion that self-organization of pyrite microcrystals occurs during microcrystal growth is further evidenced by the requirement of the energetics of Brownian motion to overcome the gravitational forces on relatively dense pyrite microcrystals. It seems unlikely that self-organization will occur by this process at microcrystal sizes much greater than 1 μm, for example. It is noteworthy in this context that

Früh (1885) reported observing Brownian motion in the microcrystals he had released from framboids.

This is consistent with the DVLO theory described earlier (section 13.1.5). The van der Waals attractive force is a function of the cube of the reciprocal intracrystalline distance, whereas the electrostatic repulsive force decreases exponentially as the intracrystalline distance decreases. The net effect (Figure 13.3) is that there is a spatial region when the net forces between the microcrystals are adequate to keep them together but not so great as to prevent rotation.

The mechanical stability of the aggregates of pyrite microcrystals that constitute framboids is easily overlooked. Individual framboids can be readily separated from the rock or sediment matrix and they preserve their integrity. They will disintegrate if placed between two glass slides which are not only pushed together but rotated with respect to one another. This suggests that the framboids will usually withstand direction stress but may disintegrate under relatively mild shear stress. This feature of framboids is independent of whether or not they appear to contain any adhesive material, such as organic matter. Thus, the final stage in framboid formation is the fixation of microcrystals into position, probably mainly by the development of the strong inter-crystalline adhesive forces on drying and frictional forces. The resultant aggregate is exceptionally stable, and framboids are preserved in rocks over geological time.

As with the synthesis of polystyrene aggregates within emulsion droplets, the self-assembly of pyrite microcrystals is accompanied by dehydration. The final framboid has a low water content. The water fraction remaining in the complete framboid is unknown, but it is obviously less than the fraction in the original assembly. The water content of the framboid at the point of burst nucleation of pyrite supernuclei with a critical radius of 3Å (section 11.3.3 of Chapter 11) will, of course, approach 100%. However, the process of dehydration is not similar to the evaporation that is used to aggregate uncharged polystyrene particles. Framboids form in the water column, for example, and there is no relationship between framboid formation and sediment burial depth.

13.3.1. The Development of Framboid Shapes

The development of the sub-spherical framboid shape is independent of the development of regular microarchitectures, and the microcrystals in most framboids are mainly disorganized. The reason for the separation of these two characteristics is caused by the independent processes involved in their formation. The framboid shape results from minimization of the surface free energy of the aggregate. This in turn occurs after the individual microcrystals have aggregated under the influence of inter-crystalline attractive forces. The development of the regular microarchitecture is caused by entropic maximization (e.g., de Nijs et al., 2015). It involves rotation of individual microcrystals within aggregates under the influence of surface interactions and Brownian motion (section 6.4 of Chapter 6).

The development of the external sub-spherical form is due to a minimization of the total surface energies of the aggregates. Although the minimum surface energy is a sphere, few of the framboids are actually perfectly spherical, and most are struggling to reach the perfect, least energetic, equilibrium form. The poly-hedral framboids result from different ambitions. In this case, the equilibrium form is determined by the microstructure itself and the bodies are effectively supercrystals, displaying the equilibrium form suitable for the internal geom-etry, as is the case of true crystals.

The sphericity of the aggregate is facilitated by a confined volume, such as a sediment pore, which allows concentration of the pyrite microcrystals. The con-tinued growth of the microcrystals results in a decrease in the inter-crystalline pore spaces and a decrease in the water content of the framboid. The effect is to increase the capillary forces between microcrystals in the framboid, not only expelling water, but also helping to drive the self-assembly of the microcrystals.

Interestingly, the sphericity of the synthetic polystyrene aggregates is highly sensitive to the evaporation rate of the emulsion: increased evaporation rates lead to buckled and deformed spheroidal aggregates as the droplet interface moves faster than the microparticles can consolidate (Wang et al., 2018). The implication for framboids is that intensive aggregation, as might be provided by magnetic forces for example, will not lead to spherical aggregates. There is some support for this in that, as described in section 8.2.2 of Chapter 8, many of the reported greigite "framboids" actually show irregular clumps of nanocrystalline greigite.

In the Wang et al. (2018) experiments, spheroidal aggregates form initially, which subsequently develop icosahedral symmetry. It seems that the slow con-tinuous pulling together of the polystyrene particles ultimately results in an ico-sahedral form to the aggregate. It appears, then, that the icosahedral framboids represent the ultimate stage in continuous slow particle self-assembly. In the case of syntheses by solvent evaporation, this is imposed by continued gentle evapo-ration. In the case of pyrite framboids, it is likely to be due to the combination of crystal growth and DVLO surface forces.

A further complication of pyrite framboid formation is that individual particles are not spheres, as in many of the experimental syntheses, but poly-hedra. The habits displayed by the microcrystals that constitute the single domains are mostly truncated octahedra, simple octahedra, and occasionally, cuboctahedra. Simple cubes appear to be rare. In most cases, the microcrystals are arranged in an almost uniform topological orientation by shared edges and/

or faces in cubic close packing (ccp). However, in some cases, the regular alignment of microcrystals is not perfect: some microcrystals are more or less misoriented from the orientation in which the majority are arranged. Such instances are mainly observed in microarchitectures made up of octahedral microcrystals, which are morphologically less spherical and therefore involve relatively larger void space in the packing than truncated octahedral microcrystals (Torquato and Jiao, 2009).

Chen et al. (2014) reported the results of a computer-based study of the packing of more than 55,000 convex polyhedra. They describe the prediction of object structures based solely on the shape of the polyhedral particles, constituting the object as the holy grail of packing studies. Chen et al. (2014) showed, however, that packing density is highly sensitive to small changes in shape. For example, truncating the edges of a rhombic dodecahedron results in decreased packing density, but truncating the vertices hardly affects it. Some experimental evidence backs up this view. For example, polyhedral gold nanocrystals produced hexagonal packed structures, whereas gold nanocubes produced tetragonal arrays (Ming et al., 2008). The Ming et al. (2008) polyhedra appear to be cuboctahedra (their Figure 2a).

13.3.2. The Development of the Framboid Microarchitecture

The development of organized microarchitectures is related to a maximization of entropy. There is a slight entropic advantage in ccp over hexagonal close packing (Woodcock, 1997), for example, which is consistent with the dominance of ccp in framboids. The packing density in icosahedrally packed domains of cubic close packed microcrystals is lower than that in uniform ccp (Mackay, 1962), suggesting that the entropic advantage of the development of pseudo-icosahedral domains is less than that of uniform cubic close packed microcrystals but, perhaps, greater than that of hexagonal close packing.

The rearrangement of the microcrystals in the framboid is powered by Brownian motion. As shown in section 6.3.3, the crystallographic orientation of the microcrystals tends to concentrate around two opposite poles. Initially, randomly oriented pyrite microcrystals rotate to the nearest maximum entropy position. The topology of the equant, symmetrical microcrystal polyhedra is independent of the orientation of the crystallographic axes so that there is a dissonance between the packing geometry and the crystallographic orientations of the individual microcrystals. The consequence of this is that organization of the microcrystals is a secondary process that takes some time after framboid formation. The reason for the lack of success in synthesizing ordered framboids may be that the experimentalists have not waited long enough.

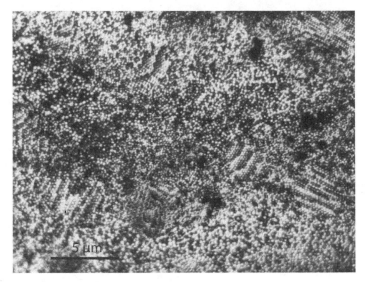

Figure 13.5. Nanoparticulate pyrite microcrystals (<100 nm in size) from a modern sediment, showing regions where the nanocrystals are organized.
Reflected light photomicrograph.

Experimental studies have demonstrated that weakly interacting spherical colloidal particles, constituted by polystyrene for example, self-organize into the cubic close packed structure. The spherical form of the resulting aggregate imposes internal elastic strain, which includes characteristic defects and enforces deviations from the cubic close packed structure. Small aggregates with less than a few hundred colloidal particles assemble into structures with varying geometry dictated by packing constraints at the spherical face boundary. Medium-sized clusters between a few hundred and about 10 000 colloidal particles favor icosahedral symmetry. These clusters can be particularly well-ordered and bear magic number particles which afford systems with closed icosahedral shells. In between the magic numbers for icosahedral clusters, decahedral clusters occur. Ultimately, $>10^5$–10^6 constituent particle clusters recover the cubic close packed structure (Wang et al., 2019).

The conclusion that self-organization of the microcrystals in ordered framboids is independent of the development of the spheroidal morphology of framboids is consistent with the regular organization observed in some non-spherical microcrystalline pyrite aggregates (Figure 13.5).

List of Symbols and Abbreviations

Symbol	Meaning	Units
$\gamma\dfrac{H_2O}{hkl}$	Surface free energy for crystal face $\{h\,k\,l\}$ in water	
$\gamma\dfrac{vac}{hkl}$	Surface free energy for crystal face $\{h\,k\,l\}$ in vacuum	
$(h\,k\,l)$	A plane of the crystal structure	
[]	A directional vector in real space	
$\{h\,k\,l\}$	A family of planes	
{i}	Activity of species i	
$\langle h\,k\,l\rangle$	A family of directions which are related by symmetry rules	
=M	Surface bound species, M	
A	Surface area of the framboid	m^2
c_0	Concentration of monomers at the framboid surface	moles μm^{-3}
c_b	Concentration of monomers in the bulk solution	moles μm^{-3}
CCC	Critical coagulation concentration	mol L^{-1}
CNT	Classical nucleation theory	
CSD	Crystal size distribution theory	
D	Diffusion coefficient	$m^2\,s^{-1}$
D_s	Whole sediment diffusion coefficient	$m^2\,s^{-1}$
D	Framboid diameter	m
DVLO	Derjaguin-Landau-Verwey-Overbeek	
e	Charge on the electron (91.6×10^{-19} coulombs)	
eV	Electron volt	
F	Faraday constant	96485 C mol^{-1}
F_{EDL}	Force per unit area due to electrostatic double layer energy	

Symbol	Meaning	Units
F_{vdW}	Force per unit area due to van der Waals energy	N
Ga	Billion years ago	
h	Distance between the crystal faces	m
H	Hamaker constant	zJ
HOMO	Highest occupied molecular orbital	eV
I	Ionic strength of solution	mol L^{-1}
IUPAC	International Union of Pure and Applied Chemistry	
J	Flux	g m^{-2} s^{-1}
J_s	Flux per unit area of the framboid	g s^{-1}
K	Equilibrium constant	
K	Absolute temperature	
k	Rate constant	
k_B	Boltzmann constant	$1.38064852 \times 10^{-23}$ m^2 kg s^{-2} K^{-1}
k_i	The rate constant for the interface reaction	
k_n	Kinetic factor in Equation 11.8	
K_{sp}	Solubility product	
LUMO	Lowest unoccupied molecular orbital	eV
Ma	Million years ago	
Myr	Million years duration	
n	Number of samples	
pH_{IEP}	Isoelectric point	pH
pH_{zpc}	Zero point of charge	pH
PT	Pressure and temperature	Pa and ºC
r	Particle radius	m
R	Gas constant	8.314 J mol^{-1} K^{-1}
r^*	Critical radius	Å
R_b	Radius of the bulk solution	m,
R_P	Rate of pyrite crystal growth	
SATP	Standard ambient temperature and pressure (IUPAC)	25ºC and 0.1 MPa
t	Time	

Symbol	Meaning	Units
T	Temperature	K
V	Volume	m^{-3}
V_{EDL}	Electrostatic double layer energy	$J\,m^{-2}$
V_{vdW}	van der Waals energy	$J\,m^{-2}$
wt %	Weight percent	
$\bar{x}$	Arithmetic (or additive) mean	
$\bar{x}^*$	Multiplicative (or geometric) mean	
$^x/$	Multiplied or divided by (cf. $\pm$)	
^xS	Sulfur with an atomic mass of x	
Z	Interaction constant in Equation 14.4	
z	Electrolyte valence	
γ	Surface free energy	
δ	Width of diffusion boundary layer	m^{-1}
$\delta^{34}S$	Deviation of $^{34}S/^{32}S$ ratio from standard	‰
ΔG_r^o	Gibbs free energy of reaction	$kJ\,mole^{-1}$
ε	Relative permittivity	dimensionless
ε_0	Permittivity in a vacuum	$F^{-1}\,m^{-1}$
ε_w	Permittivity in water	$F^{-1}\,m^{-1}$
ζ	Zeta potential	V
Θ	Tortuosity	
κ	Inverse Debye length	m^{-1}
v_0	Molar volume	m^{3}
σ	Arithmetic (or additive) standard deviation	
σ^*	Multiplicative (or geometric) standard deviation	
φ	Sediment porosity	dimensionless
Ψ_0	Stern layer potential	V
Ω	Supersaturation	
Ω^*	Critical supersaturation	
‰	Per mille	

List of Units

C	coulomb
°C	degrees centigrade
F	Farad
g	gram
J	joule
K	Kelvin
kg	kilogram
L	liter
m	meter
mol	mole
μm	micrometer (10^{-6}m)
mV	millivolt (10^{-3} V)
N	Newton
nm	nanometer (10^{-9}m)
ppm	parts per million
ppt	parts per thousand
s	second
zJ	zeptojoule (10^{-21}J)

Glossary

Geological time periods: I have given approximate ages of each geologic time period as it used in the text. The ages are indicative.

activation energy the minimum energy required to cause a reaction to occur

activity the concentration of an ion in solution corrected for interactions with other solution species

activity coefficient coefficient correcting concentration to activity of a species

advective flow transport by bulk motion

agglomerate a loose cluster of particles

aggregate a cluster of particles which are attached to each other

aldehydic carbonyl the functional group –CHO

algae aquatic plants ranging from single cells to seaweed.

amphibolite facies a mineral assemblage resulting from metamorphism corresponding to temperatures of >500°C and pressures of 1–1.2 G Pa

anaerobic life in the absence of molecular oxygen

andesite volcanic rock with an intermediate silica content

anoxic literally, free of molecular oxygen: practically <0.2 ml O_2 L^{-1}

anti-Mackay layers ideally packed, less dense, and more spherical, layers of particles in icosahedral structures near the cluster perimeter

Archimedean polyhedra the 13 convex uniform polyhedra composed of regular polygons

arithmetic mean see Equation 2.2

arithmetic standard deviation see Equation 2.4

Arrhenius activation energy the activation energy for a reaction calculated from the Arrhenius equation (q.v.)

Arrhenius equation an equation describing the dependence of the rate constant of a chemical reaction on the temperature

autolysis the destruction of cells or tissues by their own enzymes

autotroph organism that can grow on CO_2 or other simple carbon species

Berg effect supersaturation at the center of a face on a growing crystal is much lower than that at the edges and corners

biofilm an assemblage of surface-associated microbial cells that is enclosed in an extra-cellular polymeric matrix

biosphere the region of the Earth occupied by living organisms

Brownian motion the random movement of microscopic particles in a fluid resulting from the continuous bombardment from molecules of the surrounding medium

burst nucleation sudden nucleation of a phase at high supersaturations leading to multiple nuclei forming at the same time

carbonaceous chondrites primitive meteorites rich in organic compounds

Carlin-type sediment-hosted disseminated gold deposits named after the type locality in Carlin, Nevada

chalcedony cryptocrystalline variety of silica

chemostat an apparatus to carry out chemical reactions, usually involving fluids, under constant chemical and environmental conditions over extended time periods

classical nucleation theory most common model used to quantitatively study the kinetics of nucleation

cluster (chemistry) an assembly of bound molecules or atoms intermediate between a simple molecule and a nanoparticle, that is generally neutrally charged, often includes extensive metal-metal and metal-ligand-metal bonds, and can display variable compositions

cluster analysis (statistics) a multivariate method (q.v.) which aims to classify a set of measured variables into a number of different groups such that similar subjects are placed in the same group

coacervate organic-rich droplets formed via liquid-liquid phase separation

cold seep an area of the ocean floor where sulfide and hydrocarbon-rich fluid seepage occurs, often expressed in the form of a brine pool

colloform texture of successive pyrite layers, often concentric, originally thought to represent colloidal deposition.

colloids particles > ca. 100 nm and < ca. 10 μm in size

coordination number the number of atoms surrounding a central atom in a complex or crystal

critical coagulation constant the minimum concentration of counterions to induce coagulation of colloidal particles

critical radius the minimum size of a supernucleus (q.v.)

critical supersaturation the minimum supersaturation required to initiate nucleation

crystal field theory a model for the bonding interaction between transition metals and ligands which describes the effect of the attraction between the positive charge of the metal cation and negative charge on the non-bonding electrons of the ligand.

crystal size distribution theory an empirical theory based on the crystal size-frequency distribution in a batch crystallization chamber

crystallization by particle attachment crystal growth through attachment of nanoparticles rather than monomers

cubic close packing a mode of the densest packing of equal particles that produces cubic symmetry

cuboctahedron a combination of the cube and octahedron

cuboid a six-sided regular convex polyhedron with rectangular faces

Debye length the thickness of the interfacial electric double layer (q.v.)

Debye-Scherrer camera an X-ray powder diffraction instrument that collects diffraction patterns on a circular strip of photographic film

decahedron polyhedron with 10 faces

deep biosphere the part of the biosphere between 5 m and >10 km below the Earth's surface

diagenesis the physical and chemical changes occurring during the development of a sediment to sedimentary rock

diatom photosynthesizing unicellular algae with a siliceous skeleton

diffusion boundary layer the solution layer in contact with the surface where the fluxes are diffusion-controlled

diffusion constant a proportionality constant between the molar flux due to molecular diffusion and the gradient in the concentration of the species

diffusion-controlled growth crystal growth in which growth due to surface reactions is fast and the rate is determined by the diffusive flux of components to and from the crystal surface

dinoflagellate single-celled eukaryotic algae with two flagella

dislocation displacement of part of a crystal lattice

dodecahedron a polyhedron composed of 12 rhombic faces (see rhombic dodecahedron)

DVLO forces surface forces between particles according to the Derjaguin-Landau-Verwey-Overbeek model

dysoxic $0.2 < O_2 < 2.0$ ml L^{-1}

Eh potential with respect to the hydrogen electrode; a measure of the redox state of a solution

Eigen-Wilkins reaction substitution reactions (q.v.) where the overall rate is determined by the rate of water exchange

electron back-scatter diffraction a scanning electron microscope–based method of determining the microstructure and crystallography of polycrystalline materials by examining the pattern of Kikuchi bands (q.v.) generated by the interaction of electrons with the material

electron probe microanalysis an elemental analysis machine which uses wavelength dispersive X-ray analysis (q.v.)

electrostatic double layer a simplified model of a particle surface in a medium such as water which defines two layers of charge

electrostatic double layer force the repulsive force between two particles due to their surface electrostatic double layers

energy dispersive X-ray spectrometry an elemental analytical technique usually associated with electron microscopy which measures the intensities of characteristic X-rays generated by the interaction of the electron beam

epigenesis formed later than the enclosing material

epitaxial growth growth in which the crystallographic orientation of a growing crystal is determined by the structure of the substrate

equant having different diameters approximately equal, so as to be roughly cubic or spherical in shape.

equilibrium constant the ratio of the products of the ionic activities in solution raised to the power of their stoichiometric coefficients to those in equilibrium with a solid

euhedral having a regular crystal shape

eukaryote organism with cells that possess a nucleus

euxinic sulfidic water

Ewald sphere a sphere with radius equal to the reciprocal X-ray wavelength centered on the sample, used to graphically interpret the diffraction conditions

exchange reaction a reaction in which components are exchanged between substances (viz. substitution reaction)

fecal pellet organic excrement occurring in sediments, usually of a simple ovoid form less than a millimeter in length

ferrihydrites iron oxyhydroxide minerals of uncertain composition: precursors to the various FeOOH polymorphs.

ferrimagnetism type of permanent magnetism that occurs in solids in which the magnetic fields associated with individual atoms spontaneously align themselves

flagellum a thread-like structure that provides motility

foraminifera microscopic single-celled organisms characterized by streaming cytoplasm for catching food and an external shell

free energy any one of several free energies which describe the energy that is available to perform thermodynamic work at constant temperature (e.g., Gibbs free energy)

fugacity the concentration of a molecule in a gas corrected for interactions with other gaseous species

fulvic acid a soluble organic polymeric acid derived from humus

geometric mean see Equation 2.3

geometric standard deviation see Equation 2.5

Gibbs free energy of formation the change of Gibbs free energy that accompanies the formation of 1 mole of a substance from its constituent elements at 25°C and 0.1 MPa pressure

Gibbs free energy of reaction the difference in the free energies of formation of the products and reactants

greenschist facies a mineral assemblage resulting from metamorphism corresponding to temperatures of about 300°–500°C and pressures of 0.3–1 G Pa

habit the external shape of a crystal

heterogeneous nucleation nucleation where the new phase develops on a surface

heteronuclei particles on which heterogenous nucleation can occur

heterotroph an organism that requires an organic carbon source for growth

hexagonal close packing a mode of the densest packing of equal particles that produces hexagonal symmetry

high spin electronic configuration in which the orbitals are partially filled, as far as possible, with singly occupied orbitals

high-resolution transmission electron microscopy electron microscopy capable of resolution down to the Ångström level

highest occupied molecular orbital the highest energy orbital that contains electrons

hillock circular or polygonal mound formed by screw dislocations on a crystal surface

homogeneous nucleation nucleation where the new phase develops from dissolved species in solution

homonuclear bond bond between identical atoms

humic acid a natural acidic organic polymer which develops from humus

humus natural finely divided organic matter

hydrodynamic pertaining to the motion of fluids and the forces acting on solid bodies immersed in fluids

hydrophilic substances attracted to water and other polar compounds

hydrophobic substances that repel water and are attracted to non-polar (e.g., organic) compounds

hydrothermal pertaining to natural waters with temperatures above ambient.

hydrothermal bomb a container used to carry out reactions involving fluids under hydrothermal (q.v.) conditions

hydrothermal vent fissures, usually on the seafloor, from which geothermally heated water issues

hypereutrophic extremely rich in nutrients and minerals.

icosahedral packing a close packed structure in which the particles are organized into 20 differently oriented tetrahedral domains

icosahedron a polyhedron with 20 equilateral triangular faces

induction phase a period of metastable supersaturation, preceding nucleation (syn., lag phase)

inductively coupled mass spectrometer a mass spectrometer that uses an inductively coupled plasma to ionize the sample

ion activity product product of the activities of ions in solution

ion mill a machine that thins samples by firing ions at the surface

ionic strength a measure of the concentration of ions and their net electronic charges in a solution

isoelectric point the pH at which the surface carries no net electric charge in water

kaolinite a clay mineral with the chemical composition $Al_2Si_2O_5(OH)_4$

kerogen naturally occurring organic matter that is resistant to extraction using organic solvents

Kikuchi bands bands formed by diffraction of diffuse electrons originating from inelastic scattering of an incident electron beam

lag phase alternative name for induction phase (q.v.)

LaMer kinetics a model of nucleation and crystal growth which results in the formation of multiple similar microcrystals

laser-ablation inductively coupled mass spectrometry an element analytical method in which the sample is introduced to the plasma of an inductively coupled mass spectrometer (q.v.) by ablation with a focused laser beam

le Chatelier effect if equilibrium is disturbed by changing the conditions, the position of equilibrium shifts to counteract the change to re-establish an equilibrium

lipid a biomolecule that is soluble in nonpolar solvents, includes fats

lithification the processes by which a sediment becomes a rock

log-normal distribution a distribution in which the logarithm of a random variable is normally distributed

low coordinated sites edges, kinks, vertices, and defects, where surface atoms have a lower coordination number (q.v.) than the bulk

low spin electronic state in which the electrons are, as far as possible, paired in the orbitals

lowest unoccupied molecular orbital the lowest energy orbital that can accept electrons

maceration a technique to extract organic microfossils from a surrounding rock matrix using acid

machine-learning a branch application of artificial intelligence that provides systems the ability to automatically improve from experience without being explicitly programmed

Mackay layers ideally packed layers of particles in icosahedral structures near the cluster center

magic number the number of atoms or molecules forming an exceptionally stable cluster

magnetic force microscopy a method in which a sharp magnetized tip scans a sample

mass interferences ions of similar masses which interfere with analysis of the target ion

massive sulfide a rock consisting of mainly of sulfide minerals

meromictic lake in which water layers do not intermix

mesocrystal superstructures of small crystals with similar sizes and shapes with a common crystallographic orientation

metamorphism change in the structure and composition of a rock by pressure and heat

microbeam techniques analyses using microscopic beams of, usually, electrons

microcrystal crystals ca. $0.1-\lesssim 3$ μm in size

microsphere microscopic sphere with no internal structure or displaying radiating internal structure

minor element element in concentrations between 0.1 and 1 wt %—described as trace element (q.v.) in this volume

mis-indexing wrongly assigned Kikuchi bands due to pseudosymmetry (q.v.)

molarity number of moles of solute per liter of solution

mole the mass of one molecular weight in grams of a substance

mononuclear compound with a single central metal atom

multiframboid see rogenpyrit

multivariate statistics statistics where more than two variables are simultaneously analyzed

Murchison meteorite a carbonaceous chondrite (q.v.) that fell near Murchison, Victoria, in 1969 and is now one of the most studied of all meteorites

n-type semi-conductivity where the conductivity is due to free electrons in a material

NaCl structure face-centered cubic structure in which one of the two types of atom occupy the octahedral cavities formed by the arrangement of the other type of atom

nano-scale secondary ion mass spectrometry elemental analysis method with nano-scale resolution in which secondary ions are ablated from a sample with an ion beam and collected in a mass spectrometer

nanoparticles particles 1–ca. 100 nm in size

nektobenthos swimming organisms that interact with the seafloor

non-polar a molecule with a symmetric electric charge distribution

non-protonated a compound that lacks terminal H^+ ions

non-stoichiometry the deviation from simple whole number atomic ratios in a compound

normal distribution a function that represents the distribution of many random variables as a symmetrical bell-shaped graph (also Gaussian or Gauss-Laplace distribution)

nucleation first step in the formation of a new phase via assembly of atoms or molecules

nucleophilic a chemical species that donates an electron pair to form a chemical bond in a reaction

octahedron a polyhedron composed of eight faces in the shape of equilateral triangles

oriented attachment crystal growth where particles orient themselves for attachment to specific crystal faces

Ostwald ripening disappearance of small particles by dissolution and the formation of larger particles

Ostwald step rule it is not the most stable but the least stable polymorph that crystallizes first

(oxyhydr)oxide indefinite composition of oxide, hydrated oxide, hydroxide, oxyhydroxide, or mixtures of these

p-type semi-conductivity where the conductivity is due to positive holes in the electronic structure of a material

packing efficiency the fraction of the total space filled by the constituent particles (see Equation 4.2)

pelite a sedimentary rock composed of very fine clay particles

permittivity the capacity of a material to store electrical potential energy under the influence of an electric field

Platonic solid one of five convex polyhedra with congruent, regular, polygonal faces

point group a group of geometric transformations that keep at least one point fixed in space

polar a molecule with an asymmetric electric charge distribution

pole the intersection of the normal line to a crystal plane and the sphere whose center coincides with the center of the crystal

pole figure the stereographic projection of the poles used to represent the orientation of an object in space

polyframboid group of framboids clustered in spheroidal aggregates (viz. rogenpyrit)

polynuclear species with more than one central metal atom (e.g., clusters q.v.)

polysulfide sulfur moieties with the general formula S_n^{2-}

potentiometric titration a method of analysis in which the endpoint of a titration is monitored with an indicator electrode

prokaryote a single-celled organism that lacks a nucleus

protist loose group of mostly unicellular eukaryotes that are not animals, plants, or fungi

protokerogen the fraction of the insoluble organic matter in recent sediments which is resistant to acid hydrolysis

pseudo-icosahedron a 20-sided polyhedron formed from a combination of the octahedron and pyritohedron with differently sized and shaped triangular faces

pseudomorph a mineral produced by a substitution process in which the appearance and dimensions of the original mineral remain constant but the composition is replaced

pseudomorphism the process in which the appearance and dimensions of a mineral are retained by another mineral with a different composition

pseudosymmetry symmetry of a crystal which closely approaches a higher symmetry

pyritohedron a modified pentagonal dodecahedron characteristic of pyrite

radiolarian single-celled protists with intricate mineral skeletons

range rule see Equation 2.9

reactive iron shuttle the process by which iron, which is available for reaction, is transported to deeper parts of a basin

reciprocal lattice a theoretical construct used in crystallography in which regular distribution of points in which the distance between the points is proportional to the inverse of corresponding inter-planar spacings in the real lattice

redox oxidation-reduction

redoxcline a water layer with a strong redox gradient

reflected light microscope optical microscope in which objects are examined by the light reflected from their surface; the surfaces are usually polished sections

regolith the layer of unconsolidated solid material covering the bedrock

relative permittivity the ratio of the permittivity of a material in a medium to the permittivity in a vacuum (previously known as dielectric constant)

replacement the chemical process by which the composition of one mineral is substituted by another

rhombic dodecahedron a polyhedron composed of 12 rhombic faces (often shortened to dodecahedron)

rogenpyrit groups of framboids often clustered in spheroidal aggregates (viz. polyframboid)

scanning electron microscope an electron microscope in which the surface of a specimen is scanned by a beam of electrons that are reflected to form an image

screw dislocation a dislocation on the crystal surface in which the atoms are arranged in a helical pattern

secondary enrichment enrichment of a mineral deposit by later reactions with circulating groundwaters (see supergene enrichment)

seed crystal a small crystal on which larger crystals grow

self-assembly the process of particle aggregation and agglomeration

self-organization the process whereby particles in an aggregate or agglomerate achieve geometric order

semi-conduction a solid that has a conductivity between that of an insulator and that of most metals, usually due to the addition of an impurity

shear zone a planar or curviplanar zone composed of rocks that are more highly strained than adjacent rocks; deeper level equivalents to faults

single crystal X-ray diffraction a technique for determining crystal structure by exploiting the diffraction of X-rays by a single crystal

skeletal crystals a crystal that develops under conditions of rapid growth and high supersaturation, resulting in hollow faces or branched forms

soft-sediment deformation a variety of structures developing in wet sediments under the influence of gravity

sol a colloid in which the solid particles are suspended in a liquid

solubility product the product of the activities (q.v.) of the ionic components of a solid in solution raised to the power of their stoichiometric coefficients (q.v.)

space group a group of geometric transformations that are combined to describe the symmetry of the internal structure of a crystal

sphericity a measure of how closely an object approaches a perfect sphere

spheroid an object that is approximately spherical in shape

spin polarization the alignment of electron spins

spin state whether the electronic configuration is high spin (q.v.) or low spin (q.v.)

standard ambient temperature and pressure 25°C and 0.1 MPa

stereological error the error in measuring sphere diameters from 2D sections

steric hindrance retardation of a reaction rate by the shape and size of molecules, usually organic, in the system

Stern layer the internal layer of the double layer (q.v.)

stoichiometric coefficient the number of molecules of a component of a balanced reaction

stoichiometry the whole number ratio of elements in a compound (see non-stoichiometry)

substitution reaction a reaction in which one component is substituted for another—contrasting with, for example, a reaction involving exchange of electrons

sulfate-methane transition zone a zone at depth in sediments where sulfate and methane coexist

sulfate-reducing bacteria mostly anaerobic microorganisms that use SO_4^{2-} to oxidize organic matter or hydrogen reducing it to sulfide

sulfonate a compound containing the functional group $R\text{-}SO_3^-$, where R is an organic group

supergene enrichment secondary enrichment (q.v.) at the base of a surface oxidized zone in an ore deposit.

supernuclei nuclei which are stable and have the potential to grow

supersaturation the ratio of the actual concentration of ions in solution to the equilibrium solubility product of the solid

surface energy the excess energy at the surface of a material compared to the bulk

surface reaction-controlled growth crystal growth in which growth due to surface reactions is slow and rate-controlling

syngenesis formed at the same time as the enclosing material

tetrahedron a special case of the trigonal pyramid (q.v.) where the faces are all equilateral triangles

tortuosity the relative path length of the diffusion process

trace element element in concentration below 0.1 wt %; expanded to include minor elements (i.e., up to 1 wt % q.v.) in this volume

triakis octahedron a solid having 24 congruent faces meeting on the edges of a regular octahedron (see trisoctahedron)

trigonal pyramid a polyhedron consisting of four triangular faces (see tetrahedron)

trisoctahedron a solid having 24 congruent faces meeting on the edges of a regular octahedron (see triakis octahedron)

truncated describing regular polyhedra whose vertices are replaced by other faces

turbidites sediments formed by massive slope failures on the seafloor

van der Waals force the attractive intermolecular force between surfaces

varve finely banded marine or lacustrine sediments produced by seasonal environmental variations

volcanogenic sourced from volcanic activity

voltammetry an electroanalytic method in which the current is measured as the potential is varied

wavelength dispersive X-ray analysis elemental analysis method which collects the X-rays photons at element-specific wavelengths produced by a focused electron beam (see electron probe microanalysis)

wet chemical methods chemical analysis involving the dissolution of the sample

whole sediment diffusion coefficient diffusion coefficient corrected for sediment tortuosity (q.v.)

Wulff shape the equilibrium shape of crystals

X-ray powder diffraction a technique for determining crystal structure by exploiting the diffraction of X-rays by a randomly oriented powder sample of the bulk material

xylem the vascular tissue in plants which conducts water upward from the root and helps to form the woody element in the stem

zeta potential the electrical potential at the boundary between the surface static layer and the bulk solution

References

Abraitis, P. K., Pattrick, R. A. D., and Vaughan, D. J., 2004. Variations in the compositional, textural and electrical properties of natural pyrite: A review. *International Journal of Mineral Processing*, 74(1–4): 41–59.

Agaev, Y., and Emujazov, K., 1963. Some electrical properties of pyrite. *Izvestiya Akademii Nauk Turkmenskoi SSR*, F12: 104–107.

Aizcnshtat, Z., Krein, E. B., Vairavamurthy, A., and Goldstein, T. P., 1995. Role of sulfur in the transformations of sedimentary organic matter: A mechanistic overview. In: A. Vairavamurthy and M. A. A. Schoonen (Editors), *Geochemical Transformations of Sedimentary Sulfur*. American Chemical Society, Washington, DC, pp. 16–37.

Akai, J., Izumi, K., Fukuhara, H., Masuda, H., Nakano, S., Yoshimura, T., Ohfuji, H., Anawar, H. M., and Akai, K., 2004. Mineralogical and geomicrobiological investigations on groundwater arsenic enrichment in Bangladesh. *Applied Geochemistry*, 19(2): 215–230.

Alfonso, D. R., 2010. Computational investigation of FeS₂ surfaces and prediction of effects of sulfur environment on stabilities. *Journal of Physical Chemistry C*, 114(19): 8971–8980.

Allen, E. T., Crenshaw, J. L., Johnson, J., and Larsen, E. S., 1912. The mineral sulphides of iron with crystallographic study. *American Journal of Science*, 33: 169–236.

Alyanak, N., and Vogel, T. A., 1974. Framboidal chalcocite from White Pine, Michigan. *Economic Geology*, 69(5): 697–703.

Amstutz, G. C., Park, W. C., Schot, E. H., and Love, L. G., 1967. Orientation of framboidal pyrite in shale. *Mineralium Deposita*, 1: 317–321.

Andersson, S., 2005. Curved polyhedra. *Zeitschrift für anorganische und allgemeine Chemie*, 631: 499–501.

Andres, R. J., and Kasgnoc, A. D., 1998. A time-averaged inventory of subaerial volcanic sulfur emissions. *Journal of Geophysical Research Atmospheres*, 103: 25251–25261.

Annels, A. E., and Roberts, D. E., 1989. Turbidite-hosted gold mineralization at the Dolaucothi gold-mines, Dyfed, Wales, United Kingdom. *Economic Geology*, 8: 1293–1314.

Arrouvel, C., and Eon, J. G., 2019. Understanding the surfaces and crystal growth of pyrite FeS₂. *Materials Research-Ibero-American Journal of Materials*, 22(1): e20171140.

Avetisyan, K., Buchshtav, T., and Kamyshny, A., 2019. Kinetics and mechanism of polysulfides formation by a reaction between hydrogen sulfide and orthorhombic cyclooctasulfur. *Geochimica et Cosmochimica Acta*, 247: 96–105.

Aycard, M., Derenne, S., Largeau, C., Mongenot, T., Tribovillard, N., and Baudin, F., 2003. Formation pathways of proto-kerogens in Holocene sediments of the upwelling influenced Cariaco Trench, Venezuela. *Organic Geochemistry*, 34: 701–718.

Barlow, W., 1883. Probable nature of the internal symmetry of crystals. *Nature*, 29: 205–207.

Barlow, W., and Pope, W. J., 1907. CXI: The relation between the crystalline form and the chemical constitution of simple inorganic substances. *Journal of the Chemical Society, Transactions*, 91: 1150–1214.

Barnard, A. S., and Russo, S. P., 2009a. Modelling nanoscale FeS_2 formation in sulfur rich conditions. *Journal of Materials Chemistry*, 19: 3389–3394.

Barnard, A. S., and Russo, S. P., 2009b. Morphological stability of pyrite FeS_2 nanocrystals in water. *Journal of Physical Chemistry C*, 113: 5376–5380.

Baronov, A., Bufkin, K., Shaw, D. W., Johnson, B. L., and Patrick, D. L., 2015. A simple model of burst nucleation. *Physical Chemistry Chemical Physics*, 17: 20846–20852.

Bastin, E. S., 1950. *Interpretation of Ore Textures*. Geological Society of America Memoir 45. Geological Society of America, NY.

Baumgartner, R. J., Van Kranendonk, M. J., Wacey, D., Fiorentini, M. L., Saunders, M., Caruso, S., Pages, A., Homann, M., and Guagliardo, P., 2019. Nano-porous pyrite and organic matter in 3.5-billion-year-old stromatolites record primordial life. *Geology*, 47: 1039–1043.

Bebié, J., and Schoonen, M. A. A., 1999. Pyrite and phosphate in anoxia and an origin-of-life hypothesis. *Earth and Planetary Science Letters*, 171: 1–5.

Bebié, J., Schoonen, M. A. A., Fuhrmann, M., and Strongin, D. R., 1998. Surface charge development on transition metal sulfides: An electrokinetic study. *Geochimica et Cosmochimica Acta*, 62: 633–642.

Beijerinck, M. W., 1895. Ueber Spirillum desulfuricans als Ursache von Sulfatreduktion. *Centralblatt für Bakteriologie und Parasitenkunde*, 1: 1–9, 50–59, 104–114.

Berg, W. F., 1938. Crystal growth from solutions. *Proceedings of the Royal Society of London Series A—Mathematical and Physical Sciences*, 164: 0079–0095.

Bergh, S. V., 1928. Fossilifierade svavelbakterier uti alunskiffern på Kinnekulle. *Geologiska Föreningen i Stockholm Förhandlingar*, 50: 413–418.

Bergstrom, L., 1997. Hamaker constants of inorganic materials. *Advances in Colloid and Interface Science*, 70: 125–169.

Berner, R. A., 1969. Synthesis of framboidal pyrite. *Economic Geology*, 64: 383–384.

Berner, R. A., 1970. Sedimentary pyrite formation. *American Journal of Science*, 268: 1–23.

Berner, R. A., 1971. *Principles of Chemical Sedimentology*. McGraw-Hill, New York, NY.

Berner, R. A., 2006. GEOCARBSULF: A combined model for Phanerozoic atmospheric O_2 and CO_2. *Geochimica et Cosmochimica Acta*, 70: 5653–5664.

Berner, Z. A., Puchelt, H., Noltner, T., and Kramar, U., 2013. Pyrite geochemistry in the Toarcian Posidonia Shale of south-west Germany: Evidence for contrasting trace-element patterns of diagenetic and syngenetic pyrites. *Sedimentology*, 60: 548–573.

Berzelius, J. J., 1845. *Traité de Chimie*, 2. Chez Firmin Didot frères, Paris.

Beveridge, T. J., 2001. Use of the Gram stain in microbiology. *Biotechnic and Histochemistry*, 76: 111–118.

Birkholz, M., 2014. Modeling the shape of ions in pyrite-type crystals. *Crystals*, 4: 390–403.

Bond, D. P. G., and Wignall, P. B., 2010. Pyrite framboid study of marine Permian-Triassic boundary sections: A complex anoxic event and its relationship to contemporaneous mass extinction. *Geological Society of America Bulletin*, 122: 1265–1279.

Bottcher, M. E., and Lepland, A., 2000. Biogeochemistry of sulfur in a sediment core from the west-central Baltic Sea: Evidence from stable isotopes and pyrite textures. *Journal of Marine Systems*, 25: 299–312.

Bottrell, S. H., and Newton, R. J., 2004. Reconstruction of changes in global sulfur cycling from marine sulfate isotopes. *Earth Science Reviews*, 75: 59–83.

Boudreau, B. P., 1996. *Diagenetic models and their Interpretation*. Springer Verlag, Berlin.

Bragg, W. H., 1913. The reflection of X-rays by crystals. (II). *Proceedings of the Royal Society of London Series A—Containing Papers of a Mathematical and Physical Character*, 89: 246–248.

Bragg, W. L., 1914. The analysis of crystals by the X-ray spectrometer. *Proceedings of the Royal Society of London Series A—Containing Papers of a Mathematical and Physical Character*, 89: 468–489.

Bragg, W. L., 1961. The Rutherford Memorial Lecture, 1960: The development of X-ray analysis. *Proceedings of the Royal Society of London. Series A. Mathematical and Physical Sciences*, 262: 145–158.

Brimblecombe, P., 2003. The global sulfur cycle. In: D. H. Heinrich and K. T. Karl (Editors), *Treatise on Geochemistry*. Elsevier, Amsterdam. 15: 645–682.

Bruland, K. W., Middag, R., and Lohan, M. C., 2013. Controls of trace metals in seawater, In: D. H. Heinrich and K. T. Karl (Editors), *Treatise on Geochemistry*. Elsevier, Amsterdam 8: 19–51.

Burns, R. G., 1970. *Mineralogical Applications of Crystal Field Theory*. Cambridge University Press, Cambridge.

Burton, E. D., Bush, R. T., Johnston, S. G., Sullivan, L. A., and Keene, A. F., 2011. Sulfur biogeochemical cycling and novel Fe-S mineralization pathways in a tidally re-flooded wetland. *Geochimica et Cosmochimica Acta*, 75: 3434–3451.

Burton, W. K., Cabrera, N., and Frank, F. C., 1949. Role of dislocations in crystal growth. *Nature*, 163: 398–399.

Bush, R. T., Fyfe, D., and Sullivan, L. A., 2004. Occurrence and abundance of monosulfidic black ooze in coastal acid sulfate soil landscapes. *Soil Research*, 42: 609–616.

Butler, I. B., 1994. Framboid formation. PhD thesis, University of Wales, Cardiff, 434 pp.

Butler, I. B., Böttcher, M. E., Rickard, D., and Oldroyd, A., 2004. Sulfur isotope partitioning during experimental formation of pyrite via the polysulfide and hydrogen sulfide pathways: Implications for the interpretation of sedimentary and hydrothermal pyrite isotope records. *Earth and Planetary Science Letters*, 228: 495–509.

Butler, I., Grimes, S., and Rickard, D., 2000. Pyrite formation in an anoxic chemostatic reaction system. *Journal of Conference Abstracts*, 5: 274–275.

Butler, I., and Rickard, D., 2000. Framboidal pyrite formation via the oxidation of iron (II) monosulfide by hydrogen sulphide. *Geochimica et Cosmochimica Acta*, 64: 2665–2672.

Butler, I. B., and Rickard, D., 2003. An anoxic chemostat for controlled sulphide mineral synthesis. *Transactions of the Institution of Mining and Metallurgy, Section B: Applied Earth Science*, 112: B181–B182.

Canfield, D. E., 2004. The evolution of the Earth surface sulfur reservoir. *American Journal of Science*, 304: 839–861.

Canfield, D. E., Lyons, T. W., and Raiswell, R., 1996. A model for iron deposition to euxinic Black Sea sediments. *American Journal of Science*, 296: 818–834.

Canfield, D. E., Raiswell, R., and Bottrell, S., 1992. The reactivity of sedimentary iron minerals toward sulfide. *American Journal of Science*, 292: 659–683.

Carslaw, H. S., and Jaeger, J. C., 1959. *Conduction of Heat in Solids*. Clarendon Press, Oxford, UK.

Carstens, C. W., 1941. Om geokjemisk undersøkelser av malmer. *Norsk Geologiska Tidskrifter*, 21: 213–221.

Cevales, G., 1961. Erzuntersuchungen im Emissionsmikroskop. *Zeitschrift für Erzbergbau und Metallhüttenwesen*, 14: 159–210.

Chadwell, S. J., Rickard, D., and Luther, G. W., 1999. Electrochemical evidence for pentasulfide complexes with Mn^{2+}, Fe^{2+}, Co^{2+}, Ni^{2+}, Cu^{2+} and Zn^{2+}. *Aquatic Geochemistry*, 5(1): 29–57.

Chadwell, S. J., Rickard, D., and Luther, G. W., 2001. Electrochemical evidence for metal polysulfide complexes: Tetrasulfide S_4^{2-} reactions with Mn^{2+}, Fe^{2+}, Co^{2+}, Ni^{2+}, Cu^{2+} and Zn^{2+}. *Electroanalysis*, 13(1): 21–29.

Characklis, W. G., and Marshall, K. C., 1990. Biofilms: A basis for an interdisciplinary approach. In: W. G. Characklis and K. C. Marshall (Editors), *Biofilms*. Wiley-Interscience, New York, N. pp.3–15.

Chen, E. R., Klotsa, D., Engel, M., Damasceno, P. F., and Glotzer, S. C., 2014. Complexity in surfaces of densest packings for families of polyhedra. *Physical Review X* 4: 011024.

Chen, T. T., 1978. Colloform and framboidal pyrite from the Caribou deposit, New Brunswick. *Canadian Mineralogist*, 16: 9–15.

Chou, C.-L., 2012. Sulfur in coals: A review of geochemistry and origins. *International Journal of Coal Geology*, 100: 1–13.

Cölfen, H., and Antonietti, M., 2005. Mesocrystals: Inorganic superstructures made by highly parallel crystallization and controlled alignment. *Angewandte Chemie-International Edition*, 44: 5576–5591.

Cornwell, J. C., and Morse, J. W., 1987. The characterization of iron sulfide minerals in anoxic marine sediments. *Marine Chemistry*, 22: 193–206.

Cubillas, P., and Anderson, M. W., 2010. Synthesis mechanism: Crystal growth and nucleation. In: J. Čejka, A. Corma, and S. Zones (Editors), *Zeolites and Catalysis*. Wiley-VCH Verlag GmbH & Co. Kgaa., Weinheim, Germany, pp. 1–55.

Cunningham, J. A., Thomas, C. W., Bengtson, S., Marone, F., Stampanoni, M., Turner, F. R., Bailey, J. V., Raff, R. A., Raff, E. C., and Donoghue, P. C. J., 2012. Experimental taphonomy of giant sulphur bacteria: Implications for the interpretation of the embryo-like Ediacaran Doushantuo fossils. *Proceedings of the Royal Society B—Biological Sciences*, 279: 1857–1864.

D'Hondt, S., Rutherford, S., and Spivack, A. J., 2002. Metabolic activity of subsurface life in deep-sea sediments. *Science*, 295: 2067–2070.

Daubrée, M., 1875. Examples of contemporary formation of iron-pyrites in thermal springs and in sea-water. *Philosophical Magazine Series*, Series 4, 50: 562–564.

Davison, W., 1991. The solubility of iron sulphides in synthetic and natural waters at ambient temperature. *Aquatic Sciences*, 53: 309–329.

De Jonghe, A., Hart, M. B., Grimes, S. T., Mitlehner, A. G., Price, G. D., and Smart, C. W., 2011. Middle Eocene diatoms from Whitecliff Bay, Isle of Wight, England: Stratigraphy and preservation. *Proceedings of the Geologists Association*, 122: 472–483.

de Koff, J. P., Anderson, M. A., and Amrhein, C., 2008. Geochemistry of iron in the Salton Sea, California. *Hydrobiologia*, 604: 111–121.

de Nijs, B., Dussi, S., Smallenburg, F., Meeldijk, J. D., Groenendijk, D. J., Filion, L., Imhof, A., van Blaaderen, A., and Dijkstra, M., 2015. Entropy-driven formation of large icosahedral colloidal clusters by spherical confinement. *Nature Materials*, 14: 56–60.

De Yoreo, J. J., and Vekilov, P. G., 2003. Principles of crystal nucleation and growth. *Reviews in Mineralogy and Geochemistry*, 54: 57–93.

De Yoreo, J. J., Gilbert, P., Sommerdijk, N., Penn, R. L., Whitelam, S., Joester, D., Zhang, H. Z., Rimer, J. D., Navrotsky, A., Banfield, J. F., Wallace, A. F., Michel, F. M., Meldrum, F. C., Colfen, H., and Dove, P. M., 2015. Crystallization by particle attachment in synthetic, biogenic, and geologic environments. *Science*, 349: aaa6760.

Deditius, A. P., Utsunomiya, S., Reich, M., Kesler, S. E., Ewing, R. C., Hough, R., and Walshe, J., 2011. Trace metal nanoparticles in pyrite. *Ore Geology Reviews*, 42: 32–46.

Degens, E. T., Okada, H., Honjo, S., and Hathaway, J. C., 1972. Microcrystalline sphalerite in resin globules suspended in Lake Kivu, East Africa. *Mineralium Deposita*, 7: 1–12.

Delgado, A. V., Gonzalez-Caballero, E., Hunter, R. J., Koopal, L. K., and Lyklema, J., 2005. Measurement and interpretation of electrokinetic phenomena—(IUPAC technical report). *Pure and Applied Chemistry*, 77: 1753–1805.

Derjaguin, B., 1940. On the repulsive forces between charged colloid particles and on the theory of slow coagulation and stability of lyophobe sols. *Transactions of the Faraday Society*, 35: 0203–0214.

Derjaguin, B. and Landau, L.D., 1941. Theory of the stability of strongly charged lyophobic sols and of the adhesion of strongly charged particles in solutions of electrolytes. *Acta Physicochimica U.R.S.S.*, 14: 633–662.

Diehl, S., Rockefeller, G., Fryer, C. L., Riethmiller, D., and Statler, T. S., 2015. Generating optimal initial conditions for smoothed particle hydrodynamics simulations. *Publications of the Astronomical Society of Australia*, 32: E048.

Diehl, S. F., Goldhaber, M. B., and Hatch, J. R., 2004. Modes of occurrence of mercury and other trace elements in coals from the warrior field, Black Warrior Basin, Northwestern Alabama. *International Journal of Coal Geology*, 59: 193–208.

Diehl, S. F., Goldhaber, M. B., Koenig, A. E., Lowers, H. A., and Ruppert, L. F., 2012. Distribution of arsenic, selenium, and other trace elements in high pyrite Appalachian coals: Evidence for multiple episodes of pyrite formation. *International Journal of Coal Geology*, 94: 238–249.

Dmitrijeva, M., Cook, N. J., Ehrig, K., Ciobanu, C. L., Metcalfe, A. V., Kamenetsky, M., Kamenetsky, V. S., and Gilbert, S., 2020. Multivariate statistical analysis of trace elements in pyrite: Prediction, bias and artefacts in defining mineral signatures. *Minerals*, 10: 61.

Dos Santos Afonso, M., and Stumm, W., 1992. Reductive dissolution of iron(III) (hydr) oxides by hydrogen sulfide. *Langmuir*, 8: 1671–1675.

Doss, B., 1912a. Melnikovit, ein neues Eisenbisulfid, und seine Bedeuting für Genesis der Kieslagerstatten. *Zeitschrift fur Praktische Geologie*, 20: 453–483.

Doss, B., 1912b. Über die Natur und Zusammensetzung des miocänen Tonen der Gouvernements Samara auftretenden Schwefeleisens. *Neues Jahrbuch, Mineral.*, 33: 662–713.

Doye, J. P. K., and Calvo, F., 2001. Entropic effects on the size dependence of cluster structure. *Physical Review Letters*, 86: 3570–3573.

Drzaic, P. S., Marks, J., and Brauman, J. I., 1984. Electron photodetachment from gas phase molecular anions. *Gas Phase Ion Chemistry*, 3: 167–211.

Dullien, F. A. L., 1991. *Porous Media: Fluid Transport and Pore Structure*. Academic Press, New York, NY.

Eberl, D. D., Drits, V. A., and Środoń, J., 1998. Deducing growth mechanisms for minerals from the shapes of crystal size distributions. *American Journal of Science*, 298: 499–533.

Ebert, Y., Shaar, R., Emmanuel, S., Nowaczyk, N., and Stein, M., 2018. Overwriting of sedimentary magnetism by bacterially mediated mineral alteration. *Geology*, 46: 291–294.

Edwards, D., Axe, L., Parkes, J., and Rickard, D., 2006. Provenance and age of bacteria-like structures on mid-Palaeozoic plant fossils. *International Journal of Astrobiology*, 5: 109–142.

Eglinton, T. I., Irvine, J. E., Vairavamurthy, A., Zhou, W. Q., and Manowitz, B., 1994. Formation and diagenesis of macromolecular organic sulfur in Peru margin sediments. *Organic Geochemistry*, 22: 781–799.

Eigen, M., and Wilkins, R. G., 1965. The kinetics and mechanism of formation of metal complexes. *Mechanisms of Inorganic Reactions*. ACS Symposium Series. ACS, Washington, D.C. pp. 55–80.

Elderfield, H., and Schultz, A., 1996. Mid-ocean ridge hydrothermal fluxes and the chemical composition of the ocean. *Annual Review of Earth and Planetary Sciences*, 24: 191–224.

Emmings, J. F., Hennissen, J. A. I., Stephenson, M. H., Poulton, S. W., Vane, C. H., Davies, S. J., Leng, M. J., Lamb, A., and Moss-Hayes, V., 2019. Controls on amorphous organic matter type and sulphurization in a Mississippian black shale. *Review of Palaeobotany and Palynology*, 268: 1–18.

Endo, Y., 1978. Surface microtopographic study of pyrite crystals. *Bulletin of the Geological Society of Japan*, 29: 701–764.

Engelhardt, H., 2007. Are S-layers exoskeletons? The basic function of protein surface layers revisited. *Journal of Structural Biology*, 160: 115–124.

England, B. M., and Ostwald, J., 1993. Framboid-derived structures in some Tasman fold belt base-metal sulfide deposits, New South Wales, Australia. *Ore Geology Reviews*, 7: 381–412

Evans, J. E., Jungjohann, K. L., Browning, N. D., and Arslan, I., 2011. Controlled growth of nanoparticles from solution with in situ liquid transmission electron microscopy. *Nano Letters*, 11: 2809–2813.

Evitt, W. R., 1963. A discussion and proposals concerning fossil dinoflagellates, hystrichospheres, and acritarchs. 1. *Proceedings of the National Academy of Sciences of the United States of America*, 49: 158–164.

Ewald, P. P., and Friedrich, W., 1914. X-ray recording of cubic crystals, particularly pyrite. *Annalen Der Physik*, 44: 1183–1196.

Fabricius, F., 1961. Die Strukturen des "Rogenpyrits" (Kossener Schichten, rat) als Beitrag zum Problemen der "Veretzen Bakterien." *Geologische Rundschau*, 51: 647–657.

Farr, R. S., Honour, V. C., and Holness, M. B., 2017. Mean grain diameters from thin sections: matching the average to the problem. *Mineralogical Magazine*, 81: 515–530.

Farrand, M., 1970. Framboidal sulphides precipitated synthetically. *Mineralium Deposita*, 5(3): 237–247.

Faure, B., Salazar-Alvarez, G., and Bergstrom, L., 2011. Hamaker constants of iron oxide nanoparticles. *Langmuir*, 27: 8659–8664.

Favorov, V. A., Krasnikov, V. I., and Sychugov, V. S., 1974. Variations in semiconductor properties of pyrite and arsenopyrite and their determinants. *International Geology Review*, 16(4): 385–394.

Federov, E. S., 1891. The symmetry of regular systems of figures. *American Crystallographic Association Monograph* No. 7: 50–131.

Ferreira, T. O., Otero, X. L., Vidal-Torrado, P., and Macias, F., 2007. Redox processes in mangrove soils under *Rhizophora mangle* in relation to different environmental conditions. *Soil Science Society of America Journal*, 71(2): 484–491.

Finkelman, R. B., Dai, S., and French, D., 2019. The importance of minerals in coal as the hosts of chemical elements: A review. *International Journal of Coal Geology*, 212: 103251.

Fleischer, M., 1955. Minor elements in some sulfide minerals. In: A.M. Bateman (Editor), *Economic Geology Fiftieth Anniversary Volume 1905–1955, Part II*. Economic Geology Publishing Co., Urbana, Ill., pp. 970–1024.

Folk, R. L., 2005. Nannobacteria and the formation of framboidal pyrite: Textural evidence. *Journal of Earth System Science*, 114(3): 369–374.

Fornasiero, D., Eijt, V., and Ralston, J., 1992. An electrokinetic study of pyrite oxidation. *Colloids and Surfaces*, 62(1–2): 63–73.

Früh, J. J., 1885. Kritische Beiträge zur Kenntniss des Torfes. *Jahrbuch der Kaiserlich Königlichen Geologischen Reichsanstalt*, 35: 677–727.

Fuerstenau, M. C., Kuhn, M. C., and Elgillani, D. A., 1968. The role of dixantogen in the xanthate flotation of pyrite. *AIME Transactions*, 241: 148–156.

Gao, J., Yang, R., Zheng, L., Cheng, W., Chen, J., and Zhu, M., 2017. Forming mechanism analysis of the abnormally high $\delta^{34}S$ baryte deposits: A case study from the Zhenning–Ziyun large Devonian baryte deposits, Guizhou Province, China. *Geomicrobiology Journal*, 34(6): 481–488.

Garcia-Guinea, J., Martinez-Frias, J., Gonzalez-Martin, R., and Zamora, L., 1997. Framboidal pyrites in antique books. *Nature*, 388(6643): 631.

Garrels, R. M., 1960. *Mineral Equilbria*. Harper and Bros., New York, NY.

Gartman, A., and Luther, G. W., 2013. Comparison of pyrite (FeS$_2$) synthesis mechanisms to reproduce natural FeS$_2$ nanoparticles found at hydrothermal vents. *Geochimica et Cosmochimica Acta*, 120: 447–458.

Gartman, A., and Luther, G. W., 2014. Oxidation of synthesized sub-micron pyrite (FeS$_2$) in seawater. *Geochimica et Cosmochimica Acta*, 144: 96–108.

Gartman, A., Findlay, A. J., and Luther, G. W., 2014. Nanoparticulate pyrite and other nanoparticles are a widespread component of hydrothermal vent black smoker emissions. *Chemical Geology*, 366: 32–41.

Gayer, R., and Rickard, D., 1994. Colloform gold in coal from southern Wales. *Geology*, 22(1): 35–38.

Genge, M. J., Engrand, C., Gounelle, M., and Taylor, S., 2008. The classification of micrometeorites. *Meteoritics and Planetary Science*, 43(3): 497–515.

Gilbert, S. E., 2015. Development of analytical methods and standard reference materials for analysis of trace elements and isotopic ratios in sulphides. University of Tasmania, Tasmania.

Goldschmidt, V., 1920. *Atlas der Kristallformen. Tafeln*. Band VI. Markasit-Pyrit, 6. C. Winters, Heidelberg.

Goldschmidt, V., and Strock, L., 1935. Zur Geochemie des Selens II. *Nachr Akad Wiss Goettingen Math-Physik KI, IV, NF*, 1(11): 123–142.

Gong, M., Kirkeminde, A., and Ren, S., 2013. Symmetry-defying iron pyrite (FeS$_2$) nanocrystals through oriented attachment. *Scientific Reports*, 3: 2092.

Gong, Y.-M., Guang, R. S., Weldon, E. A., Du, Y.-S., and Xu, R., 2007. Pyrite framboids interpreted as microbial colonies within the Permian *Zoophycos* spreiten from southeastern Australia. *Geological Magazine*, 145: 95–103.

Graham, R. A., 1971. The Mogul base metal deposits County Tipperary, Ireland. PhD thesis, University of Western Ontario, 265 pp.

Graham, U. M., and Ohmoto, H., 1994. Experimental study of formation mechanisms of hydrothermal pyrite. *Geochimica et Cosmochimica Acta*, 58(10): 2187–2202.

Gregory, D., 2013. The trace element composition of sedimentary pyrite: Factors effecting uptake and uses of the data for determining paleo-ocean conditions. PhD thesis, University of Tasmania, 598 pp.

Gregory, D., Meffre, S., and Large, R., 2014. Comparison of metal enrichment in pyrite framboids from a metal-enriched and metal-poor estuary. *American Mineralogist*, 99(4): 633–644.

Gregory, D. D., Cracknell, M. J., Large, R. R., McGoldrick, P., Kuhn, S., Maslennikov, V. V., Baker, M. J., Fox, N., Belousov, I., Figueroa, M. C., Steadman, J. A., Fabris, A. J., and Lyons, T. W., 2019. Distinguishing ore deposit type and barren sedimentary pyrite using laser ablation-inductively coupled plasma-mass spectrometry trace element data and statistical analysis of large data sets. *Economic Geology*, 114(4): 771–786.

Gregory, D. D., Large, R. R., Bath, A. B., Steadman, J. A., Wu, S. L. N., Danyushevsky, L., Bull, S. W., Holden, P., and Ireland, T. R., 2016. Trace element content of pyrite from the Kapai Slate, St. Ives Gold District, Western Australia. *Economic Geology*, 111(6): 1297–1320.

Gregory, D. D., Large, R. R., Halpin, J. A., Baturina, E. L., Lyons, T. W., Wu, S., Danyushevsky, L., Sack, P. J., Chappaz, A., Maslennikov, V. V., and Bull, S. W., 2015. Trace element content of sedimentary pyrite in black shales. *Economic Geology*, 110(6): 1389–1410.

Gregory, D. D., Lyons, T. W., Large, R. R., Jiang, G. Q., Stepanov, A. S., Diamond, C. W., Figueroa, M. C., and Olin, P., 2017. Whole rock and discrete pyrite geochemistry as complementary tracers of ancient ocean chemistry: An example from the Neoproterozoic Doushantuo Formation, China. *Geochimica Et Cosmochimica Acta*, 216: 201–220.

Grimes, S. T., Davies, K. L., Butler, I. B., Brock, F., Edwards, D., Rickard, D., Briggs, D. E. G., and Parkes, R. J., 2002. Fossil plants from the Eocene London Clay: The use of pyrite textures to determine the mechanism of pyritization. *Journal of the Geological Society*, 159: 493–501.

Grondijs, H. F., and Schouten, C., 1937. A study of the Mount Isa ores (Queensland, Australia). *Economic Geology*, 32(4): 407–450.

Guilbaud, R., Butler, I. B., Ellam, R. M., and Rickard, D., 2010. Fe isotope exchange between $Fe(II)_{aq}$ and nanoparticulate mackinawite (FeS_m) during nanoparticle growth. *Earth and Planetary Science Letters*, 300(1–2): 174–183.

Guo, Y.-S., Furrer, J. M., Kadilak, A. L., Hinestroza, H. F., Gage, D. J., Cho, Y. K., and Shor, L. M., 2018. Bacterial extracellular polymeric substances amplify water content variability at the pore scale. *Frontiers in Environmental Science*, 6: 93.

Guy, B. M., Beukes, N. J., and Gutzmer, J., 2010. Paleoenvironmental controls on the texture and chemical composition of pyrite from nonconglomeratic sedimentary rocks of the Mesoarchean Witwatersrand Supergroup, South Africa. *South African Journal of Geology*, 113(2): 195–228.

Halbach, P., and Pracejus, B., 1993. Geology and mineralogy of massive sulfide ores from the central Okinawa Trough, Japan. *Economic Geology and the Bulletin of the Society of Economic Geologists*, 88(8): 2210–2225.

Hammarlund, E. U., Dahl, T. W., Harper, D. A. T., Bond, D. P. G., Nielsen, A. T., Bjerrum, C. J., Schovsbo, N. H., Schonlaub, H. P., Zalasiewicz, J. A., and Canfield, D. E., 2012. A sulfidic driver for the end-Ordovician mass extinction. *Earth and Planetary Science Letters*, 331: 128–139.

Hannington, M. D., 2014. Volcanogenic massive sulfide deposits. Geochemistry of mineral deposits. *Treatise on Geochemistry*, 13: 463–488.

Hannington, M. D., Galley, A. G., Gerzig, P. M., and Petersen, S., 1998. Comparison of the TAG mound and stockwork complex with Cyprus-type massive sulfide deposits. *Proceedings of the Ocean Drilling Program: Scientific Results*, 158: 389–415.

Hansen, D. J., McGuire, J. T., Mohanty, B. P., and Ziegler, B. A., 2014. Evidence of aqueous iron sulfide clusters in the vadose zone. *Vadose Zone Journal*, 13: 1–12.

Hardy, W. B.,1899. A preliminary investigation of the conditions which determine the stability of irreversible hydrosols. *Proceedings of the Royal Society of London.* 66: 110–125.

Harmandas, N. G., Navarro Fernandez, E., and Koutsoukos, P. G., 1998. Crystal growth of pyrite in aqueous solutions: Inhibition by organophosphorus compounds. *Langmuir,* 14(5): 1250–1255.

Hatton, B., and Rickard, D., 2008. Nucleic acids bind to nanoparticulate iron (II) monosulphide in aqueous solutions. *Origins of Life and Evolution of Biospheres,* 38(3): 257–270.

Hawkins, A. B., 2012. Sulphate heave: A model to explain the rapid rise of ground-bearing floor slabs. *Bulletin of Engineering Geology and the Environment,* 71: 113–117.

Hayashida, S., and Muta, K., 1952. Relation of trace element content and crystal form in pyrite. *Journal of the Mining Institute of Kyushu,* 29: 233–238.

Heath, D. F., 1967. Normal or log-normal appropriate distributions. *Nature,* 213(5081): 1159–1160.

Hellige, K., Pollok, K., Larese-Casanova, P., Behrends, T., and Peiffer, S., 2012. Pathways of ferrous iron mineral formation upon sulfidation of lepidocrocite surfaces. *Geochimica et Cosmochimica Acta,* 81: 69–81.

Hendy, S. C., and Doye, J. P. K., 2002. Surface-reconstructed icosahedral structures for lead clusters. *Physical Review B,* 66: 235402.

Herndon, E. M., Havig, J. R., Singer, D. M., McCormick, M. L., and Kump, L. R., 2018. Manganese and iron geochemistry in sediments underlying the redox-stratified Fayetteville Green Lake. *Geochimica et Cosmochimica Acta,* 231: 50–63.

Hochella, M. F., Aruguete, D., Kim, B., and Madden, A. S., 2012. Naturally occurring inorganic nanoparticles: General assessment and a global budget for one of Earth's last unexplored major geochemical components. In: A. S. Barnard and H. Guo (Editors), *Nature's Nanostructures.* Pan Stanford, Singapore, pp. 1–42.

Holmkvist, L., Ferdelman, T. G., and Jorgensen, B. B., 2011. A cryptic sulfur cycle driven by iron in the methane zone of marine sediment (Aarhus Bay, Denmark). *Geochimica et Cosmochimica Acta,* 75(12): 3581–3599.

Holser, W. T., Schidlowski, M., Mackenzie, F. T., and Maynard, J. B., 1988. Geochemical cycles of carbon and sulfur. In: C. B. Gregor, R. A. Garrels, F. T. Mackenzie, and J. B. Maynard (Editors), *Chemical Cycles in the Evolution of the Earth.* Wiley, New York, NY, pp. 105–173.

Honnorez, J., 1969. La formation actuelle d'un gisement sous-marin de sulfures fumerolliens à Vulcano (mer tyrrhénienne). Partie I. Les minéraux sulfurés des tufs immergés à faible profondeur. *Mineralium Deposita,* 4(2): 114–131.

Honnorez, J., Honnorez-Guerstein, B., Valette, J., and Wauschkuhn, A., 1973. Present-day formation of an exhalative sulfide deposit at Vulcano (Tyrrhenian Sea). 2. Active crystallization of fumarolic sulfides in the volcanic sediments of the Baja di Levante. In: G. C. Amstutz and J. A. Bernard (Editors), *Ores in Sediments.* Springer, Berlin, pp. 139–166.

Hosking, K. F. G., and Camm, G. S., 1980. Ocurrences of pyrite framboids and polyframboids in west Cornwall. *Journal of the Camborne School of Mines,* 80: 33–42.

Howarth, R. W., 1979. Pyrite: Its rapid formation in a salt-marsh and its importance in ecosystem metabolism. *Science,* 203(4375): 49–51.

Hu, S. Y., Evans, K., Craw, D., Rempel, K., Bourdet, J., Dick, J., and Grice, K., 2015. Raman characterization of carbonaceous material in the Macraes orogenic gold deposit and metasedimentary host rocks, New Zealand. *Ore Geology Reviews,* 70: 80–95.

Hu, S. Y., Evans, K., Rempel, K., Guagliardo, P., Kilburn, M., Craw, D., Grice, K., and Dick, J., 2018. Sequestration of Zn into mixed pyrite-zinc sulfide framboids: A key to Zn cycling in the ocean? *Geochimica et Cosmochimica Acta*, 241: 95–107.

Huber, C., and Wächtershauser, G., 1997. Activated acetic acid by carbon fixation on (Fe,Ni)S under primordial conditions. *Science*, 276(5310): 245–247.

Huerta-Diaz, M. A., and Morse, J. W., 1992. Pyritization of trace-metals in anoxic marine-sediments. *Geochimica et Cosmochimica Acta*, 56(7): 2681–2702.

Huerta-Diaz, M. A., Tessier, A., and Carignan, R., 1998. Geochemistry of trace metals associated with reduced sulfur in freshwater sediments. *Applied Geochemistry*, 13(2): 213–233.

Hung, A., Muscat, J., Yarovsky, I., and Russo, S. P., 2002. Density-functional theory studies of pyrite FeS_2(100) and (110) surfaces. *Surface Science*, 513(3): 511–524.

Hunt, W. F., 1915. The origin of the sulphur deposits of Sicily. *Economic Geology*, 10(6): 543–579.

Hunter, R. J., 1988. *Zeta Potential in Colloid Science*. Academic Press, New York, NY.

Israelachvili, J. N., 2011. Electrostatic forces between surfaces in liquids. In: J. N. Israelachvili (Editor), *Intermolecular and Surface Forces* (Third Edition). Academic Press, San Diego, CA, Chapter 14, pp. 291–340.

Issatshenko, B. L., 1912. On the deposit of iron sulfide in bacteria. *Bulletin du Jardin Imperial Botanique de St. Petersbourg*, 12: 1–6.

Ixer, R. K., and Vaughan, D. J., 1993. Lead-zinc-fluorite-baryte deposits of the Pennines, North Wales and Mendips. In: R. A. D. Pattrick and D. A. Polya (Editors), *Mineralization in the British Isles*. Chapman & Hall., London, UK, pp. 355–418.

Izumi, K., Endo, K., Kemp, D. B., and Inui, M., 2018. Oceanic redox conditions through the late Pliensbachian to early Toarcian on the northwestern Panthalassa margin: Insights from pyrite and geochemical data. *Palaeogeography Palaeoclimatology Palaeoecology*, 493: 1–10.

Jahn, A., and Nielsen, P. H., 1995. Extraction of extracellular polymeric substances (EPS) from biofilms using a cation exchange resin. *IAWQ International Conference and Workshop on Biofilm Structure, Growth and Dynamics*, Noordwijkerhout, Netherlands, pp. 157–164.

Jedwab, J., 1971. La magnétite de la météorite d'Orgueil vue au microscope électronique a balage. *Icarus*, 15: 319–340.

Jiang, W. T., Horng, C. S., Roberts, A. P., and Peacor, D. R., 2001. Contradictory magnetic polarities in sediments and variable timing of neoformation of authigenic greigite. *Earth and Planetary Science Letters*, 193(1–2): 1–12.

Jorgensen, B. B., 1977. Sulfur cycle of a coastal marine sediment (Limfjorden, Denmark). *Limnology and Oceanography*, 22(5): 814–832.

Kafantaris, F-C. A., and Gregory G., 2019. Kinetics of the nucleophilic dissolution of hydrophobic and hydrophilic elemental sulfur sols by sulfide. *Geochimica et Cosmochimica Acta*, 269: 554–565.

Kalliokoski, J., and Cathles, L., 1969. Morphology, mode of formation and diagenetic changes in framboids. *Bulletin of the Geological Society of Finland*, 41: 125–133.

Kamyshny, A., Goifman, A., Gun, J., Rizkov, D., and Lev, O., 2004. Equilibrium distribution of polysulfide ions in aqueous solutions at 25 °C: A new approach for the study of polysulfides equilibria. *Environmental Science & Technology*, 38(24): 6633–6644.

Kar, S., and Chaudhuri, S., 2004. Solvothermal synthesis of nanocrystalline FeS_2 with different morphologies. *Chemical Physics Letters*, 398(1–3): 22–26.

Kashchiev, D., 2011. Note: On the critical supersaturation for nucleation. *Journal of Chemical Physics*, 134: 196102.

Kato, G., 1967. Biogenic pyrite from a Miocene formation of Shimane peninsula, southwest Japan. *Memoirs of the Faculty of Science, Kyushu University, Series D, Geology*, XVIII(2): 313–330.

Kaye, T. G., Gaugler, G., and Sawlowicz, Z., 2008. Dinosaurian soft tissues interpreted as bacterial biofilms. *Plos One*, 3(7): e2808.

Khalid, S., Ahmed, E., Khan, Y., Riaz, K. N., and Malik, M. A., 2018. Nanocrystalline pyrite for photovoltaic applications. *Chemistryselect*, 3(23): 6488–6524.

Kimura, Y., Sato, T., Nakamura, N., Nozawa, J., Nakamura, T., Tsukamoto, K., and Yamamoto, K., 2013. Vortex magnetic structure in framboidal magnetite reveals existence of water droplets in an ancient asteroid. *Nature Communications*, 4, 2649.

Kitchaev, D. A., and Ceder, G., 2016. Evaluating structure selection in the hydrothermal growth of FeS₂ pyrite and marcasite. *Nature Communications*, 7, 13799.

Kleinjan, W. E., de Keizer, A., and Janssen, A. J. H., 2005. Equilibrium of the reaction between dissolved sodium sulfide and biologically produced sulfur. *Colloids and Surfaces B - Biointerfaces*, 43(3–4): 228–237.

Knight, J. B., 1931. The gastropods of the St. Louis, Missouri, Pennsylvanian outlier: The Subulitidae. *Journal of Paleontology*, 5(3): 177–229.

Koenigsberger, J., 1901. Ueber einer Apparat zur Erkennung und Messung optischer Anisotropie undurchsichtiger Substanzer und dessen Verwendung. *Centralblatt für Mineralogie, Geologie und Paläontologie*: 565–597.

Kohn, M. J., Riciputi, L. R., Stakes, D., and Orange, D. L., 1998. Sulfur isotope variability in biogenic pyrite: Reflections of heterogeneous bacterial colonization? *American Mineralogist*, 83(11–12): 1454–1468.

Kolker, A., 2012. Minor element distribution in iron disulfides in coal: A geochemical review. *International Journal of Coal Geology*, 94: 32–43.

Kortenski, J., and Kostova, I., 1996. Occurrence and morphology of pyrite in Bulgarian coals. *International Journal of Coal Geology*, 29(4): 273–290.

Kožíšek, Z., Sato, K., Ueno, S., and Demo, P., 2011. Formation of crystal nuclei near critical supersaturation in small volumes. *Journal of Chemical Physics*, 134: 094508.

Kribek, B., 1975. Origin of framboidal pyrite as a surface effect of sulfur grains. *Mineralium Deposita*, 10(4): 389–396.

Kucha, H., Schroll, E., Raith, J. G., and Halas, S., 2010. Microbial sphalerite formation in carbonate-hosted Zn-Pb ores, Bleiberg, Austria: Micro- to nanotextural and sulfur isotope evidence. *Economic Geology*, 105(5): 1005–1023.

Kullerud, G., and Yoder, H. S., 1959. Pyrite stability relations in the Fe-S system. *Economic Geology*, 54: 533–572.

Kumar, N., Pacheco, J. L., Noel, V., Dublet, G., and Brown, G. E., 2018. Sulfidation mechanisms of Fe(III)-(oxyhydr)oxide nanoparticles: A spectroscopic study. *Environmental Science-Nano*, 5(4): 1012–1026.

Kuo, K. H., 2002. Mackay, anti-Mackay, double-Mackay, pseudo-Mackay, and related icosahedral shell clusters. *Structural Chemistry*, 13(3–4): 221–230.

Labrenz, M., and Banfield, J. F., 2004. Sulfate-reducing bacteria-dominated biofilms that precipitate ZnS in a subsurface circumneutral-pH mine drainage system. *Microbial Ecology*, 47(3): 205–217.

Labrenz, M., Druschel, G. K., Thomsen-Ebert, T., Gilbert, B., Welch, S. A., Kemner, K. M., Logan, G. A., Summons, R. E., De Stasio, G., Bond, P. L., Lai, B., Kelly, S. D., and

Banfield, J. F., 2000. Formation of sphalerite (ZnS) deposits in natural biofilms of sulfate-reducing bacteria. *Science*, 290(5497): 1744–1747.

LaMer, V. K., 1952. Nucleation in phase transitions. *Industrial and Engineering Chemistry*, 44(6): 1270–1277.

LaMer, V. K., and Dinegar, R. H., 1950. Theory, production and mechanism of formation of monodispersed hydrosols. *Journal of the American Chemical Society*, 72(11): 4847–4854.

Large, D. J., Fortey, N. J., Milodowski, A. E., Christy, A. G., and Dodd, J., 2001. Petrographic observations of iron, copper, and zinc sulfides in freshwater canal sediment. *Journal of Sedimentary Research*, 71(1): 61–69.

Large, D. J., Sawlowicz, Z., and Spratt, J., 1999. A cobaltite-framboidal pyrite association from the Kupferschiefer: Possible implications for trace element behaviour during the earliest stages of diagenesis. *Mineralogical Magazine*, 63(3): 353–361.

Large, R. R., Halpin, J. A., Danyushevsky, L. V., Maslennikov, V. V., Bull, S. W., Long, J. A., Gregory, D. D., Lounejeva, E., Lyons, T. W., Sack, P. J., McGoldrick, P. J., and Calver, C. R., 2014. Trace element content of sedimentary pyrite as a new proxy for deep-time ocean-atmosphere evolution. *Earth and Planetary Science Letters*, 389: 209–220.

Large, R. R., Halpin, J. A., Lounejeva, E., Danyushevsky, L. V., Maslennikov, V. V., Gregory, D., Sack, P. J., Haines, P. W., Long, J. A., Makoundi, C., and Stepanov, A. S., 2015. Cycles of nutrient trace elements in the Phanerozoic ocean. *Gondwana Research*, 28(4): 1282–1293.

Large, R. R., Steadman, J. A., Mukherjee, I., Corkrey, R., Sack, P., and Ireland, T. R., 2020, Trends in Ocean S-Isotopes may be influenced by major LIP Events. In: R. E. Ernst, A. J. Dickson and A. Bekker (Editors). *Large Igneous Provinces. American Geophysical Union. Carbon in Earth's Interior.* John Wiley & Sons, Inc. Hoboken, N.J., pp. 341–376.

Lasaga, A. C., 1998. *Kinetic Theory in the Earth Sciences.* Princeton University Press, Princeton, NJ.

Lash, G. G., 2017. A multiproxy analysis of the Frasnian-Famennian transition in western New York State, USA. *Palaeogeography Palaeoclimatology Palaeoecology*, 473: 108–122.

Lauf, R. J., Harris, L. A., and Rawlston, S. S., 1982. Pyrite framboids as the source of magnetite spheres in fly-ash. *Environmental Science & Technology*, 16(4): 218–220.

Lett, R. E. W., and Fletcher, W. K., 1980. Syngenetic sulfide minerals in a copper-rich bog. *Mineralium Deposita*, 15(1): 61–67.

Lewan, M. D., 1998. Sulphur-radical control on petroleum formation rates. *Nature*, 391(6663): 164–166.

Li, F., Josephson, D. P., and Stein, A., 2011. Colloidal assembly: The road from particles to colloidal molecules and crystals. *Angewandte Chemie—International Edition*, 50(2): 360–388.

Li, W., Döblinger, M., Vaneski, A., Rogach, A. L., Jäckel, F., and Feldmann, J., 2011. Pyrite nanocrystals: Shape-controlled synthesis and tunable optical properties via reversible self-assembly. *Journal of Materials Chemistry*, 21(44): 17946–17952.

Li, X., Kwak, T. A. P., and Brown, R. W., 1998. Wallrock alteration in the Bendigo gold ore field, Victoria, Australia: Uses in exploration. *Ore Geology Reviews*, 13(1–5): 381–406.

Li, Y. Q., Chen, J. H., Chen, Y., and Guo, J., 2011. Density functional theory study of influence of impurity on electronic properties and reactivity of pyrite. *Transactions of Nonferrous Metals Society of China*, 21(8): 1887–1895.

Liao, W., Bond, D. P. G., Wang, Y. B., He, L., Yang, H., Weng, Z. T., and Li, G. S., 2017. An extensive anoxic event in the Triassic of the South China Block: A pyrite framboid study from Dajiang and its implications for the cause(s) of oxygen depletion. *Palaeogeography Palaeoclimatology Palaeoecology*, 486: 86–95.

Libbey, R. B., and Williams-Jones, A. E., 2016. Relating sulfide mineral zonation and trace element chemistry to subsurface processes in the Reykjanes geothermal system, Iceland. *Journal of Volcanology and Geothermal Research*, 310: 225–241.

Limpert, E., and Stahel, W. A., 2011. Problems with using the normal distribution—and ways to improve quality and efficiency of data analysis. *PLoS ONE*, 6(7): e21403.

Limpert, E., Stahel, W. A., and Abbt, M., 2001. Log-normal distributions across the sciences: Keys and clues. *Bioscience*, 51(5): 341–352.

Lin, Q., Wang, J. S., Algeo, T. J., Sun, F., and Lin, R. X., 2016. Enhanced framboidal pyrite formation related to anaerobic oxidation of methane in the sulfate-methane transition zone of the northern South China Sea. *Marine Geology*, 379: 100–108.

Liu, L., Ireland, T., Holden, P., and Mavrogenes, J., 2020. The sign of $\Delta^{33}S$ is independent of pyrite morphology. *Chemical Geology*, 532: 119369.

Liu, S. L., Li, M. M., Li, S., Li, H. L., and Yan, L., 2013. Synthesis and adsorption/photocatalysis performance of pyrite FeS_2. *Applied Surface Science*, 268: 213–217.

Love, L. G., 1957. Micro-organisms and the presence of syngenetic pyrite. *Quarterly Journal of the Geological Society, London*, 113: 428–440.

Love, L. G., 1962. Biogenic primary sulfide of the Permian Kupferschiefer and Marl Slate. *Economic Geology*, 57: 350–366.

Love, L. G., 1965. Micro-organic material with diagenetic pyrite from the Lower Proterozoic Mount Isa shale and a Carboniferous shale. *Proceedings of the Yorkshire Geological Society*, 35:187–202.

Love, L. G., 1971. Early diagenetic polyframboidal pyrite, primary and redeposited, from the Wenlockian Denbigh Grit Group, Conway, North Wales. U.K. *Journal of Sedimentary Petrology*, 41: 1038–1044.

Love, L. G., Al-Kaisy, A. T. H., and Brockley H., 1971. Mineral and organic material in matrices and coatings of framboidal pyrite from Pennsylvanian sediments, England. *Journal of Sedimentary Petrology*, 54: 869–876.

Love, L. G., and Amstutz, G. C., 1966. Review of microscopic pyrite from the Devonian Chattanooga Shale and Rammelsberg Banderz. *Fortschritte der Mineralogie*, 43: 277–309.

Love, L. G., and Amstutz, G. C., 1969. Framboidal pyrite from two andesites. *Neues Jahrbuch für Mineralogie, Geologie und Paläontologie*, 3: 97–108.

Love, L. G., Curtis, C. D., and Brockley, H., 1971. Framboidal pyrite: Morphology revealed by electron microscopy of external surfaces. *Fortschritte der Mineralogie*, 48(2): 259–264.

Love, L. G., and Zimmermann, D. O., 1961. Bedded pyrite and micro-organisms from the Mount Isa shale. *Economic Geology*, 56(5): 873–896.

Lucas, J. M., Tuan, C. -C., Lounis, S. D., Britt, D. K., Qiao, R., Yang, W., Lanzara, A., and Alivisatos, A. P., 2013. Ligand-controlled colloidal synthesis and electronic structure characterization of cubic iron pyrite (FeS_2) nanocrystals. *Chemistry of Materials*, 25(9): 1615–1620.

Luther, G. W., 1991. Pyrite synthesis via polysulfide compounds. *Geochimica et Cosmochimica Acta*, 55(10): 2839–2849.

Luther, G. W., Glazer, B., Ma, S., Trouwborst, R., Shultz, B. R., Druschel, G., and Kraiya, C., 2003. Iron and sulfur chemistry in a stratified lake: Evidence for iron-rich sulfide complexes. *Aquatic Geochemistry*, 9(2): 87–110.

Luther, G. W., Glazer, B. T., Ma, S. F., Trouwborst, R. E., Moore, T. S., Metzger, E., Kraiya, C., Waite, T. J., Druschel, G., Sundby, B., Taillefert, M., Nuzzio, D. B., Shank, T. M., Lewis, B. L., and Brendel, P. J., 2008. Use of voltammetric solid-state (micro)electrodes for studying biogeochemical processes: Laboratory measurements to real time measurements with an in situ electrochemical analyzer (ISEA). *Marine Chemistry*, 108(3–4): 221–235.

Luther, G. W., Theberge, S. M., Rickard, D. T., and Oldroyd, A., 1996. Determination of metal (bi)sulfide stability constants of Mn^{2+}, Fe^{2+}, Co^{2+}, Ni^{2+}, Cu^{2+} and Zn^{2+} by voltammetric methods. *Environmental Science & Technology*, 30(2): 671–679.

Lyons, T. W., and Berner, R. A., 1992. Carbon-sulfur-iron systematics of the uppermost deep-water sediments of the Black Sea. *Chemical Geology*, 99(1–3): 1–27.

Lyons, T. W., and Severmann, S., 2006. A critical look at iron paleoredox proxies: New insights from modern euxinic marine basins. *Geochimica et Cosmochimica Acta*, 70(23): 5698–5722.

Mackay, A. L., 1962. A dense non-crystallographic packing of equal spheres. *Acta Crystallographica*, 15(SEP): 916–919.

Maclean, L. C. W., Tyliszczak, T., Gilbert, P., Zhou, D., Pray, T. J., Onstott, T. C., and Southam, G., 2008. A high-resolution chemical and structural study of framboidal pyrite formed within a low-temperature bacterial biofilm. *Geobiology*, 6(5): 471–480.

Maier, S., and Murray, R. G. E., 1965. Fine structure of *Thioploca ingrica* and a comparison with Beggiatoa. *Canadian Journal of Microbiology*, 11(4): 645–655.

Marsh, B. D., 1988. Crystal size distribution (CSD) in rocks and the kinetics and dynamics of crystallization—I. Theory. *Contributions to Mineralogy and Petrology*, 99(3): 277–291.

Martinez-Sierra, J. G., San Bias, O. G., Gayon, J. M. M., and Alonso, J. I. G., 2015. Sulfur analysis by inductively coupled plasma-mass spectrometry: A review. *Spectrochimica Acta Part B-Atomic Spectroscopy*, 108: 35–52.

Martinez-Yanez, M., Nunez-Useche, F., Martinez, R. L., and Gardner, R. D., 2017. Paleoenvironmental conditions across the Jurassic Cretaceous boundary in central-eastern Mexico. *Journal of South American Earth Sciences*, 77: 261–275.

Marynowski, L., Rakocinski, M., Borcuch, E., Kremer, B., Schubert, B. A., and Jahren, A. H., 2011. Molecular and petrographic indicators of redox conditions and bacterial communities after the F/F mass extinction (Kowala, Holy Cross Mountains, Poland). *Palaeogeography Palaeoclimatology Palaeoecology*, 306(1–2): 1–14.

Matamoros-Veloza, A., Cespedes, O., Johnson, B. R. G., Stawski, T. M., Terranova, U., de Leeuw, N. H., and Benning, L. G., 2018. A highly reactive precursor in the iron sulfide system. *Nature Communications*, 9: 3125.

McKay, J. L., and Longstaffe, F. J., 2003. Sulphur isotope geochemistry of pyrite from the Upper Cretaceous Marshybank Formation, Western Interior Basin. *Sedimentary Geology*, 157(3–4): 175–195.

McKibben, M. A., and Riciputi, L. R., 1998. Sulfur isotopes by ion microprobe. In: M. A. McKibben, W. C. Shanks III, and W. I. Ridley (Editors), *Applications of Microanalytical Technques to Understanding Mineralizing Processes*. Society of Economic Geologists, Denver, CO, pp. 121–139.

McNeil, D. H., 1990. Stratigraphy and paleoecology of the Eocene *Stellarima* assemblage zone (pyrite diatom steinkerns) in the Beaufort-Mackenzie basin, Arctic Canada. *Bulletin of Canadian Petroleum Geology*, 38(1): 17–27.

Merinero, R., Cardenes, V., Lunar, R., Boone, M. N., and Cnudde, V., 2017. Representative size distributions of framboidal, euhedral, and sunflower pyrite from high-resolution X-ray tomography and scanning electron microscopy analyses. *American Mineralogist*, 102(3): 620–631.

Merinero, R., Lunar, R., Martinez-Fiias, J., Somoza, L., and Diaz-del-Rio, V., 2008. Iron oxyhydroxide and sulphide mineralization in hydrocarbon seep-related carbonate submarine chimneys, Gulf of Cadiz (SW Iberian Peninsula). *Marine and Petroleum Geology*, 25(8): 706–713.

Merinero, R., Ortega, L., Lunar, R., Pina, R., and Cardenes, V., 2019. Framboidal chalco-pyrite and bornite constrain redox conditions during formation of their host rocks in the copper stratabound mineralization of Picachos, north-central Chile. *Ore Geology Reviews*, 112.

Middelburg, J. J., Delange, G. J., Vandersloot, H. A., Vanemburg, P. R., and Sophiah, S., 1988. Particulate manganese and iron framboids in Kau Bay, Halmahera (eastern Indonesia). *Marine Chemistry*, 23(3–4): 353–364.

Migula, W., 1900. *Specielle Systematik der Bakterien*, 2. Gustav Fischer, Jena.

Millero, F. J., 1986. The thermodynamics and kinetics of the hydrogen sulfide system in natural waters. *Marine Chemistry*, 18(2–4): 121–147.

Ming, T., Kou, X. S., Chen, H. J., Wang, T., Tam, H. L., Cheah, K. W., Chen, J. Y., and Wang, J. F., 2008. Ordered gold nanostructure assemblies formed by droplet evapora-tion. *Angewandte Chemie-International Edition*, 47(50): 9685–9690.

Miyake, N., Wallis, M. K., and Wickramasinghe, N. C., 2012. Discovery of framboidal magnetites in the Murchison meteorite. *European Planetary Science Congress* 7: EPSC2012–921.

Moreau, J. W., Webb, R. I., and Banfield, J. F., 2004. Ultrastructure, aggregation-state, and crystal growth of biogenic nanocrystalline sphalerite and wurtzite. *American Mineralogist*, 89(7): 950–960.

Moreau, J. W., Weber, P. K., Martin, M. C., Gilbert, B., Hutcheon, I. D., and Banfield, J. F., 2007. Extracellular proteins limit the dispersal of biogenic nanoparticles. *Science*, 316(5831): 1600–1603.

Morrissey, C. J., 1972. A quasi-framboidal form of syn-sedimentary pyrite. *Transactions of the Institute of Mining and Metallurgy*, 81: B55–B56.

Morse, J. W., and Wang, Q. W., 1997. Pyrite formation under conditions approximating those in anoxic sediments. 2. Influence of precursor iron minerals and organic matter. *Marine Chemistry*, 57(3–4): 187–193.

Mucke, A., Badejoko, T. A., and Akande, S. O., 1999. Petrographic-microchemical studies and origin of the Agbaja Phanerozoic Ironstone Formation, Nupe Basin, Nigeria: A product of a ferruginized ooidal kaolin precursor not identical to the Minette-type. *Mineralium Deposita*, 34(3): 284–296.

Muir, M. D., 1981. The micro-fossils from the Proterozoic Urquhart Shale, Mount-Isa, Queensland, and their significance in relation to the depositional environment, dia-genesis, and mineralization. *Mineralium Deposita*, 16(1): 51–58.

Mukherjee, I., and Large, R., 2017. Application of pyrite trace element chemistry to explo-ration for SEDEX style Zn-Pb deposits: McArthur Basin, Northern Territory, Australia. *Ore Geology Reviews*, 81: 1249–1270.

Muramoto, J. A., Honjo, S., Fry, B., Hay, B. J., Howarth, R. W., and Cisne, J. L., 1991. Sulfur, iron and organic-carbon fluxes in the Black Sea: Sulfur isotopic evidence for origin of sulfur fluxes. *Deep-Sea Research Part A - Oceanographic Research Papers*, 38: S1151–1187.

Murowchick, J. B., and Barnes, H. L., 1986. Marcasite precipitation from hydrothermal solutions. *Geochimica et Cosmochimica Acta*, 50(12): 2615–2629.

Murowchick, J. B., and Barnes, H. L., 1987. Effects of temperature and degree of supersaturation on pyrite morphology. *American Mineralogist*, 72(11–12): 1241–1250.

Murphy, R., and Strongin, D. R., 2009. Surface reactivity of pyrite and related sulfides. *Surface Science Reports*, 64(1): 1–45.

Naik, S., and Caruntu, G., 2017. Assemblies and superstructures of inorganic colloidal nanocrystals. In: S. Hunyadi Murph, G. Larsen, and K. Coopersmith (Editors), *Anisotropic and Shape-Selective Nanomaterials: Nanostructure Science and Technology*. Springer, Cham, Switzerland, pp. 293–335.

Nazir, R., Zaffar, M. R., and Amin, I., 2019. Bacterial biofilms: The remarkable heterogeneous biological communities and nitrogen fixing microorganisms in lakes. In: S. A. Bandh, S. Shafi, and N. Shameem (Editors), *Freshwater Microbiology*. Academic Press, Cambridge, MA. Chapter 8, pp. 307–340.

Neuhaus, A., 1940. Über die Erzführung des Kupfermergels der Haaseler und der Gröditzer Mulde in Schlesien. *Zeitschrift für angewandte Mineralogie*, 2: 340–343.

Noel, V., Marchand, C., Juillot, F., Ona-Nguema, G., Viollier, E., Marakovic, G., Olivi, L., Delbes, L., Gelebart, F., and Morin, G., 2014. EXAFS analysis of iron cycling in mangrove sediments downstream a lateritized ultramafic watershed (Vavouto Bay, New Caledonia). *Geochimica et Cosmochimica Acta*, 136: 211–228.

Nolze, G., Winkelmann, A., and Boyle, A. P., 2016. Pattern matching approach to pseudosymmetry problems in electron backscatter diffraction. *Ultramicroscopy*, 160: 146–154.

Nuhfer, E. B., 1979. Temporal and lateral Variations in the Geochemistry, Mineralogy, and Microscopy of Seston collected in Automated Samplers, PhD dissertation, University of New Mexico, 396 pp.

Nuhfer, E. B., and Pavlovic, A. S., 1979. Association of kaolinite with pyritic framboids: Discussion on "Interstitial networks of kaolinite within pyrite framboids in the Meigs Creek coal of Ohio" (Scheihing, Gluskoter & Finkelman, *Journal of Sedimentary Petrology*, 48, 1978, pp. 723–732). *Journal of Sedimentary Petrology*, 49: 321–323.

Ohfuji, H., 2004. *Framboids*. PhD thesis, Cardiff University, Cardiff, 246 pp.

Ohfuji, H., and Akai, J., 2002. Icosahedral domain structure of framboidal pyrite. *American Mineralogist*, 87(1): 176–180.

Ohfuji, H., Boyle, A. P., Prior, D. J., and Rickard, D., 2005. Structure of framboidal pyrite: An electron backscatter diffraction study. *American Mineralogist*, 90(11–12): 1693–1704.

Ohfuji, H., and Rickard, D., 2005. Experimental syntheses of framboids: A review. *Earth-Science Reviews*, 71(3–4): 147–170.

Ohfuji, H., and Rickard, D., 2006. High resolution transmission electron microscopic study of synthetic nanocrystalline mackinawite. *Earth and Planetary Science Letters*, 241(1–2): 227–233.

Ohfuji, H., Rickard, D., Light, M. E., and Hursthouse, M. B., 2006. Structure of framboidal pyrite: A single crystal X-ray diffraction study. *European Journal of Mineralogy*, 18(1): 93–98.

Ostwald, J., and England, B. M., 1977. Notes on framboidal pyrite from Allandale New South Wales, Australia. *Mineralium Deposita*, 12(1): 111–116.

Ostwald, J., and England, B. M., 1979. The relationship between euhedral and framboidal pyrite in base-metal sulfide ores. *Mineralogical Magazine*, 43: 297–300.

Oszczepalski, S., 1999. Origin of the Kupferschiefer polymetallic mineralization in Poland. *Mineralium Deposita*, 34(5–6): 599–613.

Pacton, M., Gorin, G. E., and Vasconcelos, C., 2011. Amorphous organic matter: Experimental data on formation and the role of microbes. *Review of Palaeobotany and Palynology*, 166(3–4): 253–267.

Painter, M. G. M., Golding, S. D., Hannan, K. W., and Neudert, M. K., 1999. Sedimentologic, petrographic and sulfur isotope constraints on fine-grained pyrite formation at Mount Isa Mine and environs, northwest Queensland, Australia. *Economic Geology*, 94: 883–912.

Pan, Y., 2001. Inherited correlation in crystal size distribution. *Geology*, 29(3): 227–230.

Pankow, J. F., and Morgan, J. J., 1979. Dissolution of tetragonal ferrous sulfide (mackinawite) in anoxic aqueous systems.1. Dissolution rate as a function of pH, temperature, and ionic-strength. *Environmental Science & Technology*, 13(10): 1248–1255.

Papunen, H., 1966. Framboidal texture of the pyritic layer found in a peat bog in SE-Finland. *Comptes Rendus de la Société géologique de Finlande*, 38: 117–125.

Pearce, C. I., Pattrick, R. A. D., and Vaughan, D. J., 2006. Electrical and magnetic properties of sulfides. In: D. J. Vaughan (Editor), *Sulfide Mineralogy and Geochemistry: Reviews in Mineralogy & Geochemistry*. 61: 127–180.

Peiffer, S., Behrends, T., Hellige, K., Larese-Casanova, P., Wan, M., and Pollok, K., 2015. Pyrite formation and mineral transformation pathways upon sulfidation of ferric hydroxides depend on mineral type and sulfide concentration. *Chemical Geology*, 400: 44–55.

Perry, K. A., and Pedersen, T. F., 1993. Sulfur speciation and pyrite formation in meromictic ex-fjords. *Geochimica et Cosmochimica Acta*, 57(18): 4405–4418.

Peterson, B. T., and Depaolo, D. J., 2007. Mass and composition of the continental crust estimated using the CRUST2.0 model. *AGU Fall Meeting Abstracts*, 1: 1161.

Pichler, T., Giggenbach, W. F., McInnnes, B. I. A., Buhl, D., and Duck, B., 1999. Fe sulfide formation due to seawater-gas-sediment interaction in a shallow-water hydrothermal system at Lihir Island, Papua New Guinea. *Economic Geology*, 94(2): 281–287.

Pina, C. M., and Putnis, A., 2002. The kinetics of nucleation of solid solutions from aqueous solutions: A new model for calculating non-equilibrium distribution coefficients. *Geochimica et Cosmochimica Acta*, 66(2): 185–192.

Potonié, H., 1908. A classification of Caustobiolith. *Sitzungsberichte Der Koniglich Preussischen Akademie Der Wissenschaften*. Part II: 154–165.

Privman, V., Goia, D. V., Park, J., and Matijević, E., 1999. Mechanism of formation of monodispersed colloids by aggregation of nanosize precursors. *Journal of Colloid and Interface Science*, 213(1): 36–45.

Prol-Ledesma, R. M., Canet, C., Villanueva-Estrada, R. E., and Ortega-Osorio, A., 2010. Morphology of pyrite in particulate matter from shallow submarine hydrothermal vents. *American Mineralogist*, 95(10): 1500–1507.

Putnis, A., 2010. Effects of kinetics and mechanisms of crystal growth on ion-partitioning in solid solution-aqueous solution (SS-AS) systems. *European Mineralogical Union Notes in Mineralogy*. 10: 43–64.

Pyzik, A. J., and Sommer, S. E., 1981. Sedimentary iron monosulfides: Kinetics and mechanism of formation. *Geochimica et Cosmochimica Acta*, 45(5): 687–698.

Qi, S. Y., Yang, S. H., Chen, J., Niu, T. Q., Yang, Y. F., and Xin, B. P., 2019. High-yield extracellular biosynthesis of ZnS quantum dots through a unique molecular mediation mechanism by the peculiar extracellular proteins secreted by a mixed sulfate reducing bacteria. *ACS Applied Materials & Interfaces*, 11(11): 10442–10451.

Querol, X., Chinchon, S., and Lopez-Soler, A., 1989. Iron sulfide precipitation sequence in Albian coals from the Maestrazgo Basin, southeastern Iberian Range, northeastern Spain. *International Journal of Coal Geology*, 11(2): 171–189.

Raiswell, R., and Anderson, T. F., 2005. Reactive iron enrichment in sediments deposited beneath euxinic bottom waters: Constraints on supply by shelf recycling. *Geological Society, London, Special Publication*. 248: 179–194.

Raiswell, R., and Berner, R. A., 1985. Pyrite formation in euxinic and semi-euxinic sediments. *American Journal of Science*, 285(8): 710–724.

Raiswell, R., Whaler, K., Dean, S., Coleman, M. L., and Briggs, D. E. G., 1993. A simple 3-dimensional model of diffusion-with-precipitation applied to localized pyrite formation in framboids, fossils and detrital iron minerals. *Marine Geology*, 113(1–2): 89–100.

Randolph, A. D., and Larson, M. A., 1988. *Theory of Particulate Processes*. Academic Press, San Diego, CA.

Raybould, J. G., 1973. Framboidal pyrite associated with lead-zinc mineralization in mid-Wales. *Lithos*, 6: 175–182.

Read, R. A., 1968. Deformation and metamorphism of the San Dionisio pyritic ore body, Rio Tinto, Spain. PhD thesis, Imperial College, London.

Reedman, A. J., Colman, T. B., Campbell, S. D. G., and Howells, M. F., 1985. Volcanogenic mineralization related to the Snowdon volcanic group (Ordovician), Gwynedd, north Wales. *Journal of the Geological Society*, 142(SEP): 875–888.

Rhodes, J. M., Jones, C. A., Thal, L. B., and Macdonald, J. E., 2017. Phase-controlled colloidal syntheses of iron sulfide nanocrystals via sulfur precursor reactivity and direct pyrite precipitation. *Chemistry of Materials*, 29(19): 8521–8530.

Rhumbler, L., 1892. Eisenkiesablagerungen im verwesenden Weichkörper von Foraminiferen, die sogenannten Keimkugeln Max Schultze's u. A. *Nachrichten von der Königlich. Gesellschaft der Wissenschaften und der Georg-Augusts-Universität zu Göttingenaus dem Jahre 1892*, 1892: 419–428.

Rickard, D., 1966. An examination of the protective film produced in microbiological corrosion of steel plates by the sulphate -reducing anaerobe Desulphovibrio desulphuricans. *Imperial College*, London.

Rickard, D., 1968a. The geological and microbiological formation of iron sulfides. PhD thesis, Imperial College, London.

Rickard, D., 1968b. The microbiological formation of iron sulfides. *Stockholm Contributions in Geology*, 20: 49–66.

Rickard, D., 1969. The chemistry of iron sulphide formation at low temperatures. *Stockholm Contrib. Geol.*, 20(4): 67–95.

Rickard, D., 1970. The origin of framboids. *Lithos*, 3: 269–293.

Rickard, D., 1974. Kinetics and mechanism of the sulfidation of goethite. *American Journal of Science*, 274(8): 941–952.

Rickard, D., 1975. Kinetics and mechanism of pyrite formation at low-temperatures. *American Journal of Science*, 275(6): 636–652.

Rickard, D., 1995. Kinetics of FeS precipitation: Part 1. Competing reaction mechanisms. *Geochimica et Cosmochimica Acta*, 59(21): 4367–4379.

Rickard, D., 1997. Kinetics of pyrite formation by the H_2S oxidation of iron (II) monosulfide in aqueous solutions between 25 and 125 degrees C: The rate equation. *Geochimica et Cosmochimica Acta*, 61(1): 115–134.

Rickard, D., 2006. The solubility of FeS. *Geochimica et Cosmochimica Acta*, 70(23): 5779–5789.

Rickard, D., 2012a. Fossil bacteria: Evidence for the evolution of the sulfur biome. *Developments in Sedimentology*, 65: 633–683.

Rickard, D., 2012b. Sedimentary pyrite. *Developments in Sedimentology*, 65: 233–285.

Rickard, D., 2012c. Metal sequestration by sedimentary iron sulfides. *Developments in Sedimentology*, 65: 287–317.

Rickard, D., 2012d. Sedimentary sulfides. *Developments In Sedimentology*, 65: 543–604.

Rickard, D., 2012e. The evolution of the sedimentary sulfur cycle. *Developments in Sedimentology*, 65: 685–766.

Rickard, D., 2012f. The geochemistry of sulfidic sedimentary rocks. *Developments in Sedimentology*, 65: 605–632.

Rickard, D., 2013. The sedimentary sulfur system: Biogeochemistry and evolution through geologic time. In: D. H. Heinrich and K. T. Karl (Editors), *Treatise on Geochemistry*. Elsevier, Amsterdam, 9: 267–326.

Rickard, D., 2015. *Pyrite: A Natural History of Fool's Gold*. Oxford University Press, New York, NY.

Rickard, D., 2019a. How long does it take a pyrite framboid to form? *Earth and Planetary Science Letters*, 513: 64–68.

Rickard, D., 2019b. Sedimentary pyrite framboid size-frequency distributions: A meta-analysis. *Palaeogeography, Palaeoclimatology, Palaeoecology*, 522: 62–75.

Rickard, D., Butler, I. B., and Oldroyd, A., 2001. A novel iron sulphide mineral switch and its implications for Earth and planetary science. *Earth and Planetary Science Letters*, 189(1–2): 85–91.

Rickard, D., Griffith, A., Oldroyd, A., Butler, I. B., Lopez-Capel, E., Manning, D. A. C., and Apperley, D. C., 2006. The composition of nanoparticulate mackinawite, tetragonal iron(II) monosulfide. *Chemical Geology*, 235(3–4): 286–298.

Rickard, D., Grimes, S., Butler, I., Oldroyd, A., and Davies, K. L., 2007. Botanical constraints on pyrite formation. *Chemical Geology*, 236(3–4): 228–246.

Rickard, D., and Luther, G., 1997a. Kinetics of pyrite formation by the H_2S oxidation of iron(II) monosulfide in aqueous solutions between 25 and 125 degrees C: The mechanism. *Geochimica Et Cosmochimica Acta*, 61(1): 135–147.

Rickard, D., and Luther, G. W., 2006. Metal sulfide complexes and clusters. *Reviews in Mineralogy & Geochemistry*, 61: 421–504.

Rickard, D., and Luther, G. W., 2007. Chemistry of iron sulfides. *Chemical Reviews*, 107(2): 514–562.

Rickard, D., and Morse, J. W., 2005. Acid volatile sulfide (AVS). *Marine Chemistry*, 97(3–4): 141–197.

Rickard, D., and Zweifel, H., 1975. Genesis of Precambrian sulfide ores, Skellefte District, Sweden. *Economic Geology*, 70(2): 255–274.

Rinklebe, J., and Shaheen, S. M., 2017. Redox chemistry of nickel in soils and sediments: A review. *Chemosphere*, 179: 265–278.

Roberts, A. P., and Turner, G. M., 1993. Diagenetic formation of ferrimagnetic iron sulfide minerals in rapidly deposited marine-sediments, South Island, New Zealand. *Earth and Planetary Science Letters*, 115(1–4): 257–273.

Rosso, K. M., and Vaughan, D. J., 2006a. Reactivity of sulfide mineral surfaces. *Reviews in Mineralogy & Geochemistry*, 61: 557–607.

Rosso, K. M., and Vaughan, D. J., 2006b. Sulfide mineral surfaces. *Reviews in Mineralogy & Geochemistry*, 61: 505–556.

Rudnick, R. L., and Gao, S., 2003. Composition of the continental crust. In: H. D. Holland and K. K. Turekian (Editors), *Treatise on Geochemistry*. Pergamon, Oxford, UK, pp. 1–64.

Rugge, A., and Tolbert, S. H., 2002. Effect of electrostatic interactions on crystallization in binary colloidal films. *Langmuir*, 18(18): 7057–7065.

Rust, G. W., 1935. Colloidal primary copper ores at Cornwall mines, southeastern Missouri. *Journal of Geology*, 43(4): 398–426.

Ruths, M., and Israelachvili, J. N., 2011. Surface forces and nanorheology of molecularly thin films. In: B. Bhushan (Editor), *Nanotribology and Nanomechanics*. Springer-Verlag, Heidleberg, pp. 107–202.

Sassano, G. P., and Schrijver, K., 1989. Framboidal pyrite: Early-diagenetic, late-diagenetic, and hydrothermal occurrences from the Acton Vale quarry, Cambro-Ordovician, Quebec. *American Journal of Science*, 289(2): 167–179.

Sawłowicz, Z., 1990. Primary copper sulfides from the Kupferschiefer, Poland. *Mineralium Deposita*, 25(4): 262–271.

Sawłowicz, Z., 1993. Pyrite framboids and their development - a new conceptual mechanism. *Geologische Rundschau*, 82(1): 148–156.

Sawłowicz, Z., 2000. *Framboids: From Their Origin to Application*. Prace Mineralogiczne, 88. Polska Akademia Nauk, Krakow, Poland.

Schallreuter, R., 1984. Framboidal pyrite in deep-sea sediments. *Initial Reports of the Deep Sea Drilling Project*, 75: 875–891.

Scheihing, M. H., Gluskoter, H. J., and Finkelman, R. B., 1978. Interstitial networks of kaolinite within pyrite framboids in Meigs Creek Coal of Ohio. *Journal of Sedimentary Petrology*, 48(3): 723–732.

Schieber, J., and Schimmelmann, A., 2007. High resolution study of pyrite framboid distribution in Santa Barbara basin sediments and implications for water-column oxygenation. *Pacific Climate Workshop*, Pacific Grove, CA., pp. 31–32.

Schlosser, C., Streu, P., Frank, M., Lavik, G., Croot, P. L., Dengler, M., and Achterberg, E. P., 2018. H₂S events in the Peruvian oxygen minimum zone facilitate enhanced dissolved Fe concentrations. *Scientific Reports*, 8. 12642.

Schneiderhöhn, H., 1922. *Anleitung zur mikroskopischen Bestimmung und Untersuchung von Erzen und aufbereitungs Produkten besonders im auffallenden Licht*. Guertler, Berlin.

Schneiderhöhn, H., 1923. Chalkographische Untersuchung des Mansfelder Kupferschiefers. *Neues Jahrbuch fur Mineralogie, Geologie und Palaontologie*, 157: 1–38.

Schoenflies, A., 1915. XVI. Über Krystallstruktur (II). *Zeitschrift für Kristallographie - Crystalline Materials*, 55: 321–352.

Schoepfer, S. D., Tobin, T. S., Witts, J. D., and Newton, R. J., 2017. Intermittent euxinia in the high-latitude James Ross Basin during the latest Cretaceous and earliest Paleocene. *Palaeogeography Palaeoclimatology Palaeoecology*, 477: 40–54.

Schoonen, M. A. A., and Barnes, H. L., 1991a. Mechanisms of pyrite and marcasite formation from solution: III. Hydrothermal processes. *Geochimica et Cosmochimica Acta*, 55(12): 3491–3504.

Schoonen, M. A. A., and Barnes, H. L., 1991b. Reactions forming pyrite and marcasite from solution. 1. Nucleation of FeS_2 below 100°C. *Geochimica et Cosmochimica Acta*, 55(6): 1495–1504.

Schouten, C., 1946. The role of sulphur bacteria in the formation of the so-called sedimentary copper ores and pyritic ore bodies. *Economic Geology*, 41(5): 517–538.

Schulz, H. N., Brinkhoff, T., Ferdelman, T. G., Marine, M. H., Teske, A., and Jorgensen, B. B., 1999. Dense populations of a giant sulfur bacterium in Namibian shelf sediments. *Science*, 284(5413): 493–495.

Schulze, H., 1882. Schwefelarsen in wässriger Lösung. *Journal für Praktische Chemie*. 25: 431–452.

Scott, R. J., Meffre, S., Woodhead, J., Gilbert, S. E., Berry, R. F., and Emsbo, P., 2009. Development of framboidal pyrite during diagenesis, low-grade regional metamorphism, and hydrothermal alteration. *Economic Geology*, 104(8): 1143–1168.

Seager, A. F., 1952. Screw dislocations in pyrite. *Nature*, 170(4323): 425.

Sgavetti, M., Pompilio, L., Roveri, M., Manzi, V., Valentino, G. M., Lugli, S., Carli, C., Amici, S., Marchese, F., and Lacava, T., 2009. Two geologic systems providing terrestrial analogues for the exploration of sulfate deposits on Mars: Initial spectral characterization. *Planetary and Space Science*, 57(5–6): 614–627.

Short, M. N., 1931. *Microscopic Determination of the Ore Minerals*. Bulletin 825. United States Geological Survey, Washington, DC.

Shotyk, W., 1988. Review of the inorganic geochemistry of peats and peatland waters. *Earth-Science Reviews*, 25(2): 95–176.

Shtukenberg, A. G., Zhu, Z., An, Z., Bhandari, M., Song, P., Kahr, B., and Ward, M. D., 2013. Illusory spirals and loops in crystal growth. *Proceedings of the National Academy of Sciences of the United States of America*, 110(43): 17195–17198.

Sinninghe Damsté, J. S., and de Leeuw, J. W., 1990. Analysis, structure and geochemical significance of organically-bound sulfur in the geosphere: State-of-the-art and future-research. *Organic Geochemistry*, 16(4–6): 1077–1101.

Skei, J. M., 1988. Formation of framboidal iron sulfide in the water of a permanently anoxic fjord: Framvaren, south Norway. *Marine Chemistry*, 23(3–4): 345–352.

Skipchenko, N. S., and Berber'yan, T. K., 1976. The structure of framboidal pyrite. *International Geology Review*, 18: 1435–1469.

Smolarek, J., Trela, W., Bond, D. P. G., and Marynowski, L., 2017. Lower Wenlock black shales in the northern Holy Cross Mountains, Poland: Sedimentary and geochemical controls on the Ireviken Event in a deep marine setting. *Geological Magazine*, 154(2): 247–264.

Soliman, M. F., and El Goresy, A., 2012. Framboidal and idiomorphic pyrite in the upper Maastrichtian sedimentary rocks at Gabal Oweina, Nile Valley, Egypt: Formation processes, oxidation products and genetic implications to the origin of framboidal pyrite. *Geochimica et Cosmochimica Acta*, 90: 195–220.

Špillar, V., and Dolejš, D., 2013. Calculation of time-dependent nucleation and growth rates from quantitative textural data: Inversion of crystal size distribution. *Journal of Petrology*, 54(5): 913–931.

Stakes, D. S., Orange, D., Paduan, J. B., Salamy, K. A., and Maher, N., 1999. Cold-seeps and authigenic carbonate formation in Monterey Bay, California. *Marine Geology*, 159(1–4): 93–109.

Steinike, K., 1963. A further remark on biogenic sulfides: Inorganic pyrite spheres. *Economic Geology and the Bulletin of the Society of Economic Geologists*, 94: 998–1000.

Steudel, R., 1975. Properties of sulfur-sulfur bonds. *Angewandte Chemie - International Edition*, 14(10): 655–664.

Sturm, E. V., and Cölfen, H., 2017. Mesocrystals: Past, presence [sic], future. *Crystals*, 7: 207.

Suits, N. S., and Wilkin, R. T., 1998. Pyrite formation in the water column and sediments of a meromictic lake. *Geology*, 26(12): 1099–1102.

Suk, D., Peacor, D. R., and Vandervoo, R., 1990. Replacement of pyrite framboids by magnetite in limestone and implications for paleomagnetism. *Nature*, 345(6276): 611–613.

Sun, Y. D., Wignall, P. B., Joachimski, M. M., Bond, D. P. G., Grasby, S. E., Sun, S., Yan, C. B., Wang, L. N., Chen, Y. L., and Lai, X. L., 2015. High amplitude redox changes in the late Early Triassic of South China and the Smithian-Spathian extinction. *Palaeogeography Palaeoclimatology Palaeoecology*, 427: 62–78.

Sunagawa, I., 1957. Variation in crystal habit of pyrite. *Reports of the Geological Survey of Japan*, 175: 1–41.

Sunagawa, I., 1987. Morphology of minerals. In: I. Sunagawa (Editor), *Morphology of Crystals, Part B*. D. Reidel, Dortrecht, pp. 509–587.

Sunagawa, I., 1993. Crystal growth in the mineral kingdom. *Journal of Crystal Growth*, 128(1): 397–402.

Sunagawa, I., Endo, Y., and Nakai, N., 1971. Hydrothermal synthesis of framboidal pyrite. *Mining Geology*, 2: 10–14.

Sunagawa, I., and Takahishi, K., 1955. Preliminary report on the relation between the o(111) face of pyrite crystals and its minor content of arsenic. *Geological Society of Japan Bulletin*, 6: 1–10.

Sweeney, R. E., and Kaplan, I. R., 1973. Pyrite framboid formation: Laboratory synthesis and marine sediments. *Economic Geology and the Bulletin of the Society of Economic Geologists*, 68: 618–634.

Szczepanik, P., Gize, A., and Sawłowicz, Z., 2017. Pyritization of dinoflagellate cysts: A case study from the Polish Middle Jurassic (Bathonian). *Review of Palaeobotany and Palynology*, 247: 1–12.

Takahashi, S., Yamasaki, S. I., Ogawa, K., Kaiho, K., and Tsuchiya, N., 2015. Redox conditions in the end-Early Triassic Panthalassa. *Palaeogeography Palaeoclimatology Palaeoecology*, 432: 15–28.

Tamaru, Y., Takani, Y., Yoshida, T., and Sakamoto, T., 2005. Crucial role of extracellular polysaccharides in desiccation and freezing tolerance in the terrestrial cyanobacterium *Nostoc commune*. *Applied and Environmental Microbiology*, 71(11): 7327–7333.

Tardani, D., Reich, M., Deditius, A. P., Chryssoulis, S., Sanchez-Alfaro, P., Wrage, J., and Roberts, M. P., 2017. Copper-arsenic decoupling in an active geothermal system: A link between pyrite and fluid composition. *Geochimica et Cosmochimica Acta*, 204: 179–204.

Taylor, G. R., 1982. A mechanism for framboid formation as illustrated by a volcanic exhalative sediment. *Mineralium Deposita*, 17(1): 23–36.

Theberge, S., and Luther, G. W., 1997. Determination of the electrochemical properties of a soluble aqueous FeS species present in sulfide solutions. *Aquatic Geochemistry*, 3: 191–211.

Thiel, J., Byrne, J. M., Kappler, A., Schink, B., and Pester, M., 2019. Pyrite formation from FeS and H₂S is mediated through microbial redox activity. *Proceedings of the National Academy of Sciences*, 116(14): 6897–6902.

Thiessen, R., 1920. Occurrence and origin of finely disseminated sulfur compounds in coal. *Transactions of the American Institute of Mining and Metallurgical Engineers*, 63: 913–926.

Tokody, L., 1931. Pyritformen und -fundorte. *Zeitschrift fur Kristallographie, Kristallgeometrie, Kristallphysik, Kristallchemie*, 80: 255–348.

Torquato, S., and Jiao, Y., 2009. Dense packings of polyhedra: Platonic and Archimedean solids. *Physical Review E*, 80: 041104.

Trefalt, G., Szilagyi, I., Tellez, G., and Borkovec, M., 2017. Colloidal stability in asymmetric electrolytes: Modifications of the Schulze-Hardy Rule. *Langmuir*, 33(7): 1695–1704.

Tribovillard, N., Bialkowski, A., Tyson, R. V., Lallier-Verges, E., and Deconinck, J. F., 2001. Organic facies variation in the late Kimmeridgian of the Boulonnais area (northernmost France). *Marine and Petroleum Geology*, 18(3): 371–389.

Tribovillard, N., Lyons, T. W., Riboulleau, A., and Bout-Roumazeilles, V., 2008. A possible capture of molybdenum during early diagenesis of dysoxic sediments. *Bulletin de la Société Géologique de France*, 179(1): 3–12.

Turchyn, A. V., and Schrag, D. P., 2004. Oxygen isotope constraints on the sulfur cycle over the past 10 million years. *Science*, 303(5666): 2004–2007.

Vallentyne, J. R., 1962. Concerning Love, microfossils and pyrite spherules. *Transactions of the New York Academy of Science*, Ser. 2, 25: 177–189.

Vallentyne, J. R., 1963. Isolation of pyrite spherules from recent sediments. *Limnology and Oceanography*, 8(1): 16–29.

van Bemmelen, J. M., 1866. Boden Unterssuchungen in den Niederlanden. *Die landwirthschaftlichen Versuchs-Stationen*, 8: 255–306.

Verwey, E. J. W., and Overbeek, J. Th. G., 1948. *Theory of the stability of lyophobic colloids*. Elsevier. New York.

Vietti, L. A., Bailey, J. V., Fox, D. L., and Rogers, R. R., 2015. Rapid formation of framboidal sulfides on bone surfaces from a simulated marine carcass fall. *Palaios*, 30(4): 327–334.

Vilinska, A., and Rao, K. H., 2011. Surface thermodynamics and extended DLVO theory of *Leptospirillum ferrooxidans* cells' adhesion on sulfide minerals. *Minerals & Metallurgical Processing*, 28(3): 151–158.

Voigt, B., Moore, W., Manno, M., Walter, J., Jeremiason, J. D., Aydil, E. S., and Leighton, C., 2019. Transport evidence for sulfur vacancies as the origin of unintentional n-type doping in pyrite FeS₂. *ACS Applied Materials & Interfaces*, 11(17): 15552–15563.

Wacey, D., Kilburn, M. R., Saunders, M., Cliff, J. B., Kong, C., Liu, A. G., Matthews, J. J., and Brasier, M. D., 2015. Uncovering framboidal pyrite biogenicity using nano-scale CNorg mapping. *Geology*, 43(1): 27–30.

Wadell, H., 1935. Volume, shape, and roundness of quartz particles. *Journal of Geology*, 43(3): 250–280.

Wan, M., Schroder, C., and Peiffer, S., 2017. Fe(III): S(-II) concentration ratio controls the pathway and the kinetics of pyrite formation during sulfidation of ferric hydroxides. *Geochimica et Cosmochimica Acta*, 217: 334–348.

Wan, M. L., Shchukarev, A., Lohmayer, R., Planer-Friedrich, B., and Peiffer, S., 2014. Occurrence of surface polysulfides during the interaction between ferric (hydr)oxides and aqueous sulfide. *Environmental Science & Technology*, 48(9): 5076–5084.

Wang, D., Wang, Q., and Wang, T., 2010. Shape controlled growth of pyrite FeS$_2$ crystallites via a polymer-assisted hydrothermal route. *CrystEngComm*, 12(11): 3797–3805.

Wang, F., Richards, V. N., Shields, S. P., and Buhro, W. E., 2014. Kinetics and mechanisms of aggregative nanocrystal growth. *Chemistry of Materials*, 26(1): 5–21.

Wang, J. W., Mbah, C. F., Przybilla, T., Englisch, S., Spiecker, E., Engel, M., and Vogel, N., 2019. Free energy landscape of colloidal clusters in spherical confinement. *ACS Nano*, 13(8): 9005–9015.

Wang, J. W., Mbah, C. F., Przybilla, T., Zubiri, B. A., Spiecker, E., Engel, M., and Vogel, N., 2018. Magic number colloidal clusters as minimum free energy structures. *Nature Communications*, 9(1): 5259.

Wang, J. W., Sultan, U., Goerlitzer, E. S. A., Mbah, C. F., Engel, M., and Vogel, N., 2019b. Structural color of colloidal clusters as a tool to investigate structure and dynamics. *Advanced Functional Materials*, 30: 1907730.

Wang, L., Shi, X. Y., and Jiang, G. Q., 2012. Pyrite morphology and redox fluctuations recorded in the Ediacaran Doushantuo Formation. *Palaeogeography Palaeoclimatology Palaeoecology*, 333: 218–227.

Wang, Q., and Morse, J. W., 1996. Pyrite formation under conditions approximating those in anoxic sediments: I. Pathway and morphology. *Marine Chemistry*, 52(2): 99–121.

Weerasooriya, R., and Tobschall, H. J., 2005. Pyrite-water interactions: Effects of pH and pFe on surface charge. *Colloids and Surfaces A - Physicochemical and Engineering Aspects*, 264(1–3): 68–74.

Wei, H. Y., Algeo, T. J., Yu, H., Wang, J. G., Guo, C., and Shi, G., 2015. Episodic euxinia in the Changhsingian (late Permian) of South China: Evidence from framboidal pyrite and geochemical data. *Sedimentary Geology*, 319: 78–97.

Wei, H. Y., Wei, X. M., Qiu, Z., Song, H. Y., and Shi, G., 2016. Redox conditions across the G-L boundary in South China: Evidence from pyrite morphology and sulfur isotopic compositions. *Chemical Geology*, 440: 1–14.

Wicksell, S. D., 1925. The corpuscle problem: A mathematical study of a biometric problem. *Biometrika*, 17(1/2): 84–99.

Wiese, R. G., and Fyfe, W. S., 1986. Occurrences of iron sulfides in Ohio coals. *International Journal of Coal Geology*, 6(3): 251–276.

Wignall, P. B., and Newton, R., 1998. Pyrite framboid diameter as a measure of oxygen deficiency in ancient mudrocks. *American Journal of Science*, 298(7): 537–552.

Wignall, P. B., Bond, D. P. G., Kuwahara, K., Kakuwa, Y., Newton, R. J., and Poulton, S. W., 2010. An 80 million year oceanic redox history from Permian to Jurassic pelagic sediments of the Mino-Tamba terrane, SW Japan, and the origin of four mass extinctions. *Global and Planetary Change*, 71(1–2): 109–123.

Wikjord, A. G., Rummery, T. E., and Doern, F. E., 1976. Crystallization of pyrite from deoxygenated aqueous sulfide solutions at elevated temperature and pressure. *Canadian Mineralogist*, 14: 571–573.

Wilkin, R. T., and Arthur, M. A., 2001. Variations in pyrite texture, sulfur isotope composition, and iron systematics in the Black Sea: Evidence for Late Pleistocene to Holocene excursions of the O_2-H_2S redox transition. *Geochimica et Cosmochimica Acta*, 65(9): 1399–1416.

Wilkin, R. T., and Barnes, H. L., 1996. Pyrite formation by reactions of iron monosulfides with dissolved inorganic and organic sulfur species. *Geochimica et Cosmochimica Acta*, 60(21): 4167–4179.

Wilkin, R. T., and Barnes, H. L., 1997a. Formation processes of framboidal pyrite. *Geochimica et Cosmochimica Acta*, 61(2): 323–339.

Wilkin, R. T., and Barnes, H. L., 1997b. Pyrite formation in an anoxic estuarine basin. *American Journal of Science*, 297(6): 620–650.

Wilkin, R. T., Barnes, H. L., and Brantley, S. L., 1996. The size distribution of framboidal pyrite in modern sediments: An indicator of redox conditions. *Geochimica et Cosmochimica Acta*, 60(20): 3897–3912.

Wilson, N. S. F., and Zentilli, M., 1999. The role of organic matter in the genesis of the El Soldado volcanic-hosted manto-type Cu deposit, Chile. *Economic Geology and the Bulletin of the Society of Economic Geologists*, 94(7): 1115–1135.

Wolthers, M., Butler, I. B., Rickard, D., and Mason, P. R. D., 2006. Arsenic uptake by pyrite at ambient environmental conditions: A continuous-flow experiment. *ACS Symposium Series*, 915: 60–76.

Wolthers, M., van Der Gaast, S. J., and Rickard, D., 2003. The structure of disordered mackinawite. *American Mineralogist*, 88: 2007–2015.

Woodcock, L. V., 1997. Entropy difference between the face-centred cubic and hexagonal close-packed crystal structures. *Nature*, 385(6612): 141–143.

Wu, R., Zheng, Y .F., Zhang, X. G., Sun, Y. F., Xu, J. B., and Jian, J. K., 2004. Hydrothermal synthesis and crystal structure of pyrite. *Journal of Crystal Growth*, 266(4): 523–527.

Xian, H., Zhu, J., Liang, X., and He, H., 2016. Morphology controllable syntheses of micro- and nano-iron pyrite mono- and poly-crystals: A review. *RSC Advances*, 6(38): 31988–31999.

Xian, H., Zhu, J., Tang, H., Liang, X., He, H., and Xi, Y., 2016. Aggregative growth of quasi-octahedral iron pyrite mesocrystals in a polyol solution through oriented attachment. *CrystEngComm*, 18(46): 8823–8828.

Xu, R., Tang, H., Li, Y., and Guo, Z., 2019. The significance of the framboidal magnetite in Murchison meteoite [*sic*]. 50th Lunar and Planetary Science Conference 2019, Houston, Texas. 2186.

Yao, W. S., and Millero, F. J., 1996. Oxidation of hydrogen sulfide by hydrous Fe(III) oxides in seawater. *Marine Chemistry*, 52: 1–16.

Yoon, S.-J., Yáñez, C., Bruns, M. A., Martínez-Villegas, N., and Martínez, C. E., 2012. Natural zinc enrichment in peatlands: Biogeochemistry of ZnS formation. *Geochimica et Cosmochimica Acta*, 84: 165–176.

Youngson, J. H., 1995. Sulphur mobility and sulphur mineral precipitation during early Miocene: Recent uplift and sedimentation in Central Otago, New Zealand. *New Zealand Journal of Geology and Geophysics*, 38(4): 407–417.

Yuan, B. X., Luan, W. L., Tu, S. T., and Wu, J., 2015. One-step synthesis of pure pyrite FeS_2 with different morphologies in water. *New Journal of Chemistry*, 39(5): 3571–3577.

Yuan, W., Liu, G. D., Stebbins, A., Xu, L. M., Niu, X. B., Luo, W. B., and Li, C. Z., 2017. Reconstruction of redox conditions during deposition of organic-rich shales of the Upper Triassic Yanchang Formation, Ordos Basin, China. *Palaeogeography Palaeoclimatology Palaeoecology*, 486: 158–170.

Yücel, M., Gartman, A., Chan, C. S., and Luther, G. W., 2011. Hydrothermal vents as a kinetically stable source of iron-sulphide-bearing nanoparticles to the ocean. *Nature Geoscience*, 4(6): 367–371.

Zhang, H. Z., Gilbert, B., Huang, F., and Banfield, J. F., 2003. Water-driven structure transformation in nanoparticles at room temperature. *Nature*, 424(6952): 1025–1029.

Zhang, J., Liang, H., He, X., Yang, Y., and Chen, B., 2011. Sulfur isotopes of framboidal pyrite in the Permian-Triassic boundary clay at Meishan section. *Acta Geologica Sinica - English Edition*, 85(3): 694–701.

Zhang, L., Chang, S., Khan, M. Z., Feng, Q. L., Danelian, T., Clausen, S., Tribovillard, N., and Steiner, M., 2018. The link between metazoan diversity and paleo-oxygenation in the early Cambrian: An integrated palaeontological and geochemical record from the eastern Three Gorges Region of South China. *Palaeogeography Palaeoclimatology Palaeoecology*, 495: 24–41.

Zhang, M., Konishi, H., Xu, H. F., Sun, X. M., Lu, H. F., Wu, D. D., and Wu, N. Y., 2014. Morphology and formation mechanism of pyrite induced by the anaerobic oxidation of methane from the continental slope of the NE South China Sea. *Journal of Asian Earth Sciences*, 92: 293–301.

Zhang, X., 2015. Synthesis, characterization and electric transport properties of thin film iron pyrite for photovoltaic applications. PhD thesis, University of Minnesota, 217 pp.

Zhang, X., Li, M. Q., Walter, J., O'Brien, L., Manno, M. A., Voigt, B., Mork, F., Baryshev, S. V., Kakalios, J., Aydil, E. S., and Leighton, C., 2017. Potential resolution to the doping puzzle in iron pyrite: Carrier type determination by Hall effect and thermopower. *Physical Review Materials*, 1: 015402.

Zhang, Y. N., Law, M., and Wu, R. Q., 2015. Atomistic modeling of sulfur vacancy diffusion near iron pyrite surfaces. *Journal of Physical Chemistry C*, 119(44): 24859–24864.

Zhao, J., Liang, J. L., Long, X. P., Li, J., Xiang, Q. R., Zhang, J. C., and Hao, J. L., 2018. Genesis and evolution of framboidal pyrite and its implications for the ore-forming process of Carlin-style gold deposits, southwestern China. *Ore Geology Reviews*, 102: 426–436.

Zheng, H. M., Smith, R. K., Jun, Y. W., Kisielowski, C., Dahmen, U., and Alivisatos, A. P., 2009b. Observation of single colloidal platinum nanocrystal growth trajectories. *Science*, 324(5932): 1309–1312.

Zheng, X., Wang, Z., Wang, L., Xu, Y., and Liu, J., 2017. Mineralogical and geochemical compositions of the Lopingian coals and carbonaceous rocks in the Shugentian coalfield, Yunnan, China: With emphasis on Fe-bearing minerals in a continental-marine transitional environment. *Minerals*, 7(9): 170.

Zhou, C. M., and Jiang, S. Y., 2009. Palaeoceanographic redox environments for the lower Cambrian Hetang Formation in South China: Evidence from pyrite framboids, redox sensitive trace elements, and sponge biota occurrence. *Palaeogeography Palaeoclimatology Palaeoecology*, 271(3–4): 279–286.

Zhu, L., Richardson, B. J., and Yu, Q., 2015. Anisotropic growth of iron pyrite FeS_2 nanocrystals via oriented attachment. *Chemistry of Materials*, 27(9): 3516–3525.

Zies, E. G., Allen, E. T., and Merwin, H. E., 1916. Some reactions involved in secondary copper enrichment. *Economic Geology*, 11: 407–503.

Zimmerman, R. A., and Amstutz, G. C., 1973. Relations of sections of cubes, octahedra and pyritohedra. *Neues Jahrbuch für Mineralogie - Abhandlungen*, 120(1): 15–30.

Index

Tables and figures are indicated by *t* and *f* following the page number